T0350471

Real Science
What it is, and what it means

Scientists and 'anti-scientists' alike need a more realistic image of science. The traditional mode of research, *academic* science, is not just a 'method': it is a distinctive culture, whose members win esteem and employment by making public their findings. Fierce competition for credibility is strictly regulated by established practices such as peer review. Highly specialized international communities of independent experts form spontaneously and generate the type of knowledge we call 'scientific' – systematic, theoretical, empirically tested, quantitative, and so on. Ziman shows that these familiar, 'philosophical' features of scientific knowledge are inseparable from the ordinary cognitive capabilities and peculiar social relationships of its producers. This wide-angled close-up of the natural and human sciences recognizes their unique value, whilst revealing the limits of their rationality, reliability, and universal applicability. It also shows how, for better or worse, the new 'post-academic' research culture of teamwork, accountability, etc. is changing these supposedly eternal philosophical characteristics.

JOHN ZIMAN is well known internationally for his many scholarly and popular books on condensed-matter physics and on science, technology and society. He was born in 1925, and was brought up in New Zealand. He took his DPhil at Oxford and lectured at Cambridge before becoming Professor of Theoretical Physics at Bristol in 1964. His research on the electrical properties of metals earned his election to the Royal Society in 1967. After voluntary early retirement from Bristol in 1982 he devoted himself to the systematic analysis and public exposition of various aspects of the social relations of science and technology, on which he is a recognized world authority. He was for many years chairman of the Council for Science and Society, and between 1986 and 1991 he headed the Science Policy Support Group. He is currently Convenor of the Epistemology Group.

To Joan: IN HEART, MIND AND SPIRIT.

Real Science

What it is, and what it means

JOHN ZIMAN

CAMBRIDGE
UNIVERSITY PRESS

CAMBRIDGE UNIVERSITY PRESS
Cambridge, New York, Melbourne, Madrid, Cape Town, Singapore, São Paulo

Cambridge University Press
The Edinburgh Building, Cambridge CB2 8RU, UK

Published in the United States of America by Cambridge University Press, New York

www.cambridge.org
Information on this title: www.cambridge.org/9780521772297

First published 2000
Reprinted 2001
First paperback edition 2002

A catalogue record for this publication is available from the British Library

Library of Congress Cataloguing in Publication data

Ziman, J. M. (John M.), 1925–
    Real science : what it is, and what it means / John Ziman.
        p.   cm.
    Includes bibliographical references and index.
    ISBN 0 521 77229 X
    1. Science–Philosophy.   2. Science–Methodology.   I. Title.
Q175.Z547   2000
501–dc21   99–38905 CIP

ISBN 978-0-521-77229-7 hardback
ISBN 978-0-521-89310-7 paperback

Transferred to digital printing 2008

Contents

Preface

The seeds of this book were sown forty years ago. I was always infatuated with science and beguiled by philosophy. They seemed made for each other – and for me. But the better I came to know science, the more I realized that the philosophers were not telling it like it is. Then, sometime around 1959, I was asked to review Michael Polanyi's *Personal Knowledge*[1] and Karl Popper's *The Logic of Scientific Discovery*[2]. Each of these great books says important things about science; but in both I noticed a whole pack of dogs that didn't bark. What about the web of lectures, examinations, seminars, conferences, papers, citations, referee reports, books, personal references, job interviews, appointments, prizes, etc. in which my scientific life was entangled? Surely these must have some influence on the work I was doing. So in radio talks and articles I began to say strange things, such as 'Science is *social*' and 'Research is a *profession*'[3].

Those were rash words for a young and aspiring physicist without official credentials in philosophy or sociology. Nevertheless, the heterodoxy was overlooked and my academic career prospered. The books in which I developed this theme – *Public Knowledge*[4], *Reliable Knowledge*[5] and *An Introduction to Science Studies*[6] – were also very well received, and are still read and cited. Indeed, many of the notions that germinated in these books have since been planted out more formally by other scholars. And just as I foresaw, sociology has superseded philosophy at the theoretical core of 'science studies'.

This metascientific revolution has certainly opened science to much more searching enquiry. But the spirit in which this enquiry has been conducted has actually widened the gulf between those who do science and those who observe their doings. What is more, as I pointed out in detail in *Prometheus Bound*[7] and *Of One Mind* [8], science itself is changing rapidly, as a profession and as an institution. What is happening? Where are we going? Now, more than ever, scientists, science users and science watchers need a clear vision of how it really, really works and what it can really, really do. But just when they ought to be getting sympathetic, well-informed advice from their metascientific colleagues, they are being offered little but deconstruction and doubt.

It seems to me, nevertheless, that a much more substantial model of science can be discerned within the booming buzzing confusion of contemporary science studies. So, what can I now do to bring this model out into the open, to help science understand itself? In the end, it comes down to the same basic question: is science to be *believed* – and if so, in what sense? This is a much more subtle question than it used to seem. For all their labours, the philosophers have failed to come up with any simple, generally agreed principle on which belief in science might be safely grounded. But sociological critiques are vacuous without reference to the specific contexts in which beliefs are held or made. When I said that science is *social*, I meant that this context includes the whole network of social and epistemic practices where scientific beliefs actually emerge and are sustained.

The trouble is that this network is not regulated by any single prince or principle. To appreciate the significance of scientific knowledge, one must understand the nature of science as a complex whole. That is why I decided, about five years ago, to start again from scratch and work systematically through the whole argument. As in all my writings *about* science, I wanted to show that this argument did not require much scientific knowledge as such, but could be presented perfectly clearly in the everyday language of the common reader. The line of reasoning of this book is lengthy, and visits many different academic sites, but it is not at all technical or intellectually convoluted.

But I also wanted to show that this line of reasoning is no longer a personal fancy. The naturalistic account of science that I am presenting in this book is accepted – albeit tacitly – by numerous reputable scholars. This is clear from the the size of the bibliography, even though this does not pretend to cover all the standard metascientific literature. Indeed, this book would never end if I had set out to expound and/or refute all that has been written about the nature of science, most of which is irrelevant or tangential to my main argument.

But that still left me with a practical problem. I wanted to say what I thought in my own words; so how should these be linked to the words and thoughts of so many other authors? A mosaic of verbatim quotations would have been unreadably ponderous. Even the scholarly practice of citing every author by name in the text – e.g. 'As Tadpole & Taper (1843) have shown (see also Disraeli 1844)....' – would have interfered with the flow of ideas, and repelled the non-academic readers to whom this book is mainly addressed. On the other hand, the lazy custom of mentioning by name just a few of the usual suspects – Karl Popper, Thomas Kuhn, Robert Merton, Donald Campbell, and so on – does less than justice to many less eminent scholars whose ideas were no less perceptive and original in their time. Or, to put it another way, what really matters is the idea itself, not whether it is conventionally associated with some famous name.

What I have done, therefore, is to indicate all such linkages in the main text by inconspicuous superscripts referring to endnotes on each chapter. To avoid loading these notes down with formal bibliographic information, which tends to be very repetitive and difficult to scan, I have compacted this into a comprehensive alphabetical list of references, accessible directly from the notes by the author's name and date. Moreover, since each entry on this list bears a coded reference back to the various notes where it is cited, the bibliography also operates as an author index for the book as a whole.

But what else beside such bibliographic pointers should these notes contain? In principle I am permissive in such matters, and have always enjoyed reading (and even composing) the addenda, qualificanda, divertenda, detractenda, joculanda, etc. with which a gristly book can be made more palatable. In practice, however, notes easily get out of hand. As footnotes they clutter the printed page, and as endnotes they are out of context. A more austere academic tradition would restrict notes to their ostensible function of relating the contents of the cited work specifically to the place in the text where they are cited. Thus, in addition to recording intellectual priorities, they should give the reader an indication of the attitude of the cited author to the point being made, whether of enthusiastic agreement, qualified acceptance or downright opposition.

But as I have observed ruefully when my own work has been cited by others, that can seldom be done accurately or equitably in a few words. Suppose, for example, that I want to cite Kuhn (1963) on 'paradigms'. Yes, he should certainly get abundant credit for first formulating this invaluable concept – but did he mean exactly the same by it as I am now proposing? Shouldn't I at least hint at differences through verbatim quotes?

What about his later responses to the critical literature that it provoked? How have other scholars interpreted this concept – and so on? Before you know where you are, your note has become a small essay. Practise the same scrupulosity for hundreds of other equally worthy authors, and your whole book has again drowned in its own notes.

Presenting references satisfactorily is a hard problem in a non-academic book. So I have cut the Gordian knot, and included nothing in the notes beyond the formal bibliographic citation. This could refer to anything from a favourable treatise to a scornful aside. But at least it indicates a linkage, a certain congruence of interest, a wave of recognition to a fellow pilgrim on the path through the forest, an invitation to further discourse on a theme of mutual intellectual concern. Academic reviewers will never, of course, forgive this breach of ponderous current practice; but those for whom this book is really written may well be grateful for all those unwritten pages that they don't have to buy or pretend to read.

Finally, I ought to acknowledge the help of all those kind people who have contributed to the creation of this book. Ah, but they are too numerous to list individually. As I said, I began thinking and talking about these matters years ago, and have discussed them personally, pro and con, back and forth, with a great many other scholars with similar interests. In fact, this list would include about half the authors I have cited in the bibliography – although I guess that some of these would not wish to have it thought that they had actually *helped* to bring *these* ideas to birth! Let me just say 'Thank you all!', for the courtesy, conviviality, collegiality and straightforward friendship that has graced these innumerable conversations and communications.

John Ziman
Oakley, August 1998

1

A peculiar institution

1.1 Defending a legend

Science is under attack. People are losing confidence in its powers. Pseudo-scientific beliefs thrive. Anti-science speakers win public debates. Industrial firms misuse technology. Legislators curb experiments. Governments slash research funding. Even fellow scholars are becoming sceptical of its claims[1].

And yet, opinion surveys regularly report large majorities in its favour. Science education expands at all levels. Writers and broadcasters enrich public understanding. Exciting discoveries and useful inventions flow out of the research laboratories. Vast research instruments are built at public expense. Science has never been so popular or influential.

This is not a contradiction. Science has always been under attack. It is still a newcomer to large areas of our culture. As it extends and becomes more deeply embedded, it touches upon issues where its competence is more doubtful, and opens itself more to well-based criticism. The claims of science are often highly questionable. Strenuous debate on particular points is not a symptom of disease: it signifies mental health and moral vigour.

Blanket hostility to 'science' is another matter. Taken literally, that would make no more sense than hostility to 'law', or 'art', or even to 'life' itself. What such an attitude really indicates is that certain general features of science are thought to be objectionable in principle, or unacceptable in practice. These features are deemed to be so essential to science as such that it is rejected as a whole – typically in favour of some other supposedly holistic system.

The arguments favouring 'anti-science' attitudes may well be misinformed, misconceived and mischievous. Nevertheless, they carry surprising weight in society at large. Those of us who do not share these attitudes have a duty to combat them. But what are the grounds on which science should be defended?

Many supporters of science simply challenge the various specific objections put forward by various schools of anti-science. In doing so, however, they usually assume that the general features in dispute are, indeed, essential to science. They may agree, for example, that scientific knowledge is arcane and elitist, and then try to show that this need not be a serious disadvantage in practice[2]. The danger of this type of defence is that it accepts without question an analysis which may itself be deeply flawed. In many cases, the objectionable feature is incorrectly attributed to 'science', or is far from essential to it. Dogged defence of every feature of 'the Legend'[3] – the stereotype of science that idealizes its every aspect – is almost as damaging as the attack it is supposed to be fending off.

1.2 Science as it is and does

In the long run, science has to survive on its merits. It must be cherished for what it *is*, and what it can *do*. The moral basis for the defence of science must be a clear understanding of its nature and of its powers. One might have thought that this understanding was already widely shared, especially amongst working scientists. Unfortunately, this is not the case. Most people who have thought about this at all are aware that the notion of an all-conquering intellectual 'method' *is* just a legend. This legend has been shot full of holes, but they do not know how it can be repaired or replaced. They are full of doubt about past certainties, but full of uncertainty about what they ought now to believe.

A more up-to-date and convincing 'theory of science'[4] is required for a variety of other reasons. The place of science in society is not just a matter of personal preference or cultural tradition: it is a line item in the national budget. There are increasing tensions in the relationships between scientific and other forms of knowledge and action, such as technology, medicine, law and politics. Scientists are asked by their students whether they are being prepared for a vocation or for a profession. People are expected to make rational decisions arising from, and affecting, radical changes in the way in which science is organized and performed.

The uncertainties and confusions have not been dispelled by the sociologists who have displaced the philosophers from the centre of 'science studies'. On the contrary, the various schools of sociological 'relativism' and 'constructivism' that have emerged in the past twenty years[5] often seem to be hostile to science, and eager to belittle its capabilities. In their enthusiasm to expose scientific pretensions to objectivity and truth, they exaggerate the genuine uncertainties and perplexities of scientific research[6], and propagate an equally false and damaging stereotype of pervasive cynicism and doubt. Despite repeated assertions that they too love science, and don't really dispute its practical claims, they thus confirm the natural scientists in their mistrust of the social sciences, and often seem to ally themselves with anti-scientific populism.

It must be emphasized, however, that this sceptical stance does not go unchallenged in the world of science studies. Many *metascientists* – that is, philosophers, sociologists, political scientists, economists, anthropologists and other scholars who study science as a human activity – have pointed out the weaknesses of this stance. They try to understand just what scientists think they are doing when they engage in research, and how much weight should be given to their results. They are interested in the way that scientists work, as individuals and as groups, and how this affects their findings. They may not accept scientific knowledge as uniquely true and real, but they do treat it as a peculiar human product worthy of special study.

These expeditions into the unknown heart of science branch out in all directions[7]. Metascientists make their observations through the intellectual instruments of many different disciplines, and analyse what they see along many different dimensions. The study of each such aspect has become a research specialty in its own right, with results that are often scarcely intelligible outside that specialty. We know much more about science nowadays than can be put together into a comprehensive, coherent image.

Metascientific pluralism is a wise recoil from overambitious attempts to encompass a complex human enterprise in a single formula. Nevertheless, these various modern accounts of science are not all disconnected. They start from the outside to explore the same range of ideas, activities and institutions, and they often come back with similar findings. In each case – that is the way that scholars work – they tend to put a personal interpretative spin on these findings. But one often discovers that essentially the same tale is being told by travellers who set out with quite different intellectual goals[8].

These findings are consistent with a relatively straightforward, if sketchy, overall picture of what science is and does. In effect, a sociological dimension is introduced, not to replace the traditional philosophical dimension but to enlarge it. Ideas are seen as *cultural* elements as well as *cognitive* entities. *Individual* acts of observation and explanation are seen to gain their scientific meaning from *collective* processes of communication and public criticism[9]. The notion of a scientific 'method' is thus seen to extend outside the laboratory to a whole range of social practices. And so on.

This new picture of science is somewhat more complicated than the outmoded stereotype. It is not so sharply defined. It does not claim total competence. It treats human knowledge as a product of the natural world. It does not pretend to be impregnable against thorough-going scepticism or cynicism. It calls for more modesty and tolerance than scientists have customarily cultivated about themselves and their calling. But it does provide a stout intellectual and moral defence for science at the level of ordinary human affairs – the level at which nothing is absolute or eternal, but where we often forget that life is short, and feel passionately about pasts that we have not personally experienced, or plan conscientiously for the future welfare of people whom we shall never know.

1.3 A peculiar social institution

The most tangible aspect of science is that it is a *social institution*. It involves large numbers of specific people regularly performing specific actions which are consciously coordinated into larger schemes. Although research scientists often have a great deal of freedom in what they do and how they do it, their individual thoughts and actions only have *scientific* meaning in these larger schemes. Like many facts of life, this is so obvious that it was for long overlooked!

Science is one of a number of somewhat similar institutions, such as organized religion, law, the humanities and the fine arts. These institutions differ from one another in interesting ways. But what they all do – among many other things – is to produce quantities of *knowledge*. The peculiarity of science is that knowledge as such is deemed to be its principal product and purpose. This not only shapes its internal structure and its place in society. It also strongly colours the type of knowledge that it actually produces.

The sociological dimension is thus fundamental to our picture. But the self-styled 'sociologists of scientific knowledge'[10] have become attached to a principle of 'symmetry' as between different forms of knowledge, and are mainly attracted to the features that science shares with other forms of social life. They have therefore largely ignored the procedures, practices, social roles, etc. that actually distinguish science from other institutions. Attention to these distinctive features does not mean that science is sacred. Scientific life would not be human if it were not permeated with folly, incompetence, self-interest, moral myopia, bureaucracy, anarchy and so on. It is no longer news that even the most high-minded institutions are depressingly alike in some of their less admirable characteristics. But it is only when we have understood the *differences* that make scientific knowledge unusual that we can appreciate the *similarities* that make it ordinary.

This may seem a rather obvious point, but it needs firm emphasis[11]. Sociologists who deliberately 'bracket out' the distinctive institutional characteristics of science inevitably arrive at an extreme version of cultural relativism [8.13, 10.4]. This, in turn, generates a sceptical quagmire that blocks every path towards revision of the traditional Legend. They really have no reason to deny the plain evidence of our senses that science *does* have a number of unmistakable social features which should surely figure in our picture of it.

1.4 A body of knowledge

Science generates *knowledge*. The actual observations, data, concepts, diagrams, theories, etc., etc. that make up this knowledge often appear in tangible forms, as written texts, maps, computer files and so on. Some of it is also very well founded, and no more questionable in practice than the warmth of the sun or the solidity of the ground under one's feet. But there are many forms of knowledge[12], so what makes any particular form of it *scientific*? And if it *is* scientific, how firmly should we *believe* in it?

Until recently, the answers to such questions were considered entirely a matter for philosophy. Scientific knowledge was thought to be no more than a carefully edited version of aggregated reports of innumerable independent explorations of the natural world. What made these explorations 'scientific' was their particular subject matter, and the particular way in which they were carried out. The main project of philosophers of science was to define the general principles of *demarcation* between scientific and non-scientific knowledge[13]. They could then show – it was hoped – that knowledge fully satisfying these principles was – or would be – worthy of complete belief. In the heyday of the Legend it was even argued that the idea of 'non-scientific knowledge' was a contradiction in terms, as if there were no other reality than the world revealed by science.

The failure of this project[14] has not taken these questions off the meta-scientific agenda. Our picture of science is still heavily impregnated with *epistemology* – that is, the 'theory' of knowledge. What is now clear is that fundamental epistemological issues cannot be resolved by an appeal to abstract general principles. For example, as we have already noted, scientific activity involves *social* factors operating far outside the normal scope of philosophy[15].

Metascientists are also beginning to realize that it is not feasible to separate 'knowledge' from acts of 'knowing'. Scientific knowledge is not just a disembodied stream of data or the books on a library shelf. It is generated and received, regenerated or revised, communicated and interpreted, by human *minds*. Human mental capabilities are remarkable, but also limited. They are also closely adapted to the *cultures* in which they operate. Many of the characteristic features of science are shaped by the psychological machinery that scientists employ, individually and collectively, in their study of the world. In other words, *cognition* is the vital link between the social and epistemic dimensions of science.

The appearance of cognitive factors in our picture is a decisive break with the Legend. Philosophers of science have always steered clear of 'psychologism', for fear that it would rob science of its much-prized objectivity. Personal judgements of fact or meaning might well be required to make discoveries, but they were bound to introduce irrational elements which would have to be systematically excluded from the final analysis.

Fortunately, modern cognitive science is not completely clouded over with subjectivity. Human minds are all different, but they are built to the same general plan, and acquire common standards from the scientific culture of their research discipline. For many purposes they are just as alike as many artificial instruments of perception, calculation and communication. In practice, the *social stability* of scientific knowledge is a reasonable indicator of its objectivity.

1.5 Naturalism in the study of Nature

Our new picture of science thus draws on a very wide range of academic disciplines. Conventional philosophical questions about what is to be believed have to be combined with the sociological analysis of communities of believers. Perception, cognition and language all play their part. Even the humanistic concept of *empathy* – the capacity to enter into the thoughts and feelings of another person – has its place in the social and behavioural sciences [5.11]. A philosophy *of* science does not have to encompass all that might be required of a general philosophy *for* science[16], but it still involves many elements drawn from wider accounts of the human condition.

The involvement of so many disciplines does not merely complicate the picture. It also means that we are taking a *naturalistic* point of view[17]. By including 'scientific' concepts in our overall picture of science itself, we are assuming that it too is 'natural', in the sense of being susceptible to description and explanation by the same methods, and according to the same criteria, as other features of the natural world – including human society.

To be consistent with other forms of knowledge, epistemological naturalism has to be *evolutionary*[18]. Modern science is seen as the heir to an unbroken lineage of knowledge-acquiring organic forms, stretching back to the beginnings of life on earth [9.7, 10.3]. This is a useful unifying principle for what sometimes seems no more than 'a cluster of symbols, languages, orientations, institutions and practices, ways of seeing etc.'[19]. It recognizes that many of the peculiarities of science are historical survivals rather than current necessities, and accepts that the institution as a whole is bound to change over time.

Epistemological naturalism also emphasizes the *dynamism* of science. Even the knowledge it generates is continually changing. The noun 'science' is closely identified with the verb 'to research', indicating that it is an active *process*. At any given moment, this process involves the coordinated actions of many quasi-permanent entities, such as research scientists, research instruments, research institutions, research journals etc. By its very nature, science is a complex *system*[20]. It cannot be understood without an explanation of the way that its various elements *interact*.

A naturalistic 'picture' of a dynamic system is a *model* [6.10]. Although this word means no more than a simplified representation of a complex entity, and is often used very loosely to mean any abstract theory, it conveys intuitive notions of internal structures and mechanisms. In ordinary scientific usage, a theoretical model can be taken apart conceptually, and then put together again to make a working whole. Meaningful theoretical questions can then be asked about the functioning of the various parts, and the consequences of specific changes in their make up or interactions.

Such questions are not just theoretical; they arise continually out of practical issues in the real world – issues that are often in dispute between the defenders of science and its opponents. Has our understanding of nature been influenced by the gender-specific cognitive capabilities of predominantly male researchers? Does peer review quench scientific creativity? What scientific weight should be attached to a single, carefully recorded, but unconfirmed observation? To deal with questions of this kind we need more than a new 'theory' or 'picture'. This book is about the new *model* of science that is required to replace the stereotype of the Legend.

1.6 Keeping it simple

Naturalism, as such, is not enough to hold our model together. It merely affirms that scientific research is not essentially distinct from the many other ways in which we humans typically get to know about 'life, the universe, and everything'. In principle, we should be taking a holistic view that covers 'the whole picture'. In practice, each discipline may still look at only one particular aspect of this picture, and report what it sees in its own particular language.

The fact is, quite simply, that the barriers of comprehension between these languages are so high that a *transdisciplinary* viewpoint is required to transcend them. It is all too easy to be mentally trapped in a particular discipline, unable to cross the conceptual Divide into other modes of thought. This is a familiar situation in the natural sciences. What, for example, is a 'gene'? Is it a heritable trait, as seen by the geneticist? Is it a segment of DNA, as seen by the molecular biologist? Is it a protein factory, as seen by the biochemist? Is it a developmental switch, as seen by the embryologist? Is it even, perhaps, an active, utterly selfish being, as depicted by some evolutionists? It takes the general standpoint of the biologist to see these diverse concepts as different aspects of the same entity.

Similarly, if we were to begin our metascientific explorations deep in the realm of sociology, and insist, for example, that science has to be thought of primarily as a *heterogeneous actor network*[21], we would find it very difficult to accept that it is also, in some ways, a *sequence of refutable conjectures*[22], or a *bundle of research traditions*[23], or a *problem-solving, computational algorithm*[24]. Corresponding difficulties would arise if we were start from inside any other well-established discipline. Yet valuable insights have come from each of these specialized points of view. The trouble is that, although the academic languages in which they are expressed are not necessarily 'incommensurable'[25], they have evolved independently to answer very different types of question. A great deal of intellectual boundary work is needed to translate ideas directly from one such specialized language to another and to make them consistent and coherent.

It is much more profitable to *start* looking at science from a standpoint where it can be seen and depicted – however indistinctly – as a whole. The great merit of naturalism is that it automatically takes just such a standpoint. Educated citizens of economically advanced countries know that there is 'a thing called science'[26], and can say quite a lot about it. In response to detailed questioning, they might say that it is 'a body of knowledge', or 'an organised social activity', or 'a way of life for certain people', or 'a heavy burden on the taxpayer', or 'a power for good and/or evil'. Although these answers would be very diverse, and often contradictory, they would all be based upon a shared understanding of a simple truth – that these are indeed only different aspects of a *single entity*, of whose existence they are as sure as of death and taxes.

This 'natural ontological attitude'[27] is largely tacit. Its power resides in everyday usage. It implies, and is implied by, the way in which we ordinarily talk. A familiar word such as 'scientist', 'experiment', 'research', 'apparatus', 'scientific paper', etc. can carry a whole raft of formal meanings. It may well be defined quite differently in different contexts. But in each context it is understood to refer to the same discernible element, or feature of the world. Such words thus act as mental bridges re-uniting the various aspects or dimensions into which a natural entity may have been analysed. In saying, for example, that 'science involves experiments', we are not really bothered by the fact that the philosopher's concept of an experiment as, say, 'an attempt to refute a hypothesis', seems quite remote from the economist's characterization of it as 'a speculative investment whose ultimate rent may be difficult to appropriate'. We simply rely on our practical knowledge that these are just two different ways of looking closely at the same type of activity.

Many scholars abhor the fuzziness and ambiguity of the 'natural language' used by 'lay persons'. Their ideal would seem to be a precise, unambiguous, quite general *metascientific language* into which all the results obtained by different disciplines could be accurately translated, without fear or favour. But this is an unattainable goal [6.6], even for the representation of a much less complex entity than modern science.

Failing that, they prefer to concentrate on enlarging and perfecting their own particular domain, perhaps hoping that its specialized viewpoint could eventually be widened to take in the whole scene.

This is a perfectly reasonable preference. In science studies, as in other scholarly enterprises, progress is made largely by narrowly focussed research [8.3]. I am not in any way suggesting that this research is invalid or inappropriate. Nor am I insisting that its results are irrelevant unless they can be expressed in non-specialist terms. The true strength of a discipline often resides in a highly specialized framework of concepts [8.4] which can only be mastered by a lengthy effort.

But the purpose of this book is not to review in detail the work of scholars in these various specialties: it is to derive from their work a model of science that can be understood and accepted by a much wider public. This model has to be presented in terms that are equally widely shared. To start with, these terms must already be 'common knowledge' – that is, they cannot be much more sophisticated than the words and concepts that people ordinarily use when talking about science.

At first sight, this would seem to make everything unacceptably vague. Take, for example, the very word 'knowledge'. What does it mean to say that we 'know' something? Does it convey broadly that we are 'familiar with it', or 'have been informed of it', or does it imply well-founded conviction, if not complete and justifiable certainty? As any good dictionary will show, ordinary English usage covers a whole range of meanings, often in closed circles of reciprocal definition. But what would be gained here by trying to define 'knowledge' more precisely? Not only would it pre-empt the whole issue of scientific credibility; it would also rob us of a general word for a familiar human capability.

The incorporation of bodily experiences into mental traces is a primary feature of our very existence. By referring to it in ordinary lay terms we show that we understand that 'knowing' is a 'fact of life', and as such is a major functional module in our model of science. As a natural process it certainly demand systematic analysis. But the main argument would not be made more definite if we took this module apart and reduced it formally to more basic elements which were less familiar but really just as vague [10.8]. Admittedly, it doesn't have an adjectival form; but in using the philosopher's word 'epistemic' I simply mean 'relating to knowledge' in the same everyday sense.

Let me emphasize, however, that the new model is not already latent in 'folk discourse', just waiting to be developed like a photographic image. It is not even widely accepted in the world of science studies, although I believe that it is now beginning to take shape in the minds of certain other scholars there. To put this model together I have had to enter quite deeply into various specialized disciplines, select what seemed to be relevant conceptual material, translate this material roughly into lay terms, and trim the various bits and pieces into an untidy fit.

No earnest scholar in any of these disciplines will be satisfied by such a patchwork of heterogeneous concepts. I must surely have missed out, misunderstood, or misrepresented innumerable points that are deemed to be crucial in each particular domain. But this book does not have to be 'academically correct'. It is addressed to a more general audience, including a great many people who have had first-hand experience of scientific life and work. They will reject elaborate or esoteric interpretations of scientific activity that seem remote from this experience. Their acceptance is the real test both of the adequacy of the model and of the intelligibility of my account of it[28].

To sum up: I could not say what I wanted to say in this book *except* in the most direct and simple 'lay language'. This is not just because most people would not otherwise be able to understand it, nor because scientists would not otherwise recognize themselves in it. It is also because this is the only language in which 'science' stands for a many-sided natural entity, and in which there is a consistent terminology for describing all its aspects.

2

Basically, it's purely academic

2.1 Framing the indefinable

We encounter science as a *natural kind*[1], not as an abstract *category*. In other words, like a chair, or a tiger, or a city, we recognize it when we come across it, without having to refer to an explicit formula. Indeed, such a formula is not feasible. It would not only have to be elaborate enough to indicate that science has many different aspects – institutional, mental, material, and so on. It would also have to be broad enough to extend over many different instances of scientific activity, from classifying beetles to theorizing about black holes, from recording folk tales to mapping the human genome, from ancient Chinese medicine to modern Japanese pharmacology, from explaining earthquakes to failing to explain inflation.

A catalogue of all these aspects and instances would obviously be quite unmanageable. It would merely demonstrate that science is too diverse, too protean, to be captured in full by a *definition*. Moreover, any such definition would pre-empt the outcome of our enquiry. By telling us in advance what science *is*, it would effectively determine what we would later surely find. We may be very well informed about science, and have a very good idea of various features that are typical of it, but we must be careful not to insist that any of these features are invariable or definitive.

How can we make a model of something so indefinable? The first thing to do is to *frame* it[2]. In effect, we must enclose it in notional boundaries, and limit the analysis to a carefully chosen *exemplar* or *ideal type*. A model based on a detailed study of this exemplar may not depict correctly all the various instances of the entity under consideration, but because it represents an *actual* instance it can be made to 'work' in a self-consistent manner. This, after all, is just how natural historians have always dealt with the variability of individual organisms, even when they belong to a single species.

2.2 Narrowing the frame

The scope of 'science' is immense. In the time dimension, it includes instances stretching far back into history, or even prehistory. We may well suppose that the development of agriculture was facilitated by 'scientific' observations of the germination of seeds, and or that the legends governing metallurgical practices were akin to 'scientific' theories. Modern science evolved from the ideas and techniques of the great literate civilizations of Eurasia, whose achievements still amaze us.

But the intellectual and social distance between ancient Athens and modern Cambridge is too great to be bridged by a single model. This applies even to the seventeenth century 'natural philosophers', who invented so many of our present-day scientific usages[3]. Case studies drawn from such periods are instructive mainly by comparison with more recent models. The history of science inspires many valuable meta-scientific insights but our exemplar must be drawn from a narrower, more recent frame.

'Science' also extends far into everyday life. The farmer adding growth hormones to pigfeed is, in a sense, 'being scientific'. So is the police officer collecting blood samples, the sailor heeding a weather forecast, the rich man downsizing his firm in the name of supply-side economics, and the poor Third World woman half-starving her baby on artificial milk. But everybody would agree that there is much more to science than the use of very sophisticated techniques or hyper-rational argument[4] in daily life.

In common parlance, the word 'science' often includes medicine, engineering and other practical technologies. This is what is usually meant, for example, in statements deploring the deficiencies of 'public understanding of science'[5]. These activities are certainly closely associated with science, and permeated with scientific knowledge. Even in routine practice, they often use the same elaborate instruments, and rely upon the same sophisticated theories. Indeed, it is a notable feature of the past few decades that many of the boundaries between the natural sciences and their associated technologies have been dissolving before our eyes[6].

Yet medical and engineering practitioners firmly insist that they are professionally and intellectually distinct from scientists[7]. In other words, their model of a science differs in some fundamental way from their model of a technology. This is obviously a very important point of principle, with many practical consequences. It is essential, therefore, that we should not prejudge this issue by including any unmistakable technologies in the frame enclosing our exemplar of science. Although the notion of a monolithic, over-arching *technoscience* [2.8, 3.11] is politically potent[8], it is too coarse-grained for our present study.

2.3 Research as inquiry

Technology is science in *application*: science in action is *research*. This marks out a frame in the dimension of practice. We can zoom in on our exemplar through successive layers of terms such as 'investigation', 'exploration', 'analysis' and 'explanation'. The history of science can be presented as progressively detailed and systematic *inquiry*, normally directed towards increasingly sophisticated and powerful means for *solving problems* [4.4 – 4.11, 7.9, 8.1, 8.7, 10.3].

It is popularly supposed that science can be distinguished from other modes of systematic inquiry by a distinctive *method*. This is not what is observed. The *techniques* used in scientific research are extraordinarily diverse, from counting sheep and watching birds to detecting quasars and creating quarks. The epistemic *methodologies* of research are equally varied, from mental introspection to electronic computation, from quantitative measurement to speculative inference.

These diverse methods do not fall into an obvious pattern. The significance of such features as they share can be determined only by reference to a more general model. To fix on any one of them as exemplary would once again prejudge the issue. For this reason, the frame should not exclude, on principle, any of the many techniques and methodologies that scientists actually use. In practice, that means that our model will have to be capable of exhibiting the whole range of what counts nowadays as 'good science'.

Good science produces knowledge. But research is not just *discovery* [8.9]. It is conscious *action* to acquire a particular kind of knowledge for some particular purpose. Even in its most exploratory mode, scientific research is always carried out according to a conscious plan[9]. This plan may be very flexible. It may only last for a week, or a day, or half an hour. Or it may require a billion dollar instrument taking years to design, build and operate. Scientists, loving research, romanticize it as a form of play. In reality, science makes little progress by inspired improvisation or artistic doodling.

What, then, is the purpose of research? The notion that it can be defined simply as 'solving problems'[10] is enshrined in the funding of research in terms of *projects* [8.2][11]. Researchers are expected to present detailed proposals for particular investigations. To promise a specific *outcome* would be self-contradictory. But a project proposal would seem pointless without an indication of the question or questions that might be answered by what might be discovered. Hence, the presumed purpose of research is to solve problems that can be formulated in advance.

Indeed, the societal function of science has always been thought of primarily in terms of the practical human needs that it might serve[12]. The contributions of science to the health, wealth and welfare – and war-making capacity – of mankind are legendary. Nowadays, this function is operationalized. Projects are typical instruments of *science policy*[13]. They are handles by which governments, industrial firms, medical charities and other institutions endeavour to catch hold of science and bend it to their ends.

2.4 Science in the instrumental mode

The *instrumental* attitude to science [7.2] is summed up by the acronym 'R&D' – a hybrid of scientific *Research* and technological *Development*. This locates science at the 'upstream' end of a one-way process by which useful discoveries and inventions eventually flow down into the home, the shop, the hospital and the workplace. The *linear model of technological innovation* is obviously over-simplified, and calls for considerable elaboration and modification[14], but it underlies what most politicians, business people, civil servants and journalists say about science.

Socio-economic theory sanctifies science as a wealth-creating component of R&D or demonizes it as an active principle in the techno-scientific 'military-industrial complex'[15]. In either case, the supposed role of research is to produce, by any feasible means, whatever knowledge is required, or seems likely to be required, to satisfy an actual, or envisaged, material need. In effect, each field of science is treated as an optional facility bolted on to the front end of a practical technology to improve its inventive power.

Industrial R&D and other forms of *applied science* do indeed constitute the greater part of all modern scientific activity [4.10]. Some estimates of the proportion run as high as ninety per cent[16], but this is a notional figure. There is no generally agreed definition for scientific research as a line item in the financial accounts of government departments and commercial firms. In the laboratory, factory, hospital or field station, science is inseparably intermingled with technological development, design, demonstration and practice.

The *instrumental mode* dominates all other forms of science in its use of human and material resources, and in its direct impact on society. It governs the conception that many well-informed and influential people actually have of science. Surely this is the ideal type that our model should represent.

The trouble is that such a model would be very schematic and banal. In its details, it would be pulled apart by the instrumental demands of the diverse interests it has to serve [7.6]. Institutionally, for example, an R&D organization has to conform to the managerial practices of its 'owners', whether these are commercial or governmental, civil or military, private or public, entrepreneurially independent or corporately bureaucratic. The sociology of science is thus limited to the particularities of *R&D management*, with its concerns about project planning, budgeting efficiency, career ladders, administrative responsibilities, market information and so forth. These are, of course, important issues in management studies, but they are not peculiar to scientific work.

Again, from a philosophical point of view, issues of validity, reliability, objectivity and so on are reduced to a single question: 'Does it work?' But this *pragmatism* [7.2] has to be applied under very varied circumstances, and is not founded on a coherent set of general principles. What would the answers to this question have in common, as between, say, a novel surgical procedure and a new type of microchip, an experimental fusion facility and a genetic test for schizophrenia, an economic indicator and a remote-sensing satellite? In each case, there might well be a good answer, but it would be based on a particular body of scientific, technological and societal knowledge, much of which would be very debatable.

Much wise thinking has been devoted to such cases. But this wisdom is not coherent or self-contained, for it almost always involves personal, political and cultural *values*. For this reason, the metascientific spotlight has shifted to *ethical* issues[17]. The interests and motives of the 'owners' and sponsors of R&D are questioned. The formulation of 'needs' is scrutinized. The sensitivity of scientists to moral dilemmas is queried. Such issues are so diverse and disputable that science figures in them mainly as a 'black box'. It is treated as a powerful but mysterious machine whose use has to be carefully watched, and brought under social control. But the way in which this machine works is of little interest, provided that what it produces is morally acceptable to society.

This concentration on ethical issues is, of course, appropriate and desirable [7.3–7.10]. Social antipathy to science normally arises out of perceived threats to treasured values, and should be accepted or contested as such. But more general questions of validity also enter into such debates. Are established scientific theories perfectly reliable? What weight should be given to a highly unorthodox scientific opinion? Can the uncertainties of scientific prediction be quantified? Does 'more research' always produce better understanding? Could all that we ever needed to know be deduced from a single principle? Questions such as these are often at the centre of apparently practical disputes, and cannot be answered without an appeal to our deeper beliefs about the nature of science and of scientific knowledge.

What happens, unfortunately, is that one side falls back on the Legend – the romantic philosophical conception of science as a 'method' of guaranteed, unassailable competence[18]. The other side then claims support from populist sociological works that caricature and debunk normal scientific activity[19]. The result is a stalemate. The champions of each side talk past each other, as if in different worlds. In other words, a purely instrumental model of science is an empty shell. Its intellectual vigour, like its spiritual health, does not have sources within itself, and has to be sustained from elsewhere.

2.5 Basic research as a policy category

Scientific research is not, in fact, entirely instrumental. At least 10% of scientific activity is what is often called 'pure science' – or even just 'science' – to mark it off from applied science, technological development and other high-tech work. At first sight, this seems to solve our problem. 'Pure' science, surely, is the ideal type. Here, almost by definition, is the natural frame we have been seeking. Analysis of exemplars chosen from within this frame should provide a naturalistic model of science free of alien elements.

Unfortunately, this is a circular argument. It proposes no independent criterion of purity. Thus, anyone who *claimed* to be a 'pure scientist' would apparently be qualified to judge what should count as 'pure science'. To escape tautology, we need to be able to define our frame by some other means than the say-so of those who may have an interest in being inside or outside it. Further analysis of the whole concept is necessary.

Other terms, such as 'open-ended', 'curiosity-driven', 'blue skies', 'basic', 'foundational' or 'fundamental', are widely used to make essentially the same distinction. Celebratory rhetoric falls back on stock phrases, such as 'honest seekers after truth', who are 'pushing back the frontiers of knowledge', etc. These terms are not synonymous, and are used under somewhat different circumstances to emphasize somewhat different aspects of non-instrumental scientific activity. Between them, however, they indicate features around which a frame might be constructed.

For example, official documents customarily refer to one of the components of R&D as *basic research*. This usage, along with the associated notion of *science policy*, did not become common until the 1970s. But what does it mean? The standard 'Frascati' definition[20] is peculiarly negative:

> 'Basic research is experimental or theoretical work undertaken primarily to acquire new knowledge of the underlying foundation of phenomena and observable facts, *without any particular application in view*'.

Various attempts to elaborate on this definition all retain a version of the clause I have italicized. In more Basic English: 'Basic research is what you are doing when you don't know what you are doing it for!'

The desire to distance basic research from all instrumental purposes is understandable. But even if such a residual category is logically acceptable, it is very fuzzy. What, for example, are the practical applications that 'might be in view'? Are they restricted to the 'objectives' stated in a formal project proposal – 'The purpose of this research is to discover a more alluring cheese for baiting a bigger and better mousetrap . . .' and so on? Or might they be inferred from the nature of the problem which the research is designed to solve: – 'The problem of why cheese attracts mice to mousetraps has long puzzled rodentologists . . .' etc.? Should one discount a completely unrealistic objective, such as using quarks as a source of electrical power[21]?

The trouble with formal research objectives [8.2] is that nobody expects them to be met to the letter. They are addressed to particular interests. If you are approaching The Kindly Killer Kompany (KKK) for a contract, indicate that a patent is in the offing. For the Small Animal Protection Society (SAPS) you suggest that the research could lead to the development of an anti-mousetrap olfactory vaccine. The Fundamental Biology Research Council (FBRC) would expect you to show just how deeply the same project might be expected to advance our understanding of zoo-physio-micro-molecular-ethological phenomena. And so on.

The Russian dolls of patronage in modern science confound the issue. OK, so Dr X, an enthusiastic zoo-physio-micro-molecular-ethologist, has conceived this exciting FBRC project as a personal contribution to *basic* knowledge in his subdiscipline. But he is working under Professor Y, who heads an interdisciplinary group specifically devoted to *oriented basic* research on the behaviour of rodents in man-made environments. This group is partly financed by SAPS, whose Chairman, Lord Z, thinks of it as *applied* research, along the way to the development of a novel generic technology. Little do they know that most of the funding of the SAPS comes by a roundabout route from the KKK, who have already gone beyond the stage of *pre-market* research on the design of an even kindlier mousetrap based on just such a principle. How many of these boxes should we open, or close, in our hunt for a label?

In the end, the notion of pure science cannot be defined in policy terms. Policy [8.7] is all about future action. Policy talk is so steeped in practical intentionality that it cannot attach any precise meaning to a non-instrumental activity. Policymakers try to define basic research by exclusion, and then have to invent elastic concepts such as 'potential applicability' to bridge the gap that they have created[22].

Indeed, even the classification of activities on the applied side of this gap is thoroughly confused[23]. The term 'strategic research'[24], for example, is often used to justify rather general research that might later turn out to be of service in reaching certain specific practical goals. But it is seldom difficult to imagine an *eventual* application for almost any competent research in almost any field of biology, chemistry or the life sciences[25]. Interpreted in that spirit, the Frascati formula would limit basic research to remote disciplines such as pure mathematics, high energy physics and cosmology. Nobody supposes that 'pure' science ought to be framed so very narrowly.

2.6 Fundamental knowledge as an epistemic category

Pure science is often said to be concerned mainly with *fundamental* problems. The Frascati formula describes *basic* research as being undertaken 'primarily to acquire new knowledge of the *underlying foundation* of phenomena and observable facts'. Metaphorically speaking, scientific knowledge is likened to a many-storeyed building. It is supposed that the products of research can be arranged in a layered structure, where the 'deeper' layers provide support for the layers that are 'above' them.

This gravitational metaphor[26] expresses a familiar feature of scientific progress. Certain bodies of knowledge have been found to stand in a one-way relationship to other bodies of knowledge, which it is said they can *explain* or to which they can be *reduced*. This finding has been elevated into a general epistemic principle. A widespread belief in *reductionism* is typical of modern science.

The grounds for this belief will be explored in a later chapter [10.8]. We must be careful, therefore, not to prejudice the results of this exploration by tacitly taking them for granted. Indeed, if we were to do so, we would almost inevitably be drawn to the same implausible conclusion as in the previous section. If reductionism really rules, then the purest, most exemplary, forms of science would again seem to be elementary particle physics, cosmology and pure mathematics, since these are generally supposed to be the most 'fundamental' in this sense.

In reality, science is nowhere near any such grand reductionist goal. Nevertheless, fundamentalism is regarded as a desirable feature of research. General *theories* are favoured because they seem more fundamental than specific *facts*. *Invisible* entities, such as quarks, molecules and genes, are thought to be particularly fundamental because they operate behind the scenes[27]. Research focussing on *naturally occurring* objects, such as stars, rocks and organisms, which might reveal such hidden mechanisms, is considered to be more fundamental than the study of *artificial* systems, such as magnetically confined plasmas, anti-sense DNA or insurance companies, whose existence can be explained in terms of human agency.

More generally, scientists often describe a piece of research as fundamental when it is particularly esteemed, either for greatly improving human understanding of puzzling phenomena, or for opening the way into a totally unexplored field. They thus rate it as one of the highest *epistemic* qualities of good science. It is for their fundamental contributions to knowledge that great scientists are most admired. But this is a very exclusive criterion that only applies to a very small proportion of basic research. An *exemplar*, or *ideal* type, should not be confused with an *exemplary*, or *idolized* type. When people refer to scientists 'such as Einstein (or Darwin, or Pasteur)' they are not really saying much about scientists in general. One of the defects of the traditional Legend is that its model of science seems designed solely for the personal use of rare geniuses. It would completely sabotage our project to suppose that the practice of pure science is confined to this elite group.

More seriously, as philosophers now largely agree[28], 'being fundamental' is not an objective property of scientific activity [8.5]. Basic research cannot be recognized simply by inspection of its scientific objectives or results. At best, we can say that a piece of research is more or less fundamental *in relation to some other research*. Typically, this occurs when the behaviour of a complex entity can be explained reasonably well in terms of the properties of its components [10.1]. But these components may themselves have simpler elements. Thus, living cells are fundamental in relation to organisms, chemical molecules are fundamental in relation to cells, atoms are fundamental in relation to molecules – and so on. But neither cells nor molecules are absolutely fundamental in themselves.

We are thus dealing with a *relative* characteristic that depends entirely on the context. Biochemistry, for example, is fundamental in a biological context, but not as a branch of chemistry. In appropriate circumstances, almost any research project might turn out to be 'fundamental' relative to *some* body of knowledge. For example, apparently routine investigations of the metabolism of micro-organisms are now turning out to be fundamental to our understanding of climate change.

This context-dependence explains why the term 'basic' is used in policy language as the antithesis of 'applied'. Policymakers are primarily interested in R&D that is clearly directed towards practical applications. In a specific organizational context, the dominant mode of research is likely to be empirical rather than theoretical, in order to make contact with everyday realities. Research that explores the foundations of this activity will thus be linked only indirectly to its applications. In effect, the Frascati formula advises policymakers to treat this type of research as non-instrumental, despite its obvious strategic potential[29]. This a sound maxim in each context, but does not solve our general problem.

The notion that some forms of research are pure because they are *intrinsically* more fundamental than others is not merely elitist. It simply does not define a fixed frame. The traditional hierarchy of the sciences put the social sciences and humanities at the top, reducing them down through psychology and biology to chemistry, physics and mathematics. But now we know that mathematics cannot be reduced to logic without reference to human languages and other social institutions. The foundations of the whole edifice are not merely insecure: they are all up in the air.

2.7 Out of pure curiosity

Pure science usually appears in government R&D statistics as a residual category with a vague label such as *advancement of knowledge*. Research managers sometimes give a positive spin to its lack of practical purpose by referring to it as *open-ended*. Sociologists and economists point to – but seldom attempt to calculate – the *cultural* value of the 'useless' knowledge obtained by pure research[30]. In the past, scientists would have said that this was work for the glory of God and the benefit of mankind[31]. This is described nowadays as *blue-skies research*, which is short for 'the pursuit of knowledge for its own sake, wherever it may lead them, perhaps up in the air, out of touch with the solid earth of established theories, etc.' The inconsistency of this image with other equally apt metaphors, such as 'exploring foundations' and 'pushing back frontiers', shows just how difficult it is to define pure science in general terms.

The idea that it is *curiosity-driven* does provide pure research with a purpose that is unrelated to any particular application. It also frames it in a new dimension. We often talk about science being powered by 'human curiosity' as if this were a collective social force[32], but we are well aware that it is essentially a variable *psychological* trait of individuals. In effect, this idea suggests that the purity of scientific research is determined by neither its purpose nor its product, but by the *personality* of the researcher. It draws attention to the researcher as a *person*, rather than as a cog in a social machine.

Curiosity is, indeed, one the most notable qualities of many (but not all) outstanding scientists. Take Einstein, again, or Darwin, or Pasteur, or Marie Curie, or Dorothy Hodgkin, or . . . It can be said of almost any seriously famous scientist that he or she had an inquiring mind, and was alert to, and fascinated by, strange ideas or events. This is almost a truism. Since research is a mode of inquiry, then a strongly motivated and effective researcher is necessarily 'inquisitive'.

Once again, we are in danger of taking just one component of scientific excellence as a framing principle. Very few of the thousands of research workers who engage usefully in pure science are scientific *virtuosos*, brimming over with insatiable curiosity. But many of them have other equally valuable personal qualities, such as intelligence or persistence or imaginative insight. Scientific research is much more than the enlightened exercise of personal curiosity. 'Science uses curiosity, it needs curiosity, but curiosity did not make science'[33]. Elaborate intellectual and institutional frameworks are required to harness this individual trait to the collective production of reliable knowledge.

By definition, curiosity-driven research is not intended to be applicable. But its results are often of great practical value. Why, then, is it usually presented as a particularly pure form of basic research – the very antithesis of technological R&D? Surely the world of human artefacts and cultural institutions is full of unexplained phenomena and mysterious patterns. More curiosity may actually be required to make a practical invention than a theoretical discovery. One reason might be that *scientific* curiosity is supposedly focussed principally on the *natural* world. It is aroused by regularities, or by deviations from such regularities, amongst naturally occurring objects, and hence seeks peculiarly 'fundamental' explanations and interpretations [2.6, 8.5].

A more obvious reason is that curiosity is a peculiarly *individualistic* virtue. It goes with the romantic stereotype of the pure scientist as a brilliant nonconformist[34], a 'lonely seeker after truth'. Ideally, the pure scientist is an *amateur*, in the true sense of the word. She plays the research game obsessively, with no other concern than the hope of making a contribution to knowledge. Her purity is moral. It is associated with commitment to a transcendental goal[35] and indifference to such worldly considerations as the possibility of winning a valuable prize or making a commercial profit.

This is how scientists like to see themselves. A sense of personal commitment is still a major element in scientific work[36]. Scientists nowadays can scarcely claim that they are really gifted amateurs. They know very well that they are typical social actors performing a typical professional role. But they are the sort of people who want to be in charge of what they do. They celebrate curiosity because it implies *autonomy*. It can only be exercised by someone who is free to look around them, reflect on what seems strange, and inquire further into it. In other words, by describing pure research as 'curiosity-driven' – even as 'unfettered'[37] – they proclaim that it ought to be undertaken by researchers who formulate their own research problems [8.1, 8.7] and apply their own criteria to what counts as good science.

Basic research cannot be differentiated from other forms of research solely in terms of its psychology[38]. Nor can complete personal autonomy be its sole guiding principle. The image of the pure researcher as a perfectly isolated individual animated by an inner vision is a fantasy. At best, it is a generalized and simplified version of the stories that researchers customarily tell when trying to explain their actions. Such stories are often quite sincere. But they usually systematically ignore the social setting in which these actions are performed. This is the theme to which we now turn.

2.8 Academic science as a culture

Pure, non-instrumental science turns out to be an elusive concept. For policy purposes, it is only a residual category. As a body of knowledge, it is not especially fundamental. It is not always the product of personal curiosity. And yet we have no difficulty in talking about it, and describing some of its typical features. We have a distinct image of an ideal type, even though we cannot target its essence.

This is where naturalism comes to our aid. Pure science, like science in general, is a recognizable natural kind. Instead of trying to define it in the abstract, let us point to it as an existing entity. What we actually have in mind when we use this term is, of course, an extremely familiar and distinctive activity – *academic* science[39]. Pure research is framed by its social setting. It is the type of science that is carried out in universities. The stereotype of the pure scientist is the professor, engaged in both the pursuit of knowledge and its onward transmission.

In effect, academic science is a *culture*[40]. It is a complex way of life that has evolved in 'a group of people with shared traditions[41], which are transmitted and reinforced by members of the group'[42]. That is how it would be seen by a visiting anthropologist from Mars, and that is how it should be represented in our model.

This is well understood by official policymakers. What they sometimes call the *Science Base*[43] includes a mass of research projects, widely dispersed throughout *academia*. When asked to give an account of their support for basic science, they do not laboriously inspect their records, project by project, trying to decide whether it satisfies the Frascati criterion. They simply lump together all the money paid out for 'government funded R&D undertaken within university-level establishments', as if all the scientific activity in these institutions were of the same general type.

In addition – and this is significant – they include under the same heading 'government-funded research in other closely linked, or similar organisations'[44]. In other words, scientific policymakers recognize that this is a distinctive activity which is not confined to educational institutions. Indeed, in some countries, the bulk of academic science is located in separate *institutes* run by a *National Academy, National Centre, Research Council*, or some such bureaucratic organization, with no direct teaching responsibilities. Pockets of the same culture are also sometimes found inside large governmental and industrial R&D *laboratories*. The most notable of these, Bell Labs, was famous both for its Nobel Prizes and for its university-like attitude to research.

Like any other cultural form, academic science has a *history* of development and change. Many of its characteristic features can be traced back to the seventeenth-century Scientific Revolution, or even earlier. It emerged in essentially its modern form in Western Europe in the first half of the nineteenth century[45]. Since then it has evolved into a coherent and elaborate social activity, increasingly integrated into society at large. Indeed, science has grown and spread around the world as a characteristic *subculture* of the general culture of *modernity*. Technoscience [2.2] is an essential component of economic development. Although academic science requires a very sophisticated social environment, it is cultivated arduously in tiny plots even in the poorest and least developed countries.

2.9 Many disciplines in one science

Academic science is widely dispersed, geographically and institutionally, and does not have any system of overall control. Nevertheless, it is remarkably open and uniform in its practices and principles. A research scientist can move from university to university, or from a university to a research institute, or even from country to country[46], without serious cultural hindrance. On the other hand, mobility between fields of research is severely restricted[47], even within the same organization. One of the main features of academic science is that it is sharply differentiated and structured in terms of *disciplines* [8.4]. A professor of physics in Bristol may have more in common with a physicist in Jakarta than with the professor of chemistry in the next building.

It is worth remarking, finally, that a modern university is expected to be *multidisciplinary*. For educational reasons, it customarily covers as wide a range of disciplines as possible – from Classical Greek Literature, say, to Computational Cosmology. Bundled together arbitrarily into Schools and Faculties, academics of diverse disciplines are driven through parallel career hoops, by the same standardized performance indicators. In spite of vast differences of subject matter their research cultures are stereotyped and homogenized. They are expected to carry out original research, publish their findings in books and articles, be aware of all that is going on and become international authorities in recognized fields of knowledge, supervise the research work of graduate students and post-doctoral assistants, subject their work and career aspirations to the anonymous assessments and public critiques of their scholarly peers, serve as committee persons, editors, referees etc. in their learned societies, seek funds to support their research and the research of their colleagues, attend innumerable conferences, congresses, seminars, workshops and other meetings, give and listen to innumerable indigestible chunks of academic discourse, ask and receive polite but pointed questions concerning their own and other research claims in their specialty – and so on.

The institutional structures that anchor and rule these activities do differ considerably from country to country. British, French and American academics, for example, are indeed heirs to very different intellectual traditions, very different organizational arrangements and very different styles of work[48]. Insiders can even distinguish between apparently identical universities, such as Oxford and Cambridge[49]. But these are only variations on a theme. Such differences are insignificant when compared with life outside the ivory tower, in whatever country.

From this perspective, the differences between the faculties are also insignificant. As academic persons, the professors of physics and chemistry have almost as much in common with the professors of theology and accountancy as they do with each other. The same ideal types, the same social norms and epistemic principles, are deemed to motivate research activity, whether in the natural sciences, the social sciences or the humanities, in management studies or in clinical medicine, in ethnology or in engineering.

In other words, academic science is not restricted to the natural sciences. It includes many distinctive intellectual traditions and disciplines. In some traditions, 'scholarship'[50] – the enlightened re-formulation of existing knowledge – is preferred to 'research' – the generation of new knowledge. Some disciplines are so dominated by their educational and vocational responsibilities that they do not draw a line between original research and exemplary practice. This diversity is a feature of academic science which is sometimes overlooked by policymakers – for example, in the standardized criteria by which they try to measure research performance in quite different subjects.

The grouping of disciplines into Faculties or Schools is a necessary organizational device. But these groups do not separate neatly into just two cultures [7.10]. The Continental European usage is justified: the word 'science' should always be interpreted to cover the whole range of organized knowledge. There are, of course, many different 'sciences' – physical, biological, behavioural, social, human, medical, engineering and so on – but they are all varieties of the same cultural species. This is what I mean when I talk about academic science in the rest of this book.

3

Academic science

3.1 The republic of learning

Academic science is the stereotype of science in its purest form. When people talk about scientific research (as distinct from technology) they primarily have in mind the sort of scientific work that is done in universities. They think of it as the characteristic activity of members of a particular social group in a particular social frame.

Scientists themselves insist that they belong to a *community*, indicating that they recognize each other as people who share many values, traditions and goals[1]. But this community is essentially notional. The word is used to mean 'all those people who subscribe to certain general principles of rationality and objectivity, and have such high standards of expertise and mutual trust that they can be relied upon to work together for the benefit of humanity in the attainment of truth'[2]. On the one hand, it proclaims the unity of this group within society at large. On the other hand, it asserts that its members are individuals who are linked together voluntarily by their common attitude to learning and research.

The concept of a scientific community is part of the traditional philosophical Legend. At the same time, however, it encases science in a sociological 'black box', whose internal structure is deemed to be irrelevant to the pursuit of knowledge. Indeed, the power of the Legend lingers on, even amongst the champions of a 'sociology of scientific knowledge'[3]. In effect, they treat the wider institutional frame as the *product* of the processes at work inside the research laboratory, discounting its *influence* on those processes[4].

It is true that the scientific culture fosters rationality [6.8] and relies heavily on trust [5.6]. But the rhetoric of cooperation and fraternity has to be squared with the reality [9.1] that science is also notoriously competitive and disputatious[5]. Scientific biographies are deeply scarred by private episodes of vitriolic personal jealousy between individuals. The public history of science is a chronicle of bitter intellectual controversies between strongly partisan groups. Every research laboratory is a miniature arena of individual opportunism and social conflict[6]. The frontiers of knowledge are partitioned into a thousand specialized territories.

How are these disruptive psychological and social forces kept in check? It is impossible to believe that a community of passionate individualists would hold together spontaneously solely on the basis of a common philosophical attitude. But the first priority of science studies is not to uncover all the disreputable negotiations, opportunism, rationalizations, tinkering, appeals to authority, power ploys, etc. that actually negate the Legend[7]. Nor is it to unmask the class, corporate, governmental or religious interests [7.4–7.10] that supposedly manipulate their scientific puppets[8]. It is to understand the social process by which 'organised knowledge'[9] emerges out of this disorder[10].

In reality, science is more than a cultural community. Scientific behaviour is 'regulated by well-established, easily recognized and relatively stable norms, values and laws'. Sociologically speaking, therefore, academic science as a whole is an *institution*[11]. It does not have a written constitution, a legal identity, a chief executive officer or a corporate plan. In essence, it is a social order that relies enormously on established relationships of personal and institutional trust[12]. Nevertheless, it holds together[13] and operates as an 'implementation structure'[14], working towards the common goals of its members.

It has sometimes been argued that this 'commonwealth of science', or 'republic of learning'[15], should be taken up as a general model for a participatory political democracy[16]. This greatly exaggerates the possibility of governing a country as if it were an 'ideal speech community'[17]. The historical evidence points in the opposite direction. The modern institutional form of science reflects and harmonizes with the democratic (or perhaps oligarchic) pluralism of the society in which it typically arose and is largely embedded[18].

It would overload our inquiry to pursue it back into the past. But we know that science has changed remarkably in the course of a few centuries. Although we are analysing it as we find it today, our study must be framed in time. Unfortunately, this boundary cannot be attached to an obvious historical discontinuity. Science has evolved so smoothly, along a rising curve of size and influence[19], that it is difficult to fix on the moment when it might be said to have taken on its present form.

The historians of science have long been interested in the sources of its *epistemic* 'norms, values and laws'. There is an enormous literature showing how these can be traced back to philosophers such as Francis Bacon, René Descartes and John Locke, or linked with more diffuse religious, governmental, commercial, legal or humanistic ideas that were already prevalent in late seventeenth-century Europe. For this reason, the birth of 'modern' science is often assumed to coincide with the revolution of ideas signalled by, say, the publication of Isaac Newton's *Principia* in 1687.

In recent years, however, attention has been given to the origins of crucial *social* practices, such as the systematic publication of research findings[20], criteria for membership of scientific communities[21], the public accreditation of experimental results[22], academic employment and preferment on the basis of research reputation[23], the foundation of specialist scientific societies[24], the award of highly prestigious prizes for scientific discoveries[25], and so on. Although this literature is still not comprehensive, it does not support the view that modern science emerged fully formed at the Foundation Meeting of the Royal Society on the 28th of November 1660. Many of the structural features of science that we now take for granted, such as professionalism, subject specialization and peer review of publications, only became important much later.

In dealing with this dynamic yet deep-rooted culture, we must look back far enough to include a number of long-lived traditions that remain surprisingly alive, and yet not give too much weight to practices that are now quite archaic. A convenient cut-off date would seem to be somewhere in the first half of the nineteenth century. This was the period when a self-consciously scientific culture – applying as much to history, theology, linguistics and the other humanities as to the natural sciences – emerged in the state universities of pre-unified Germany. This culture diffused, by direct imitation or by convergent co-evolution, throughout Continental Europe and the English-speaking world. In other words, in talking about 'academic science' we are referring to a distinctive social institution that has existed in advanced countries in something like its present form for rather more than a century.

The choice of this time frame is essentially arbitrary. It does not imply that the scientific culture came into existence at the beginning of this period, that it is uniform from country to country and from subject to subject, or that it has remained static ever since. On the contrary, science is a cultural form that has always evolved and diversified as it adapts to technical progress and societal change. What we can say, however, is that academic science, as practised in the second half of the twentieth century, is a highly successful, well-founded institution, solidly based on a long record of high achievement. Its social 'norms, values and laws' have slowly developed in close partnership with its philosophical principles[26]. This stable combination of practices and principles is what people think of when they refer to 'science'. And yet, as we shall see, this robust institution, this sturdy culture, is now entering a period of rapid and profound change.

3.2 Elements of the scientific ethos

How then does science hold together as a social institution? Unlike most other professions, it does not have written regulations or a formal system of governance. But newcomers to research soon discover that they are not just learning technical skills. They are entering a self-perpetuating 'tribe', where their behaviour is governed by many unspoken rules[27]. These rules vary in detail from discipline to discipline[28], from country to country, and from decade to decade. But the sub-tribes of academia span a common culture. In 1942, Robert Merton suggested that the 'prescriptions, proscriptions, preferences and permissions' that scientists feel bound to follow could be summed up in a small number of more general *norms*[29].

These norms are usually presented as traditions rather than moral principles. They are not codified and are not enforced by specific sanctions. They are transmitted in the form of precepts and examples, and are eventually incorporated as an *ethos* into the 'scientific conscience' of individual scientists. People are not born with a 'scientific attitude'[30]. The personal detachment and public humility of the Nobel Laureate have been painfully learnt; even then, they do not always repress the self-interest and vanity that drove him originally along the road to the Prize.

Indeed, norms only affirm ideals; they do not describe realities. They function precisely to resist contrary impulses. The moral order in any social system always involves a tense balance between its norms and the corresponding *counter-norms*[31]. It would make no sense to insist, for example, on 'universalism', if 'particularism' were not a tempting alternative. As we shall see, the norms of science are hard to live up to, personally and communally. By assuming general virtue, they open opportunities for the unscrupulous[32] [9.4]. They also conflict with many 'tribal' values, such as group cohesion and loyalty[33]. Scientists such as

Galileo and Sakharov who dared to 'speak truth to power' were seen at the time as traitors rather than heroes.

The significance of the Mertonian norms is much disputed. They were originally conceived as structural elements in a theoretical model of the scientific culture. Nowadays they are often regarded as no more than useful words for moralizing about actions and ideals in scientific life[34]. In principle, however, they provide each member of the scientific community with a stable social environment. So long as everybody keeps to the rules, then their responses to events and to one another's actions are reasonably predictable. A community of otherwise independent individuals is thus enabled to organize itself spontaneously [9.8] into a well-structured institution[35].

Nevertheless, it would be wrong to regard these norms as applying to scientists in every aspect of their lives. 'Being a scientist', and 'doing research' is a social *role*, just like 'being a lawyer' and 'defending a client'. The norms only come into play in situations where this role is being performed[36]. What counts as 'scientific' is what is published, or stated formally in public, or communicated as an expert opinion. What a scientist says informally, in private, about, say, her colleagues and their work, is not subject to the same constraints[37].

Critics of science make a lot of mileage out of the manifest discrepancies between the private and public actions of scientists. They also fasten on examples of scientific behaviour that obviously deviate from the norms – fraud, plagiarism, partisan disputes over priority, and so on. These are serious matters for concern, but they are not so widespread and prevalent that they completely corrupt the whole enterprise. Indeed, the fact that such episodes are still generally regarded as both deviant and scandalous is a tribute to the continuing moral authority of the ethos that they flout. This ethos is easy to debunk as a collective 'false consciousness' that conceals from scientists the true significance of their activities[38], but it sustains their morale in what can be a very discouraging calling[39]. Any reader who believes otherwise should not bother with the rest of this book.

For our present purposes, however, the main question about the Mertonian scheme is not its normative force. It is whether it provides a row of pegs on which to hang a naturalistic account of some of the social and psychological features of academic science. These features can, of course, be analysed in a variety of ways[40]. They are by no means independent of the philosophical framework by which they are usually justified, and vary substantially from science to science[41].

The particular virtue of the Mertonian norms is that they emphasize practices and principles that impact directly on individuals and that genuinely distinguish science from other institutions and callings [10.5]. It really does seem to be both true and important that scientists do not normally behave like spies – who are trained to keep their actions secret – nor like the members of a religious sect – who must accept the authority of their guru – nor like shopkeepers – who earn their living by advertising their wares – nor like antique dealers – whose best goods are all old and second-hand – nor like soldiers – whose duty is not to reason why. The culture of science is both characterized and structured by just such differences.

In the next few sections, we look at this culture as a system of particular social practices, temporarily dissociated from the type of knowledge that they generate. These are grouped according to the various norms from which they derive. What we shall find is that academic science cannot be defined completely in terms of these norms. It also has features derived from certain other structural principles, such as career motivation and subject specialization. We shall then be in a position to look at the way in which this whole culture is now changing.

In later chapters we shall re-establish the connection between the sociology of science and its philosophy. This will take us back into the core of our model, where science as a particular mode of knowledge production shapes, and is shaped by, scientific knowledge as a particular mode of human understanding.

3.3 Communalism

Let us go through the Mertonian norms one by one. The norm of *communalism* [ch.5] requires that the fruits of academic science should be regarded as 'public knowledge'[42]. It thus covers the multitude of practices involved in the communication of research results to other scientists, to students, and to society at large. By exchanges of truthful depictions, science creates its context of serious thought[43].

The institutional scale and complexity of these practices is enormous. It involves learned journals and books, conferences and seminars, academic and commercial publishers, libraries and electronic databases. A significant proportion – at least several per cent – of the funding of science goes into supporting its communication system. The precept 'publish or perish' is not a joke. Natural scientists spend much time and scrupulous effort into 'writing up' their research results[44]. A great deal of scientific skill is employed, often without pay, in editing research journals and books. Indeed, in the social sciences and humanities many researchers work almost entirely inside the communication system, reading the texts of other scholars and writing texts for other scholars to read.

The results of original research are customarily published in the *primary literature* of science. This consists mainly of 'papers' – that is, articles in quarterly, monthly or weekly periodicals, chapters in books, or even whole books. A scientific paper is the contribution of a named individual (or group of individuals) to the communal stock of knowledge. It reports the outcome of an investigation, typically in the form of a claim to have made an observational, experimental or theoretical discovery. The evidence for this claim is presented in the form of a verbal text, backed up by numerical data, graphs, diagrams, photographs, mathematical formulae etc. Ideally, any competent reader is given sufficient information to reproduce the investigation and arrive at the same conclusion.

But the notional *archive* to which this information is communicated [9.3] is absolutely enormous. In principle, this is open to anyone who can get to an academic library and make sense of the specialized technical languages in which most of it is written. In practice, most research findings would be completely lost without a very detailed system of classification and retrieval. Scientific progress does not only depend on an elaborate machinery of library catalogues, bibliographic indexes, abstract services and other databases that can be searched for particular items of information. It also depends on an extensive *secondary* literature of review journals, monographs, handbooks, and other works where this information is collected, selected, analysed and recombined into more general pieces of knowledge[45]. It is no accident, moreover, that academic science is closely associated with higher education, where students get from textbooks a suitably simplified – often sanitized – version of what is now supposedly known to science[46].

The *formal* communication system of science is only one of the many channels through which research findings flow out of the laboratories or libraries where they are produced. It is paralleled by *informal* processes that extend far out into society[47]. Scientists chatter incessantly and compulsively amongst themselves – in the laboratory, in the hallways, around the lunch table and everywhere else they meet[48]. They also spend a great deal of time giving lectures and listening to one another at seminars, conferences and other meetings. Some of them try to improve 'public understanding of science' by writing popular books, talking on the radio and appearing on television. Scientific information is also transferred in many forms, in both directions, across the interfaces with medicine, engineering, law, management, public policy and other practical activities.

The essence of the norm of communalism is, of course, its prohibition of *secrecy*. Indeed, a security-conscious official once defined basic research by its 'publishability' – having previously ruled that research that was not basic might not be published! Quite apart from such circular nonsense, there is a convention that 'unpublished research data' and 'private communications' from other scientists carry little weight, and should be cited very sparingly. *All* the evidence needed to sustain the results of the investigation should be put before the reader.

The norm requires, moreover, that these results should be made public as quickly and comprehensively as is humanly possible. Under normal circumstances, the time that this takes is not begrudged. But the intimation of a sensational discovery can put the whole ethos under pressure. Should researchers keep quiet about their findings until they feel that they have been fully confirmed[49]? Can journal editors and referees be trusted not to exploit what they have been told in confidence? How much can be said publicly in the interval between the submission of a paper and its appearance in a journal? Academic science has developed an elaborate code of courtesies, conventions and rules – mostly unwritten and often inconsistent – to deal with such eventualities.

What is really at stake, however, is seldom secrecy as such [5.13]: it is the distinction between a *formal* and an *informal* scientific communication[50]. A scientific paper is sent in for publication as a 'contribution' to knowledge. Its eventual appearance in a learned journal (or in a book from a reputable academic publisher) registers its 'acceptance' by the scientific community. It is thus transformed into an item of *communal* knowledge, as the norm requires.

The scientific archive is not, therefore, just a repository of everything that has been written on scientific matters. Although its margins are often hotly disputed, it only includes material that has been filtered through a process of formal communication. This is how scientists proclaim *socially* the distinctiveness of their particular type of knowledge.

The norm of communalism is thus the charter for a social institution that is central to academic science[51]. This institution is not only governed by various other norms, it is also a major influence on the way that scientists deal with one another, gain their livelihood and plan their careers. For this reason, the scientific culture is very sensitive to attempts to bypass its traditional channels of communication[52]. That is why, for example, there are such strong objections to the disclosure of an important discovery in a press release before it has been accepted for publication in a scientific journal.

3.4 Universalism

Inclusion in the archive is socially advantageous, both for items of knowledge and for the scientists who produce them. These are the research claims, these are the researchers, that other scientists must take seriously. Had Einstein's first papers been published privately as pamphlets, instead of appearing in a regular scientific journal, no self-respecting physicist would have taken the least notice of them – or of their author. Conversely, the claim to have observed 'cold fusion' gained a hearing in the scientific world because it came from a professor with a long list of papers to his name, rather than from an unknown backyard inventor.

The criteria for inclusion are thus of great importance. They do not, of course, guarantee that the archive is all 'good science' in any absolute sense, or even that all the 'best' science is to be found there. But they dominate and shape academic science, as a body of knowledge and as a mode of knowledge production.

Supporters of the Legend insist that these criteria are purely rational and objective. Although individual scientists may be far from perfect in these qualities, their limitations are eventually cancelled out by collecting a diversity of opinions and rubbing them against one another until the truth emerges. But science is a turbulent enterprise [9.1], where ideas and people compete ruthlessly for attention. This operation has to be carried out by a loosely structured mechanism which cannot be cleansed of social and personal influences. The criteria for inclusion in the archive have to take explicit account of these influences if they are to counter them effectively.

The norm of *universalism* [ch.6] establishes this clearly. It requires that contributions to science should not be excluded because of race, nationality, religion, social status or other irrelevant criteria[53]. Observe that this norm applies to *persons*, not to *ideas*. It does not relate to the intellectual substance of the communication, but to the social context out of which it has come.

Thus, *gender* bias is firmly forbidden [7.5]. Papers submitted by woman scientists should be assessed by exactly the same criteria as those submitted by men, and should never be rejected on that account. Note, however, that this – like the Ten Commandments – is essentially a negative norm. By vetoing social obstacles to the rise of potential scientific merit it strongly favours the principle of a career open to the talents[54], as much for women as for men. But it says nothing about affirmative action to correct the results of past injustices. For example, academic science has no mechanism for countering ideological prejudices against feminist theories or research programmes in a predominantly male scientific community[55].

It is a commonplace that the scientific ethos comes under strain when the larger culture opposes universalism. Academic science – as distinct from purely utilitarian technoscience – cannot truly thrive in an authoritarian monoculture such as Nazism, Communism or a fundamentalist religion [10.5]. But it may also be incompatible with a condition of complete cultural pluralism. In the multicultural ideal, justice has to be done to each and every one of the diverse social factors that are systematically excluded from science by the universalist norm. This is seldom a serious problem for the natural sciences, but it raises a number of questions about the general applicability of the social and behavioural sciences across a world of many cultures.

In practice, this norm is effective mainly within the scientific community. Scientists are very conscious of their personal standing, whether or not it is acknowledged by an official title. Nevertheless, in the eyes of science, they are all equal. A full professor cannot pull rank on an assistant professor in a scientific debate. Their communications to the archive must satisfy exactly the same criteria. A dissenting voice cannot be suppressed because it comes from a researcher who has not yet obtained a PhD. Delicate courtesies are required in voluntary collaborations between researchers, and even more so in the unequal relationship between students and their supervisors[56].

However elitist and self-serving it may be to outsiders, the scientific community is enjoined to be democratic and fair to its own members[57]. Such evidence as can be found indicates that it comes nearer to this ideal than might be expected[58]. This is not to deny the many horror stories about research supervisors appropriating the ideas of their students[59], editors refusing to publish critiques of their pet theories, plum jobs for the protegés of powerful professors, the closure of 'schools of thought' and scholarly disciplines against external criticism, and so on. The point is that such events are regarded as scandalous. They are reported as deviations from the established conventions of scientific life. Obviously, people are strongly drawn towards various forms of behaviour that flout this peculiarly 'anti-tribal' norm[60]; that does not mean that the corresponding 'counter-norms'[61] are of similar moral authority.

Let me point out, again, that scientists are not to be expected to observe the norm of universalism in all their doings. In private life, they may be male chauvinist pigs. As citizens, they may be rabid nationalists. As spiritual beings, they may belong to exclusive religious sects. As academic personages – even as scientific notables – they may take pleasure in exercising arbitrary powers. But as members of the scientific community, they must repress these partialities and adopt a universalist stance. In effect, this aspect of their lives is enclosed by a frame[62], within which they are called on to perform a variety of stereotyped roles. Each of these roles – as author, editor, referee, conference speaker, chairperson, research supervisor, etc. – is governed by well-established conventions. If they are to retain any credibility as scientists, then this is how they must behave.

3.5 Disinterestedness, humility

The notion that academic scientists have to be *humble*[63] and *disinterested* [ch.7] seems to contradict all our impressions of the research world. Scientists are usually passionate advocates of their own ideas, and often extremely vain. Once again, this is a norm that really only applies to the way that they present themselves and their work in formal scientific settings. In particular, it governs the style and approach of formal scientific communications, written or spoken[64].

The neutral, impersonal, passive voice of the typical scientific paper is clearly an affectation. It is adopted to conceal the personal and social factors that may have originally motivated the research and might influence it towards a particular outcome. The author presents himself as a mere name, a disembodied instrument of factual observation or logical inference, morally detached from the events or arguments reported.

For this reason, any reminder of the actual social background of the work is kept to a minimum. The address of the author may indicate that he is an employee of a firm with known commercial interests in this field of the research; nevertheless the text will carefully avoid any implication that the investigation might be connected with those interests. Indeed, acknowledged support from a highly 'interested' source – for example, from a tobacco company in research on some aspect of lung cancer –

might make a paper unacceptable as a contribution to pure science. The scientific ethos assumes that academic research is undertaken principally by 'academics', whose livelihood does not depend directly on the material outcomes of their activities.

This detachment from the lifeworld [10.2] – the world of everyday humanity – is enhanced by a stance of *humility*. By systematically citing formal scientific sources for everything that is not entirely their own work, researchers limit immodest claims to personal originality [ch.8]. In talking about their achievements, they are expected to minimize them by comparison with those of 'the giants on whose shoulders they stand'. They thus locate themselves in an abstract world of purely scientific entities [6.11], where ordinary human interests have no place. In effect, by conforming to this norm they give up the tangible benefits of membership of a real scientific community in favour of a permanent place in the shadowy tribe of 'scientific authors'. There is even a prejudice against scientists who write successful popular books about science[65], for fear that they might use worldly acclaim to boost their standing in this tribal Hall of Fame.

The remarkable fact about the norm of disinterestedness is that it is so well observed. Scientists *do* write in this curiously impersonal way, and *do* pepper their texts with innumerable notes gratefully acknowledging their indebtedness to the work of other scientists. It is true that thesis advisers, journal editors, referees, book reviewers and other authorities police the literature and censor communications that do not conform to its conventions [3.7, 9.1, 9.5]. But apprentice researchers soon pick up these conventions, and adopt them for themselves. The sociology of the scientific community thus merges into the psychology of the individual scientist.

Needless to say, this is another example of role play. The full picture is not so flattering. In private, scientists do not repress their personal interest in getting their work accepted. This is revealed in numerous accounts of their opinions and behaviour – for example in the rivalries and controversies that often flare up in the race to make and be credited with an outstanding discovery[66]. The public expression of such feelings obviously offends against the communal norms, but tends to be treated as a discourtesy rather than a crime.

But the most malign form of egoism – dishonesty – is treated much more severely. In the long run, it is kept in check by the norms of communalism and scepticism. All scientists know that their research claims will be subject to public criticism and must be based on findings that other scientists already accept or could replicate. This very strict condition on their formal behaviour spreads into all their informal dealings as scientists. They live and work in a social environment that relies overwhelmingly on personal trust [5.7]. Although research papers are written in a barbarously impersonal style [7.1], they are never anonymous. Every published research claim puts the author's reputation for probity on the line[67].

Credibility [8.10] is so much the stock in trade of every academic scientist that even petty insincerity on scientific matters is socially unacceptable and earns black marks amongst colleagues[68]. Indeed, this convention is so deeply entrenched in the scientific culture that there is no regular communal machinery for enforcing it. Academic scientists and their institutions thus have great difficulty in dealing with cases of outright deception concerning the substance or authorship of research findings[69].

3.6 Originality

Originality [ch. 8] energizes the scientific enterprise. Although academic scientists are not always inspired by curiosity [2.7], they are trained to be self-reliant and independent, and live in a highly individualistic culture[70], where they are expected to be 'self-winding' in their choice of research questions[71], and to come up with previously unpublished answers. Their most cherished traditions celebrate and sustain this aspect of the scientific culture. This is the norm that keeps academic science progressive, and open to novelty.

Originality is required at two levels. At a mundane level, every scientific paper must contribute *something* new to the archive. It must suggest a new scientific problem, propose a new type of investigation, present new data, argue for a new theory, or offer a new explanation. Ideally, it should combine several such virtues.

Referees are primed to advise editors to reject papers that lack any of these features. At the very least, this process should establish that the work is authentic – that it is not a slavish copy of another item in the archive, whether by the same or a different author. More rigorous scrutiny should eliminate purely imitative research, or material that has obviously been recycled from previously published texts. Although copyright disputes are rare in academia, *plagiarism* is as infamous as fabrication in a scientific paper.

Paradoxically, insistence on originality entrenches orthodoxy. Conscientious referees often reject papers because their authors 'seem unaware of' (i.e. have failed to cite) relevant earlier work. To avoid such rebuffs, scientists have to be fully acquainted with all prior research in their field. From their doctoral studies onwards, they have to work hard at 'keeping up with' their subject. In addition to exchanging informal information about current developments, they must continually survey a rapidly growing and changing body of scientific literature. In practice, this usually means that they each restrict their research interests to relatively conventional problems in a limited domain – i.e. to 'normal' science within an established 'paradigm'[72].

To be fair, this norm imposes a strain on the typical professional scientist. High 'creativity' – to invoke that undefinable term – is a very rare gift, even amongst well-esteemed scientists. To make the most of what one has, it is prudent to focus narrowly on a restricted range of problems, to become an expert on all that is already known about them, and to use 'state of the art' techniques to tackle them. One cannot hope to be another Einstein: one might at least make a modest contribution to knowledge by joining the hundred or so theoretical physicists around the world who are still exploring some of the approximate solutions to a simplified version of his equations. In other words, the notoriously hyperfine *specialization* of academic science[73] is a rational personal response to the social imperative to 'be original'.

The daily routine of research is thus so adapted to the norm of originality that it is largely taken for granted. At the level of notable achievement, however – for example, where prestigious prizes and other public distinctions are at stake – originality is such a paramount consideration that it has to be *proved*. Many of the traditions and practices of academic science are designed to establish who was first with a noteworthy discovery. This has become one of the major functions of the scientific archives [9.3]. In principle, *priority* is decided by the date – even the time of day – on which a discovery claim was formally communicated to a reputable journal. In practice, however, real life is much more untidy than is allowed for by this convention. The issue may eventually depend on much more contentious considerations, such as previous informal transfers of information, past collaboration on relevant research themes, the balance of effort between co-workers, and so on[74].

In spite of the well-known fact that scientific discoveries are often made quite independently at nearly the same time[75], priority disputes between scientists are notoriously bitter and confused. Although often kept private[76], when they erupt in public they bring to the surface the psychological tensions inherent in the norm of disinterestedness and seem to flout the ethos of the collective pursuit of truth[77]. But such conflicts do not prove that scientists are peculiarly jealous and contentious as *individuals*. They indicate, rather, that science is a powerful and intensely normative *institution*, where people rightly feel that proper behaviour and outstanding achievement should be fairly recognized and rewarded[78].

3.7 Scepticism

If originality is the motor of scientific progress, *scepticism* [ch.9] is its brake. This norm triggers many academic practices, such as public debate and peer review. Scientists are unlike other people in that they are continually being presented with novel observations and theories that are relevant to their research. In making use of the work of others, they cannot depend entirely on their own unaided judgement. No one scientist has the expert knowledge required to decide which items can be relied on and which are suspect. Systematic procedures for the scrutiny of research claims are an essential feature of science.

Original research is difficult. It is not easy to get really convincing results. Scepticism towards novel findings is usually well justified. But that is not a licence for downright, comprehensive philosophical doubt. Indeed, this norm was originally designated *organized* scepticism[79], indicating that it stands in firm opposition to radical sociological relativism [10.5] and other forms of epistemic nihilism. A policy of 'informed criticism' might be a more accurate description.

In practice, this norm only operates systematically at the entry points to the archives. Reputable scientific journals only publish papers that have been subjected to *peer review* – that is, critical scrutiny by scientific experts. When a manuscript is submitted, *referees* specializing in the same research field are asked to report on it to the editor. Peer review of contributions to the primary research literature is the principal social mechanism for quality control in academic science[80].

In effect, a referee acts as a representative of the scientific community. This involves a temporary reversal of social roles, from making the case for a research claim to trying to pick holes in one. Referees may even be inclined to be too zealous, precisely because they are dealing with work in which they have an interest, carried out by friends and colleagues to whom they must not show favour[81]. Numerous anecdotes emphasize the delicacy of the relationships between authors, editors and referees. In a single week, an established academic scientist may have had to occupy each of the corners of this agonistic triangle. Peer review may thus put strains on the psyche, but is not such a socially divisive process as it might seem, since it is customarily confidential and anonymous. On the contrary, the fact that most mature scientists have experienced its traumas from both sides helps to consolidate the scientific culture and its collective authority[82].

A referee's report is nominally personal, but it is really an individual interpretation of the communal criteria for 'good science'. Originality is only one of these criteria. As well as detecting errors, omissions, *non sequiturs* and other technical defects, referees are expected to assess a paper in terms of the overall credibility and scientific significance of its findings. And that, as recruits to research soon learn, is highly disputed, and very much a matter of opinion.

Scientists are thus held accountable to their community, rather than to their superiors or for themselves[83]. Peer review keeps the official scientific literature reasonably honest and factually reliable. It favours precise, thorough and cogent argumentation and sets high benchmarks for technical performance. But it does not pretend to eliminate error, nor does it guarantee certainty or truth. On the contrary, it is often the occasion of fierce disputes that illustrate graphically the uncertainties, arbitrary assumptions and half-truths of scientific knowledge. In the *social* system of academic science, this is where the *intellectual* action is – and *vice versa*.

By comparison with peer review, the other sceptical practices of the scientific community are very unsystematic. This is not to minimize their critical function in the production of knowledge. Most scientists are continually engaged in private and public debate about their work. New ideas are usually presented informally and sharply questioned in seminars and conferences, before they are published. Learned journals often provide facilities for contentious correspondence about published work. In the social sciences and humanities, most academic books are reviewed publicly, very thoroughly, and at length, by expert authorities. The primary literature is indexed, summarized, raked over, reformulated as 'progress', searched for anomalies, transformed into lists of 'current problems' [8.1] and otherwise critically surveyed in review journals, symposia volumes and other secondary publications.

Generally speaking, although often driven by strong feelings, critical comments are phrased courteously and dispassionately. A coded language is used to express even such strong opinions as utter disbelief or suspicions of deceit. Disputes conducted in reputable journals are expected to conform to the traditions of formal communication. Consider, for example, Peter Duesberg's controversial opinion that AIDS is not caused by HIV. An overwhelming majority of the experts on this subject regarded these views as wrong-headed and completely disproved by a vast body of contrary evidence. Nevertheless, Duesberg and his small band of supporters were given journal space for their 'heresies', provided that these were presented impersonally, in a form that referees and editors could accept as contributing in some small manner to scientific understanding.

The fact is that there is no way of purging the formal archive of errors. However decisively the substance of a published paper has been confuted, it remains on record as a putative piece of knowledge. Scientists who have been proved wrong rarely recant in public; when they change their minds they usually shift to the now dominant view and present it as if they had believed it all along[84].

Science has no formal machinery for officially closing a controversy[85]. An academic debate may continue for as long as anyone cares to keep it going. But the scientific community often exercises the final sanctions of the sceptic: disregard. Observations and theories that are considered unsound are dismissed briefly in review articles, not cited in the textbooks, and eventually forgotten. The scientific archive is very largely a garbage heap of discarded notions, whose archaeological exploration[86] is the responsibility of historians and philosophers, not of working scientists.

3.8 CUDOS institutionalized

There is much more to the practice of academic science than an individual pursuit guided by a general ethos. Even the loneliest 'seeker after truth' must eventually interact with other people, if only as informed critics or supporters. Cooperation is as much the name of the game as competition. A scientist must not only know what it means to do 'good science': she must know where she stands amongst her scientific 'peers' and how to work with them. Academic science could not function, even as a loosely knit community, without a mechanism for signing up members, and acknowledging the part they play in the common cause.

This mechanism is 'CUDOS'. The initial letters of the Mertonian norms spell out the reward that academic scientists get for communicating their research results to the communal archive[87]. In brief, they make *contributions* to knowledge in the expectation of receiving *recognition* by the community. The tradition of exchanging scientific information for social esteem is more like ethnographic *potlatch* than a market transaction[88]. Nevertheless, this exchange serves the individual interests of scientists so well that they are readily induced to respect the norms of the group[89].

The CUDOS mechanism works because the rewards are graded to match the quality of the work that they recognize. Routine scientific papers are acknowledged principally by being cited occasionally in the subsequent literature. A list of such papers in a personal *curriculum vitae* should then be enough for a foothold on the academic ladder. A longer CV, including several more distinctive contributions, would be needed for a more senior post, such as associate or full professor. Yet higher standing might be signalled by a sideways move to a more distinguished university, perhaps as the outcome of a semi-public bidding game between rival institutions. Really famous scientists can join the jet set, commuting periodically across the world to perform their scholarly duties at two or more universities.

Learned societies and academies recognize notable scientific achievements by the award of medals and prizes. Office in such a society carries prestige as well as responsibilities – not every British Nobel Laureate becomes President of the Royal Society. A governmental appointment, as an expert advisor, member of a research council, chairman of a commission of enquiry, etc., can signify high scholarly reputation[90], capped by an official public honour such as a knighthood.

Indeed, even the most unworldly scholar, publicly disavowing any thought of reward for his modest contributions to knowledge[91], secretly treasures the invitation to give a keynote address or take the chair at a plenary session of a major international congress. The recompense for scientific achievement can even extend beyond the grave. The most enduring and prestigious reward is recognition by *eponymy* – *Darwinism*, *Einstein's* equation, *pasteurization*, *Creutzfeldt-Jacob* disease, *Listeria*, *Diesel* fuel, *ampères*, and so on[92].

Some forms of recognition are obviously valued as doors to well-paid employment or social influence. But they also have a deeper psychological meaning. Academic research is basically individualistic. Scientists are presumed to work very much on their own. They need personal reassurance that they have successfully lived up to the norm of originality, the most exacting requirement of the scientific life[93]. The first experience of having one's work cited favourably can feel like getting a prize for a discovery, and spur one on to further effort. On the other hand, an immensely prestigious award, such as a Nobel Prize, may actually reduce the subsequent scientific productivity of its recipient[94].

It is sometimes argued that the CUDOS system offends against the norm of universalism by generating undue social stratification within the scientific community[95]. Quasi-economic cycles[96] of 'credibility' or 'symbolic credit' [8.10] – i.e. trading on one's name for access to the resources required to do more research, and hence win more recognition – are not inherently stable. The system is very vulnerable to the 'Matthew Effect', where honours and resources accumulate around those who already have them[97]. Such a position, once achieved, is unassailable. In fact, members of the upper strata of the scientific community often retain their authority long after their scientific capabilities have declined.

Nevertheless, the 'republic of learning' could not function without clear, unambiguous criteria of personal merit based on genuine achievement[98]. That is why the tokens of esteem come best from the scientific community, usually by procedures equivalent to peer review, where the participation of competitors guarantees legitimacy and sincerity[99]. Although often scorned as a mere ritual, the validation of research findings by formal communication to the archive [9.3] also plays an essential part in these procedures[100].

3.9 Specialization

Science does not really operate as a single community of tens of thousands of individual scientists. Not only is it stratified in terms of scholarly authority; it is also subdivided into *subject specialties* [8.3 – 8.4].

Academia is carved up into *disciplines*, each of which is a recognized domain of organized teaching and research. These cut right across institutional boundaries. A major scholarly discipline such as economics has members in almost every university in the world. They are normally assembled locally into academic departments or institutes, but this is not essential. Indeed, the continuous creation of *interdisciplinary* units for research and teaching is vital for the emergence of new disciplines[101] and other modes of scholarly progress [8.8].

It is practically impossible to be an academic scientist without first locating oneself in an established discipline. Academic disciplines are surprisingly real, even though they are often differentiated very arbitrarily[102]. This might be expected in the humanities and behavioural sciences where disciplines that started from very different positions have found it difficult to agree on how they should be connected. It would be quite an intellectual exercise, for example, to assign research projects correctly as between sociology, social anthropology and social psychology. But few recognized disciplines are really 'compact' in all their defining characteristics[103]. Even in the physical sciences, there is no general agreement on the boundaries between, say, applied mathematics, theoretical physics and theoretical chemistry. These were established by historical accident and vary somewhat from university to university and from country to country.

Nevertheless, scholars often put a good deal of effort into 'boundary work'[104], especially when it comes to fending off a threat to the place of their discipline in the notional hierarchy of academic esteem[105]. A major academic discipline such as physics is much more than a conglomerate of university departments, learned societies, and scientific journals. Even though it is only loosely organized, it is a well-defined *institution*, delineated by a variety of intangible but effective social practices. In effect, each discipline is a distinct 'tribe', living out its particular version of the general scientific culture[106]. This is where academic scientists acquire the various theoretical paradigms, codes of practice and technical methods that are considered 'good science' amongst their peers[107].

An established discipline provides each of its members with a career base, a social identity, a public stage on which to perform as a researcher or teacher. Membership is not just a matter of occasional, personal preference: it is a privilege earned by laborious apprenticeship[108] and maintained by life-long commitment[109]. The organization and management of universities and other formal academic institutions is largely determined by the way that they are subdivided into faculties and departments along disciplinary boundaries.

Disciplines, faculties and departments, however, are more significant for *teaching* than for research. The subdivision of disciplines into small 'problem areas' seems to be an unavoidable feature of academic science[110]. The number of distinct 'fields' of science and scholarship runs into the thousands, all carefully differentiated in the minds of their devotees. At the core of each of these *specialties* is an 'invisible college'[111], whose members share a particular research tradition.

Problem areas do not, of course, last for ever, but they are not just temporary research sites, to be entered casually as occasion demands. Nor are they just safe refuges for timid, unimaginative souls. Many a distinguished scientist retires after spending a whole career making a string of contributions to a particular specialty. The papers listed in her CV were all 'good science', yet they could all be found on the same library shelf, classified together under a subject heading covering only a few per cent of the archival literature of a major discipline.

Extreme specialization is the typical reaction of academic scientists to the tension between the norms of originality and scepticism [3.6]. It is feasible to follow – even aspire to lead – the detailed fortunes of a world-wide 'invisible college' of a few hundred members, reading their papers before or after publication, and meeting them as colleagues and friends for a few days each year[112]. There are obvious psychological dangers in putting one's personal eggs into such a small basket[113] – especially when this basket is quite likely to be overturned by progress in other fields. But for many highly committed researchers this is as far as they want to go in their desire to overcome their isolation and acknowledge the cooperative aspect of their activities[114]. Perversely, they may even distance themselves from the other specialties and disciplines whose knowledge and expertise they know they could not themselves master, but on which they really depend for their own achievements[115].

Needless to say, scientists and science are often accused of tunnel vision. This charge is often well justified [8.1, 8.4, 9.3]. But narrowness of focus is not just a deplorable consequence of a misconceived philosophy; it is closely connected with the social phenomenon of subject specialization in the scientific community. This, again, is not favoured by the scientific ethos. On the contrary, extreme specialization is socially divisive, and thus conflicts directly with the norms of communalism and universalism. Nevertheless, it is deeply entrenched in the scientific culture, and strongly reinforced from generation to generation by the conditions under which scientists are trained for research[116].

In an economic metaphor, subject specialization represents a natural division of labour by the hidden hand of competition amongst free-wheeling individuals [8.3, 9.1]. In pursuit of personal esteem, researchers spread out over the whole domain of knowledge, each seeking a fruitful niche for their trade in credibility and research resources. But this process is amplified and stabilized by the norm of originality. Sturdy individualism in the cultivation of specialized areas of knowledge is not merely typical of academic science: it is one of its fundamental structural principles.

3.10 Avocation

One question about academic science remains unanswered: how are academic scientists supposed to make a living? Nothing that is written about the academic ethos ever asks this question[117]. The norm of disinterestedness even ignores the possibility that members of the scientific community might be *employed* to do research, and might therefore be inclined to favour interests connected with such employment [7.8].

Until, say, the middle of the nineteenth century, almost all scientists were *amateurs*. Although they were not all gentlefolk[118], very few actually earned their livelihood by doing scientific work. Even those with university appointments were not expected to do original research. Science was largely the *avocation* – the subsidiary calling – of people with other means of support.

In a sense, academic science continues this tradition. As the name suggests, this 'other means of support' is customarily a permanent post as a university teacher. The peculiar feature of academic science is that it developed as an activity engaged in principally by 'academics', whose official employment is to teach, rather than to do research. The convention is that this research is 'their own work'[119]. They are free to undertake, publish and benefit from it entirely as individuals, even though it was actually carried out in a university laboratory, in the course of their working days as university employees – perhaps even by students or assistants rather than with their own fair hands.

Everybody knows, of course, that university teachers usually owe their posts to proven research competence, and earn further promotion by subsequent research achievements. In a sense, an academic scientist is an entrepreneur who makes a living by accumulating scholarly recognition and investing it in an *academic appointment*. That appointment may or may not carry teaching or managerial responsibilities. It may not even be in a university at all, but in a quasi-academic research institute[120]. Its essential feature is that it provides a more or less sufficient amount of personal time, on a more or less adequate personal income, to exercise a personal commitment to the pursuit of knowledge.

Academic appointments are as fundamental to the research system as landed estates are to the feudal system. The name tag of a scientist at a conference states her 'affiliation' – that is, the institution where she holds an appointment – just as an aristocratic title names a place or region. Indeed, the feudal term *tenure* is used for an appointment that is effectively permanent to the official age of retirement. 'Getting tenure' is one of the defining moments in an academic career[121]. Scientists struggle with nature and with one another to achieve it. But tenure is also vital to the academic ethos, for it empowers disinterestedness and enables originality. Good scientists need its protection whilst they are still highly productive, not just as a token of previous excellence.

On the other hand, tenure nurtures inactivity, especially if granted too young. Most academic institutions therefore offer a hierarchy of successively more prestigious, better-paid, tenured posts, to recognize and motivate continued research effort. In accordance with the norm of universalism [3.4], entry into and movement up the academic hierarchy is based on research performance, as assessed by peer review. An elaborate social apparatus of graduate education, doctoral studies[122], post-doctoral fellowships, etc. has evolved to apprentice, select and recruit the most promising young people into the scientific community.

Every country, every research organization, has its own ways of appointing and promoting academic scientists and scholars. There are striking procedural differences between, say, France[123] and the United States[124] in what would seem to the outsider to be precisely similar tasks. Needless to say, there are also frequent deviations from the official procedures – sometimes so scandalous and convoluted that they are reported more naturalistically by novelists than by biographers or social scientists! But these deviations are silhouetted against a puritanically meritocratic background. The strength of academic science depends on its being 'a career open to the talents', and less prone to jobbery than most other callings[125].

The existence of academic science as a distinctive cultural form thus involves a very peculiar social arrangement[126]. Universities and other institutions provide selected individuals with personal time and other resources for an activity from which they do not, as institutions, directly profit. All that the institution seems to get in return is a share in the scholarly reputation of its more notable employees.

This is not necessarily a one-sided transaction. Universities – even nations – compete passionately for scholarly esteem. Oxford and Harvard, Copenhagen and Calcutta, Tokyo and Sydney, are world-class universities because their leading professors are world-class scientists.

France and Germany are great scientific nations because the members of their state-funded research institutes make great scientific discoveries. Indeed, academic science was initially shaped by the competitive bidding for famous scholars between the universities of rival German principalities in the early nineteenth century. International competition for Nobel Prizes continues this very human tradition[127].

These rivalries attest to the fundamental credo of academic science – that the pursuit of knowledge is of value in itself. This is often linked with belief in the ultimate utility of the knowledge thus won. But these are articles of faith, and do not depend on a demonstration that the quantifiable benefits that academic science brings to society far outweigh their out-of-pocket costs[128]. In the end, institutional support for the advancement of knowledge is seen as a moral duty. It is an ethical conundrum whether such a virtuous action is done less 'for its own sake' than for the good name it brings.

In other words academic science relies on *patronage*. It is surprising that the scientific culture – the engine of modernity – is based upon such an old-fashioned social practice[129]. One could say, of course, that this simply transfers the CUDOS system to a higher plane. Like individual scientists, institutions and nations exchange contributions to knowledge for communal recognition. But it is very questionable whether such a symbolic social interaction can be interpreted so simply as an economic transaction.

Academic science obviously owes a great deal to *private* patronage. Over the centuries, wealthy individuals have made liberal donations to science and continue to do so on a large scale. Every university has laboratories, libraries, institutes, buildings, endowed professorships, telescopes or other research facilities named after quite amazingly generous benefactors. In some cases, completely new institutes have been created, to further research into particular problems or particular fields. This is a very significant feature of academic science, even though it does not really provide more than a fraction of the total resources, including academic salaries, that institutions actually devote to research.

Nowadays, however, most of these other funds actually come from the state. *Public* patronage of academic science is not really new. In many countries, universities have always been state institutions: since their professors are civil servants, their research time and facilities are paid for out of the national budget. There is an old tradition of official national academies and research organizations staffed by full-time researchers funded by the state. Scientists have often approached their governments for grants of money for specific research projects[130]. Since the Second World War, almost all governments have increased the flow of funds through these various channels to such a level that it surpasses all other sources.

Academic science is jealous of its autonomy, and yet it is maintained completely by public and private patronage. How does it stay so independent? How is it possible for such large funds to flow into the scientific community without violating its norms and crushing its elaborate conventions? What is the unwritten condition in the implied contract between science and the state[131] that keeps them at arm's length[132]?

The unwritten condition is actually one of the key principles of academic science. It is that all patronage, private or public, is channelled through *communal* filters. Benefactions are customarily put into the hands of academic committees for detailed allocation [4.9]. National academies and research organizations are governed by councils of scientific notables. Grants are awarded, research programmes are developed and researchers appointed on the basis of peer review[133].

This principle serves, of course, to decouple the research from the personal interests of the benefactor, and give it a communal seal of objectivity. But it also resolves the dilemma that haunts every patron: once I have handed over my money, how can I trust the recipient to use it in a way that I approve? This is particularly difficult when the possible uses are incomprehensibly technical. The prudent patron of high science inevitably turns to representatives of the scientific community for trustworthy advice on such matters[134]. In spite of its individualism, science makes itself accountable as a collective enterprise to its patrons.

3.11 Science in society

Taken together, these practices combine into a powerful social mechanism. Academic scientists are under no illusions about its effectiveness in shaping their research and framing their careers[135]. And yet it does not conflict with a thoroughly *individualistic* ethos. In the heyday of academic science, an established research scientist in a tenured academic appointment could feel remarkably free to behave like the legendary 'lonely seeker after truth'. This freedom even extends to *not* playing this role, if inspiration lapses or fades with age.

Personal autonomy within the scientific community is linked with the autonomy of the scientific community in the world at large. An elaborate and extensive social activity such as as modern science could not really be floating in the air, like the island of Laputa, unattached to its life-world. But academic science can reasonably claim to be an independent social institution, in the sense of being free to manage its own internal affairs without interference by external human authorities[136].

This autonomy is customary, rather than contractual. Universities earn their keep by preparing young people for various callings. When we say that society has always supported pure research, we mean that universities have been left free to devote their surplus resources to the scholarly interests of their professional employees[137]. Academia is a self-contained world involved in highly specialized and esoteric pursuits. It protects itself from interference by stressing the difficulty of these pursuits, and their distance from everyday affairs[138]. What would be the point of tangling with a faculty of professors of mediaeval palaeography, or computational cosmology, on their home grounds? It is only when science impinges on practice [ch.7], that the principle of academic freedom has to be invoked, and then mainly as part of the general human right to 'freedom of expression'[139].

Most academic scientists actually value this separation between their research work and their 'ordinary' lives. They are trained into it, until it becomes part of their self-identity[140]. For many scholars it is personally comforting to have a specialized domain of order and rationality into which they can retreat from the chaotic and emotional world of everyday life. Some people thrive under these circumstances; others pay heavily for them in later life[141].

Nevertheless the 'ivory tower' mentality is not just a regrettable side-effect of academic specialization. In many cases it can be justified as a necessary condition for disinterestedness. But it is so deeply ingrained that it has to be considered an integral part of the academic ethos. In effect, it is a defensive ideology. To proclaim its independence from the rest of society, the scientific community deliberately surrounds itself with a protective moat. Scientists learn to talk about science as if it were a complete and self-sufficient way of life, as well as a complete and self-sufficient world view [10.5, 10.8]. When they refer to 'the real world', or 'the external world', they are as likely to mean social realities outside the walls of the research institute as material realities outside the scope of scientific theory.

Needless to say, 'science' is not really disconnected from 'society'[142].

The main bridge across the moat is *educational*. Their teaching responsibilities bring scholars and their institutions into direct contact with professional practice. Teachers of law, medicine, engineering, agriculture and many other disciplines must have had personal experience as practitioners, and often plan their research around practical problems. Even in fields as abstract as mathematical economics and theoretical chemistry, research for a PhD frequently leads to a job in government, commerce or industry.

The moat around academic science is also crossed by an *epistemic* bridge. This bridge emerges from the ivory tower through the archive [9.3]. The linear model of technological innovation [2.4] is an essential part of the Legend. In this model, academic science is conveniently located 'upstream' from applied science, so that purely scientific knowledge can 'trickle down' into technological research and development, where it is eventually put to practical use. Indeed, the norm of communalism [3.3, ch.5] requires academic scientists to facilitate this process by publishing their research results promptly, fully and freely.

The academic ethos makes much of the *outward* flow of trained people and potentially useful knowledge across these bridges. It does not mention possible *inputs* to science from society. In principle, the traffic is all one-way. The right to academic freedom is not counterpoised by an ethic of social responsibility[143]. The systematic utilization of research discoveries is not complemented by a counterflow of ideas for utilitarian research. The campaign for better 'public understanding of science'[144] is not balanced by an effort to achieve better scientific understanding of the public.

That is the official ideology. In reality, academic science has always had many two-way connections with society. It cannot dissociate itself from the diversity of institutions and enterprises with which it shares the title 'science'. Academic science is an integral part of technoscience [2.2], socially as well as epistemically[145]. Industrial R&D and academic science are culturally very different[146], but they use many of the same ideas and techniques. Thriving colonies of academic science are to be found inside many R&D organizations. Most industrial researchers have been trained in academia, which cannot quite ignore their career intentions. Indeed, it has always been feasible for quite senior scientists to move back and forth across this divide, especially in fields such as chemistry and economics where the distinction between basic and applied research has never been very sharp.

In the end, a naturalistic account of academic science would have to include this much larger complex of institutions and ways of life. In other words, it would become yet another book about 'the place of science in society'. Our purpose here is more limited. We are trying to pin down the familiar notion of 'basic science' – that is, science simply as a source of knowledge. My argument is that this notion is little more than an idealized image of 'academic science', a social institution whose features are taken to be characteristic of science in general.

The Mertonian norms are particularly useful because they stress the sociological features that academic scientists consider to be peculiar to their profession. They take for granted most of what science has in common with other cultural activities and institutions[147]. Their virtue is that they draw attention to and justify the social practices and conventions that make it distinctive. Indeed, the claim to be distinctive as a mode of knowledge production [10.6] is one of the most distinctive features of science as a social institution.

4

New modes of knowledge production

4.1 The academic mode

Science is a mode of *knowledge production*. Its social norms are inseparable from its epistemic norms – what philosophers call its *regulative
principles*[1]. Scientists' ideas about what should count as 'the truth' cannot
be disentangled from the ways they work together in pursuing it. The
philosophy of academic science is part and parcel of its culture.

The regulative principles of academic science are thus important components of its ethos. They are actually so familiar to most scientists, and
are stated so often in different forms, that it is not easy to produce a standard list. The simplest way of describing them is to say that they involve
such concepts as theory, conjecture, experiment, observation, discovery,
objectivity, inference, etc., which we shall be analysing in detail in later
chapters.

The significant point here is that although these are usually taken to
be independent philosophical concepts they can be directly linked with
sociological aspects of the academic ethos[2]. For example, the norm of
'communalism' is closely connected with the principle of *empiricism* – that
is, reliance on the results of replicable observation and experiment.
Again, social 'universalism' is related to *explanatory unification*; 'disinterestedness' is normally associated with belief in an *objective reality*; insistence on 'originality' motivates *conjectures* and *discoveries*; 'organized
scepticism' requires that these be fully *tested* and *justified* before being
accepted as established knowledge. And so on.

This correlation with the Mertonian norms accords with our naturalistic approach. Observation suggests that academic science can be represented by a 'model' in which these regulatory principles play a major role.
Like their sociological counterparts, they set ideal standards which are
generally accepted, although seldom achieved. As before, we interpret

them descriptively. They lay out a conceptual framework, summing up in a few standard terms an immense variety of epistemic practices, procedures and traditions. Words such as theory, experiment, observation, discovery etc. are essential if we want to explain what academic scientists actually do, how these activities are related to one another, and how particular meanings are attached to their outcomes.

That does not necessarily mean that science is governed by some immutable philosophical doctrine which justifies it absolutely, regardless of the human setting. As we have remarked, there is no general agreement on how the regulatory principles of academic science should be defined. They are essentially *precepts* rather than precise rules. They tell us, for example, that it is highly desirable that theories should be tested experimentally. But this is not an absolute requirement: historical disciplines such as palaeontology are not excluded from science just because they do not involve experimental research.

The regulatory principles of science change with time. It is still not clear, for example, how much weight should be attached to a computer simulation as a test of a theoretical hypothesis. They also co-evolve with the social practices to which they are linked. The philosophical status of 'experiment', for example, was greatly enhanced in the late seventeenth century by the systematic procedures developed by Robert Boyle and other Fellows of the Royal Society to ensure that what they observed was attested by credible witnesses[3].

It can be objected that the academic ethos is rather vague, and not really true to life, sociologically or philosophically. It is obviously too high-minded to be taken at face value. Part of its effectiveness as an ideology[4] is that it matches its social practices much too neatly to its proclaimed philosophical principles. This is the image that the academic community presents of itself, and its work, to the world. It describes what ought to be, not what usually is.

Nevertheless, this idealized image captures most of the characteristic features of academic science, *as practised in its heyday*. I am not romanticizing the science and scientists of a past age. All I am saying is that a model based on the academic ethos would not grossly misrepresent the way in which scientific knowledge was produced in European and American universities between, say, 1850 and 1950. These were the social and philosophical ideals that most scientists tried to live by, and at least a few achieved[5]. Historically speaking, it would not be such a bad model of the scientific enterprise during that era.

This is also the image that many people, especially scientists, still have of 'pure science' [ch.2]. Even when they are debunking it, they are usually criticizing its failure to live up to its avowed ethos, morally or intellectually. Academic science is thus something more than a communal activity that happened to develop during a particular historical period. It is our standard exemplar of an 'epistemic institution'[6]. Likewise, academic research is something more than a particular cultural form. It is our ideal type of a 'mode of knowledge production'[7].

4.2 Is science to be believed?

The big question remains: what sort of knowledge does science produce? What does it tell us about 'this sorry Scheme of Things entire'? How much does it cover? Can I trust my life to it? Is it *really* to be believed? When people talk about science, they have in mind certain standards of credibility and reliability. The response to this question is absolutely central to our image of science.

Unfortunately, this is where the science/anti-science divide is particularly sharp. On the one side, there are the many people who are still bemused by the epistemological claims of the Legend. These assert that science is magically endowed with an infallible method for achieving absolutely perfect truth, so that it is 'irrational ' to defy its intellectual authority, or even to discuss the aspects of the world to which it might not apply. On the other side are those who are repelled by such intellectual arrogance, and draw upon one or another of a variety of arguments – many of which are sound enough – to dismiss the whole question.

Philosophers since Plato have grappled strenuously with just such issues – and come away without convincing formulae for dealing with them. But they have long been aware that the extreme claims endorsed by the Legend are unproven[8], and at the very least grossly inflated. Indeed, from a sociological point of view[9] they are so clearly untenable that the whole concept of justifiable belief is downgraded to a mere matter of culturally 'relative' opinion. But that, again, ignores too easily some very obvious features of the mental and social worlds in which science flourishes.

It is clear now that we cannot construct a satisfactory model of science until this aspect of the Legend has been challenged and cut down to size. But the credibility of science depends as much on how it operates as a collective social enterprise as it does on the principles regulating the type of information that this enterprise accepts and transforms into knowledge.

This investigation cannot be limited to an abstract analysis of the regulatory principles alone. It involves what we know about human cognition, scientific work, instrumental capabilities and other naturalistic considerations[10]. In fact, it is essentially a matter of surveying more thoroughly the territory that I mapped out roughly over twenty years ago[11].

Before we set out, however, we have to be sure that we are talking about science as it really is, not about a purely hypothetical system. When I first began to think about science as a social institution, almost forty years ago[12], this was not even a problem. One could just treat the academic model as a reasonable approximation to reality. And even now, when so much has changed, this model is a natural reference system for the analysis. Although it is a theoretical construct with many idealized features, we do know that something like it once actually worked, so it must have all the elements necessary to do its job.

The fact is, however, that real science – even the sort of science carried out in universities – is deviating more and more from the long-established academic mode. This must be very obvious to any reader familiar with the present-day research scene. For some of them, the 'academic science' described in the previous chapter must have seemed worlds away. What about intellectual property rights, project proposals and grants, directed programmes, contract researchers, global networks, interdisciplinary centres and teams, research performance evaluation, and so on? Where do these new social practices and conventions fit into the academic ethos?

In effect, academic research is being complemented or even superseded by a new 'mode of knowledge production'[13]. This is not just a little local difficulty that can be dealt with under the heading 'science policy' – not even by throwing more money into the R&D system. It is more than a glitch in the administrative and funding relationships between science and society. It involves a radical structural change in many features of our model.

Naturalism demands that we discuss these changes and their likely effects. At first sight, only the social aspects of the research culture are affected. None of the pressures for change seem to be operating directly on its philosophical aspects. It is quite clear that it is becoming more difficult for scientists to conform to the Mertonian scheme in their relations with one another. Nobody suggests, however, that their work should no longer be regulated by the traditional criteria for scientific validity, etc.

And yet the academic ethos is a tightly woven complex of social and epistemic norms, all closely dependent on one another. Academic science [ch.3] emerged as a specific mode of knowledge production nearly two hundred years ago. Its epistemology had evolved in partnership with its sociology. This historical process still goes on. Any substantial changes in the social practices of scientists must affect their intellectual practices, and *vice versa*. To give obvious examples: if scientists became noticeably more secretive about their research, would not their results be treated as less trustworthy, and if they seemed to be very vulnerable to external interests, would we accept the complete objectivity of their world view?

For most people – and especially for scientists – any shift in the ground beneath their belief in science is very distressing. The philosophical principles that regulate science seem as sacrosanct and perpetual as the physical laws that regulate the universe. Surely I am not suggesting that henceforth scientists will no longer be bound by logic or constrained by manifest facts? Might I be accepting the subversive doctrine that 'anything goes'[14]? As we shall see [ch.10], nothing so wildly anarchical is proposed. But subtle changes of emphasis and interpretation are already occurring in the way we generate and treat research results. Yes, science is still to be believed, but not with quite the old fervour, or for quite the same ends.

But before trying to weigh up the epistemic implications of current structural changes in the scientific enterprise, we need to have some idea of the general nature of these changes and the forces that seem to be driving them. A very schematic account of this large subject will occupy the remainder of this chapter. In subsequent chapters we shall look in some detail at various familiar characteristics of scientific knowledge, showing how they are connected with certain traditional features of the research process. In each case, we then consider closely various moves away from the practices sanctioned by the academic ethos, and assess their epistemic effects.

These sociological changes and their philosophical correlates do not, of course, operate piecemeal. When, eventually, we have drawn them together, we shall find that our overall concept of 'science' is undergoing a radical transformation. Our exemplar is changing before our eyes into a new form – *post-academic science* [4.4] – which performs a new social role, and is regulated by a new ethos and a new philosophy of nature. I doubt whether the science/anti-science conflict can be resolved until both sides agree that this might be where we are going, if it is not already where we are at.

4.3 What is happening in science?

'The devil is in the detail!' The small print of everyday business can change out of all recognition long before this shows up in the headlines of an ethos[15]. Whether or not the abstract norms, goals, regulatory principles, categories, etc. of science have altered, there has obviously been a real cultural revolution in its working practices and institutional arrangements[16]. This revolution has been so pervasive that it is best presented in personal terms. Let us imagine a day in the life of a typical scientist of the 1990s.

We may call her Dr Mary Jones, for an academic scientist is more likely to be female than twenty-five years ago, especially in the behavioural and humanistic sciences. She is in her early forties, and although not a 'star' is generally recognized to be a very competent and productive researcher whose work is always worth reading and citing. Meet her in her office at the good, if not world-renowned, university where she works as a Senior Research Fellow.

First there is a letter from the Dean of the Faculty with the good news that her promotion to the rank of Associate Professor has been approved. What is more, this is a permanent post – for five years in the first instance, renewable at five year intervals up to the age of retirement. Of course it doesn't offer legally waterproof tenure – nobody has that nowadays – but fortunately, her field is very active, and showing no signs of drying up. The annual career appraisal interview with the Chairman of the Department will be just a formality. She has always made an effort to keep up with the latest developments, and having held a succession of short-term contract appointments working on various applied research problems she is by no means a narrow specialist. But there was always the fear of not finding an opening for her particular skills. Now, at least, she will not have to be competing with the latest crop of PhDs for a job demanding some quite novel expertise.

Indeed, the first appointment in her diary is with one of her own doctoral students. Training people to do research is rewarding work, even though she is not sure whether what she is teaching them is quite what they will need for their later employment – most likely in industry or government.

Mary looks back for a moment over her own career. She has always had a passion for her subject, but had not really set out to be a full-time researcher. She enjoys teaching, and is quite looking forward to giving some undergraduate courses in her new post. When she first got her doctorate, she could have gone in for a post in the national research organization. It would have been great to have won against such stiff competition, and a safe position as a civil servant would have left her free to devote herself completely to research. But how would that prospect have seemed by this stage in her life? Even the brightest of her student contemporaries who took that route have not been very much more productive. On the other hand, if she had taken a secure post teaching at a minor university she would have been overloaded with lectures and seminars, and probably have dropped out of research altogether. Staying on in academia has been risky, but it is the sort of work that she really enjoys, and she is lucky to have made it at last.

The research student has some exciting new results. In fact, Mary's research group is doing excellent work. You can tell that by the number of citations they are getting. That will earn a few useful performance indicator points for the Department – and the university – in the national evaluation competition next year. The group did pretty well in the last one, only two years ago . . . Oh, better not think about it. It will stop her research for months while she is collecting all the information they ask for – most of it quite irrelevant to the scientific quality of her work. The trouble is that the whole research budget of the university will depend on getting a high score.

At this moment, there is a telephone call from the government funding agency. Bad news! Their latest research grant application has failed. The project got an alpha rating for the science, but was rejected by the merit review panel. Apparently, one of the non-specialist reviewers couldn't be persuaded of its possible long-term benefits. Admittedly, it was exploratory research, but a real expert would have seen the connections with some very practical problems. Perhaps they had been too cautious in not promising an eventual pay off. Next time they would emphasize the potential applications, and put in a lot about how they were going to disseminate and exploit them commercially – even though it would still be very basic research.

The trouble is that it isn't just her own research unit that has lost this round in the funding game – if that's what it is. For the next hour, Mary is on the email, consulting with the other groups she is working with. This was to have been the first joint project in a collaboration that goes back several years. In fact, her partners are a very mixed bunch – two other university groups, a small unit in a government laboratory, some people in the R&D department of a global multinational corporation, and a real whizz-kid who has just moved out of academia to start up his own company.

Geographically, they are so dispersed that they seldom meet in person. But they are all good scientists, and somehow they discovered that they were all interested in the same bit of science. So they began using the Internet to exchange data and criticize each other's ideas. It was very exciting to work with people from several different disciplines, coming at the problem from quite different points of view. Just trying to deal with a practical problem that had come up in the industrial laboratory had opened up some fundamental questions that stretched right across the board. She would never have thought of asking questions like that on her own, and wouldn't have known where to start to answer them.

So they are getting a lot out of working together, almost like a regular research group. But it is all very informal, and relations among the principal investigators have sometimes been strained. Mary has been congratulating herself on her tact when they were putting together this grant application. Her hope was that a joint project like this would bind them a little more strongly. Now it is taking all her diplomacy to prevent the team from breaking up, as so many do after a year or so.

But perhaps that's better than research in some fields, where people get locked into multi-centred 'collaborations' that are so big and highly organized that they are almost permanent institutions[17]. It is lucky that her work didn't require access to a big research facility – well, no more than what a major research university ought to be able to provide for its academic staff. Even that has been getting difficult lately – or so the administrators said when they insisted that research groups would be charged for their computing time. Perhaps that will encourage her research assistants to be more efficient, but it does mean so many more silly forms to fill in.

Mary is really quite good at administrative work, even though she reckons she has too much to do already, what with accounting to the funding agency for grant money, negotiating with the university finance officers about indirect costs, seeking contracts to pay for post-doctoral assistants, and so on. At lunch a friend tells her that as an Associate Professor she will have to behave more like a middle manager than a scholar – and when she gets to be a Full Professor it will be even worse. They discuss the time that some of the senior academics are spending trying to link the university more closely with local industry. Their efforts don't seem to be having much success. Or is that where she should already be looking to get funds for her group . . . ?

Funding? Always that worry. As she walks back to her office, she decides that she must get a post-doctoral post for that research student. Mary has some lovely ideas for research, but they will have to be turned into attractive projects. The big international organization is inviting applicants for a directed programme on a very knotty public issue. It's obviously driven by politics, but some of the detailed problems sound as if they might almost be in her field. She could probably write a plausible proposal that was not too far from the theoretical line she is trying to develop.

Entering the departmental building, she suddenly has a thought. Now is the moment to apply for a grant to enlarge her research group into an interdisciplinary research centre[18]. What she is doing these days is not really in the main stream of her discipline. Some of her research assistants had actually come to her from quite different departments. She prides herself on having created a centre of excellence that has nearly reached the critical mass to survive on its own. Could it be described as strategic research in a field of high national priority? At the very least, she should be getting together with some of the other people she knows in other universities who are doing similar work. If they could get it designated a special research area, they wouldn't have to propose specific projects to get funds.

In her office there is a message to call a number in a television company. They want her to take part in a programme on some of the ethical issues surrounding her research. Why not? It will be very superficial, of course, and she risks being made to seem pedantic by a half-baked interviewer. But even that sort of publicity would do no harm to her career, and the university authorities are always delighted to be in the news.

Now, at last, a chance to settle down to her own work. The results they have been getting lately need to be written up for publication. For weeks now, they have been exchanging rough drafts of a paper on the Internet. There has been a lot of discussion about how some of the data should be presented, and a messy controversy over what they meant. She feels confident that they have nearly arrived at an acceptable version. Between them, they have surely covered all the points that might be taken up by referees. In fact, she wonders, with so many of the leading people in the field already involved, where the editor will go to find independent reviewers. With such a bevy of authors, she just hopes she will get some credit for having had the original idea for this research, and for having put so much into analysing the results.

But there is a problem with the section that suggests some practical applications of what they have discovered. The patent experts don't think it constitutes prior disclosure, but intellectual property like this is valuable, especially if it can be exploited in secret for a year or so. Is that why the industrial people are so keen on this work . . . ? How much needs to be said to make sense of the science, without giving too much away?

There has also been a disturbing letter from the government department that is partially sponsoring the research. Instead of giving the usual formal consent to publication, they are suggesting a revised wording of one paragraph near the end. Could it be read as criticism of some of their recent policies? Even purely scientific issues seem to be politically sensitive these days. It's not a big deal, but her professional integrity is being questioned, and she will have to tread carefully to avoid an awkward confrontation.

And so on. If this chronicle strikes you as grey and humdrum, that is how I meant it. Partly that is because I have excluded the vivid technical details that would immediately identify the discipline. Is Dr Jones a physicist or an anthropologist? Is she a basic researcher with an interest in exploiting her ideas practically, or a technological scientist with strong theoretical leanings? To tell the truth, I don't know myself. If you happen to be familiar with a particular corner of scientific life, try colouring in this picture with appropriate names of subjects, institutions and people. It will look less dreary, and even more authentic.

The institutional background is also purposely vague. In fact, it is a pastiche of practices and institutions drawn from present-day Australia, Britain, France, Germany and the United States. It could just as easily have been filled in with similar items from other advanced countries. Perhaps I should have hinted that Mary Jones left her native land as a student, and like many other modern scientists is now a happy enough citizen of another nation. I could even have added a romantic touch by relating how she had overcome the problems that many women scientists have in reconciling their careers with marriage and children; but these problems are not new and are not peculiar to the scientific life.

Of course, this is only a snapshot of a tiny corner of an immensely varied scene. An overall picture, if one could be drawn, would be very patchy. The incidence of change is far from uniform. A few academic disciplines, such as pure mathematics, have not altered dramatically. In certain countries, such as France, some segments of the traditional scientific culture have been remarkably sheltered from the forces for change. Many venerable institutions, such as the Universities of Oxford and Cambridge, have preserved many features of their ancient ways. And everywhere, in every subject, one can find individuals who manage to go on performing their roles as 'scientists' and/or 'scholars', as if in a timeless world.

Now compare my 'Mary Jones' scenario with a day in the life of 'Viktor Jakob'[19] or 'Albert Woods'[20], or even 'Jim Watson'[21]. These fictitious (or quasi-fictitious) scientists lived in a culture framed by the academic ethos. This was a distinct form of life, spanned by the practices and procedures sketched out in the previous chapter. Indeed, until about a generation ago, academic science conformed quite closely to this stereotype[22].

As this comparison clearly shows, a great deal has changed. There is still a characteristic scientific culture. Nevertheless, somehow, it has acquired a number of quite novel features – features that scarcely existed or could hardly be imagined twenty or thirty years ago. In effect, there has been a decisive break with the academic tradition in relation to conditions of employment, problem choice, criteria of success and other major structural elements.

I do not propose to go through these changes explicitly here, since they will be discussed in detail in later chapters. In particular, we shall ask whether they are consistent with the academic ethos as traditionally interpreted. Can the commercial exploitation of intellectual property rights, for instance, be reconciled with the norm of communalism? Can scientific objectivity be claimed for research that is undertaken to solve specific social problems? In the end, we have to ask what philosophical principles really regulate knowledge produced under these circumstances.

For the moment, the main point is that these new features are not just accidental or temporary[23]. They involve whole complexes of interlinked procedures and principles, policies and practices. They have emerged very rapidly within an existing social institution without seriously disrupting its operations. They are integral to a new mode of knowledge production that has established itself in universities and other epistemic institutions throughout the world. Why has science taken this historical turn?

4.4 The advent of post-academic science

In less than a generation we have witnessed a radical, irreversible, world-wide transformation in the way that science is organized, managed and performed. We have looked at this transformation only as it affects every-day scientific life. But it obviously involves major structural changes at higher levels. These changes are taking place in all epistemic institutions – universities, research institutes, government establishments and industrial laboratories[24]. They are well documented in innumerable official reports, newspaper articles, comments by scientific notables and policy-oriented research papers. Science is being redefined at every level, and in relation to other segments of society[25].

This is what I mean when I say [4.2] that academic science is giving way to post-academic science[26]. The actual changes are so pervasive, so inter-connected, and vary so greatly in formal detail from country to country, that they have seldom been treated as elements of a general social phe-nomenon[27]. Taking a naturalistic stance, however, we can accept the reality of this phenomenon and analyse its effects without having to explain why it is actually happening.

The trouble is that it is not easy to relate the transformation to a single cause. It is true that the metascientist of the 2050s may be able to point to a sea change in the general philosophy of science, feeding back into the social structure of knowledge-producing institutions and thence down to the level of 'laboratory life'. This seems to be what worries many defenders of the Legend[28]. They fear that any revision of the theoretical epistemology of science will open the gates to 'post-modern' intellectual anarchy[29] in the whole research enterprise. But there is little evidence of such a high-level effect, and even if there were we are much too near the event to evaluate it.

In any case, the general argument of this book is that the epistemology of science is linked to its sociology primarily at the level of research prac-tice. Scientists produce knowledge in accordance with the norms and principles that apply in their particular situation. Even when they become responsible for institutional, corporate or governmental policies that affect the research culture more widely, their decisions are seldom consciously influenced by general philosophical considerations. Undoubtedly there are such influences, but they operate very indirectly and incoherently, in scarcely fathomable ways.

Nevertheless, post-academic science is not, as many scientists still hope, a temporary deviation from the onward march of science as we have always known it. Nor is it just 'a new mode of knowledge production'[30]: it is a whole new way of life. It is the resultant of innumerable improvised solutions to immediate practical problems. It is the product of expediency, not of design. Yet it constitutes a more or less coherent culture[31], not because it was planned as such but because science is typically a complex, self-organizing social system [9.8] that adapts opportunistically to changing circumstances[32].

In calling this culture 'post-academic' I am not suggesting a total repudiation or reversal of traditional goals. On the contrary, this term indicates continuity as well as difference. The continuity is so obvious that many people assume that nothing has really changed. Post-academic science is born historically out of academic science, overlaps with it, preserves many of its features, performs much the same functions, and is located in much the same social space – typically universities, research institutes and other knowledge-producing institutions. But although academic and post-academic science merge into one another[33], their cultural and epistemic differences are sufficiently important to justify the new name.

4.5 An undramatic revolution

The emergence of post-academic science does not seem to have a single underlying cause. But one can point to a number of general factors that have generated and shaped the new culture. These various factors need to be considered separately, since they operate in different ways and have different social and epistemic effects. Thus, 'externalist' accounts of science focus on the political, economic and industrial pressures that act more and more forcefully on the scientific community 'from the outside'. But the 'internal' factors are probably just as important. Academic science is a dynamical system, not a passive black box. It has to adapt socially to the accumulating strains produced 'inside science' by increasingly rapid scientific and technological progress.

These factors operate on such different time scales that it is not easy to locate the moment of transition. My impression is that the concept of a 'new regime' or a 'new model' for science was already in the air towards the end of the 1960s, but that the big changes did not begin to take place on the ground until about a decade later. But this is not based on any detailed historical evidence. When I say that the transition happened 'less than a generation ago' I am only indicating that it is a 'contemporary' social phenomenon: it started during the working lifetime of many scientists who are now still active in research and it is still going on.

Indeed, the transition might not really have been quite so dramatic or abrupt as I have here suggested[34]. A more systematic account would surely refer to trends that were already under way between the World Wars, to the prophetic writings of J.D. Bernal[35] and others, and to the profound effects of war-time developments on the scale and influence of the sciences and their associated technologies[36]. But disputes over the dates of historical 'periods' are seldom very fruitful, especially when they cannot be anchored to notably dramatic events. It is precisely the undramatic character of this cultural revolution that has concealed it even from those of us who have lived through it.

4.6 Collectivization

The most potent force for change within science has the longest time scale: it is scientific progress [9.8]. Almost by definition, scientific research is the most innovative of all human activities. Its dedication to originality draws it continually into entirely novel spheres of knowledge and technique. Ever since Galileo first looked at the moon through a telescope, knowledge has advanced hand in hand with research technology[37].

More powerful instruments make it much easier to do good science [5.5]. But increasingly expensive apparatus puts additional burdens on research budgets and the social machinery that they fuel [4.7]. Instrumental sophistication also moves research towards more *collective* modes of action[38]. One naturally thinks of high energy physics or space science, where hundreds of scientists must work together for years around an enormous research instrument just to perform a single experiment. But this style of 'Big Science'[39] is only the most spectacular manifestation of a general trend.

This trend was undoubtedly accentuated by the experience of many physical scientists during the Second World War, when they were put together with engineers into large organizations to produce fantastic new weapons[40]. But we would now see it as a natural cultural development. A researcher can save a vast amount of personal labour by using a highly automated laboratory instrument such as an electron microscope or desk-top computer. But this is achieved by dividing the labour with other people – technical support staff, instrument designers, software engineers, and a variety of specialists in other scientific fields. Even the 'lonely seeker after truth' works most of the time in a laboratory or library. She only feels 'lonely' because her research apparatus is integrated into an elaborate social apparatus which contrives to be so transparent that she can look right through it directly at nature[41].

The same trend shows up clearly in the growing proportion of scientific papers with two or more authors[42]. Teamwork, networking and other modes of collaboration between specialist researchers [5.13] are not mere fads, fostered by the joys of instant global electronic communication. They are the social consequences of the accumulation of knowledge and technique. Science has progressed to a level where its outstanding problems cannot be solved by individuals working independently.

It has often been remarked, for example, that most practical problems do not emerge ready-made in the middle of existing research specialties. They are essentially *transdisciplinary* [8.8]. They materialize amidst the booming buzzing confusion of the life world and cannot even be identified until they have been viewed from many different angles. For example, what is the scientific significance of BSE, the 'mad cow' disease? Should we seek the answer in veterinary pathology, neurophysiology, microbiology, animal nutrition, epidemiology, agricultural economics or even international law? The research questions that the politicians pose to the scientific experts require joint inputs from several distinct specialties. Further progress in understanding this knotty problem will involve more than extra funding for individual researchers. It will call for a major collective effort, involving elaborate social arrangements for setting up multidisciplinary research teams, coordinating their efforts and combining their findings.

Post-academic science is not exclusively concerned with solving urgent practical problems. But even the most 'fundamental' scientific problems are turning out to be transdisciplinary. As the various sciences extend their areas of understanding, they make contact, overlap and interpenetrate along innumerable cognitive and technical channels. The traditional metaphysical question of how to 'carve nature at the joints' thus becomes less and less answerable. The same uncertainty applies, moreover, to the boundaries between theoretical principles and practical procedures, and between natural phenomena and human artifices. As research is oriented towards larger problems, whether these arise in applied or basic contexts, it must necessarily rely on the collective activities of specialists from a variety of disciplines.

This is a much more radical change than it might appear on the surface. As we have noted, academic science is a highly individualistic culture. This individualism is sustained by a powerful but highly dispersed structure of academic specialties. It is not incompatible with the growth of specialized research groups around 'principal investigators' or other acknowledged scientific leaders[43]. But truly multidisciplinary teamwork challenges the traditional structure at every turn, affecting personal autonomy, career prospects, performance criteria, leadership roles, intellectual property rights, and so on[44].

4.7 Limits to growth

Another factor in the transition to post-academic science also has a long history. Science has always been a growth industry[45]. Its intellectual dynamism – the way that scientific knowledge and technological capabilities remake themselves every twenty-five years or so – has always been accommodated by social expansion.

The advance of knowledge opens up tempting opportunities for yet further advances. The solution to every scientific problem suggests two new questions that could now be tackled. Old-established disciplines do not wither away: they split like amoebae into several new ones, or re-emerge triumphantly in novel interdisciplinary combinations[46].

A successful researcher never quite succeeds in working herself out of a job. When a professor retires, she wants two new jobs to be created, for her two most brilliant pupils to carry on her work. Academic science is framed on the assumption that tenure for the rising talent of today will not block the path of promotion for the next generation, that there will always be new posts for innovative individuals, that there will be the social space, the mobility and the resources to satisfy the ambitious intellectual entrepreneur.

But this could not go on for ever. In 1961, Derek de Solla Price published a famous graph[47], showing that scientific activity had been expanding exponentially at a very high rate for about three hundred years. He pointed out that if, say, the publication of scientific papers went on doubling every 15 years, then soon every man, woman and child in the country would have to be spending all their time doing research and writing scientific papers. This was an absurdity: it was time to ask when the expansion would stop, and what would happen to science when it met its own limits to growth.

The ceiling that it is hitting is financial. In effect, the whole enterprise has now become too large and expensive to be allowed to go on growing. Conventional econometric studies indicate that the total R&D activity of a developed country now takes somewhere between 2 and 3 per cent of the national income. A substantial fraction of this is spent on more or less basic research. The governments, foundations and firms that fund academic research are now asking what they get out of this nebulous activity, and are putting bounds on their patronage.

In less than a generation, there has been a radical change of policy towards state support of academic science. In some countries – notably Britain – the transition to 'steady state' funding[48] took place in the 1970s. In others – notably the United States – it was deferred until the 1990s. The initial budget constraints or cuts were triggered by general economic stringency, but it is unlikely that there will be a return to 'normal' conditions of very rapid and sustained growth even when the economic climate is more favourable. There is just no room for such growth in the economy or the polity. Nobody seriously envisages a situation where, say, 10% of the GDP goes on R&D, which is more than most countries are willing to spend on health or defence.

Science has gone through hard times before. But these were only blips on a continuously rising curve of financial inputs, personal employment and archival output. From now on, it seems, it will have to make do with a fixed or slowly growing envelope of resources. A sturdy social machinery has emerged to allocate these resources and ensure that they are used to good effect. Unfamiliar words like 'accountability' and 'efficiency' are heard in academia. Derek Price foresaw that the transition to a 'steady state' regime would have a significant effect on academic science. I do not think he appreciated the strains that this transition would impose on the academic ethos, nor the cultural changes that would be required to adapt to the new environment.

4.8 Exploiting knowledge

Another major factor in the transition to post-academic science is greater stress on utility [7.2]. The present era in science seems to be unusually fruitful. Perhaps every era has seemed the same to those living in and through it. Nevertheless, recent developments such as the theoretical consolidation of the biological sciences around the mechanisms of heredity and the explosion of powerful research techniques out of the electronic computer seem to offer limitless openings for conceptual and practical advance on almost every front.

This does not mean that scientific knowledge is approaching a state of 'grand theoretical unification' [6.5, 8.5, 10.8]. Even the 'theory of everything' that the physicists are seeking will really say almost nothing about anything of human interest. But remarkable progress is being made in the imaginative combination of concepts and techniques drawn from disciplines or subdisciplines that were previously thought to be on distant continents of the academic map[49]. Many specialized fields of science have also entered a phase of 'finalization' [8.6], i.e. a phase where there is a reliable background of general understanding to guide research strategically towards envisaged and desired ends[50].

These general epistemic developments will be discussed further in later chapters. The essential point here is that they encourage the application of basic scientific knowledge to practical problems. Having observed the revolutionary capabilities of this knowledge in medicine, engineering, industry, agriculture, warfare, etc., people have become very impatient with the slow rate at which it diffuses out of the academic world. Governments, commercial firms, citizen groups and the general public are all demanding much more systematic arrangements for identifying, stimulating and exploiting potentially useful knowledge.

This is not the place to analyse the social and human effects of technological innovation. Nor can we pause to ask whether this power is in benevolent hands. The plain historical fact is that science has been so influential through its technological applications that everybody expects it to make greater contributions than ever before to whatever they happen to consider beneficial or profitable. This applies right across the political spectrum, from the military who want to harness science for war-making to the environmentalists who want to use it for the conservation of nature.

In effect, post-academic science is under pressure to give more obvious value for money. Many features of the new mode of knowledge production have arisen 'in the context of application'[51] – that is, in the course of research on technological, environmental, medical or societal problems. More generally, science is being pressed into the service of the nation as the driving force in a national R&D system, a wealth-creating techno-scientific motor for the whole economy[52].

This pressure is not completely new. As we saw earlier, academic science is far from oblivious to practical human needs. Universities have long been active sites of research in engineering, medicine and agriculture, and in the applicable social sciences such as law and economics. But the traditional functions of this research are to illuminate the background to practical problems, and to provide active practitioners with the knowledge needed to solve them.

The novel factor is the requirement that research should be explicitly targeted at recognizably practical problems. Post-academic scientists are expected to be continually conscious of the potential applications of their work. This does not mean that projects are always chosen on the basis of their immediate applicability. Nor does it imply that the promise of short-term profits takes priority over a prospect of long-run capital gains. The old cliché about very fundamental 'strategic' research still holds good: what is the use of a new-born baby?

Nevertheless, a norm of utility is being injected into every joint of the research culture[53]. Discoveries are evaluated commercially before they have been validated scientifically – witness the brouhaha over 'cold fusion'. Apparently 'useless' sciences such as astrophysics or classical archaeology have to justify themselves by emphasizing their 'cultural' value[54] to the public. Scientists themselves are seldom in a good position to assess the utility of their work, so expert peer review is enlarged into 'merit review' by non-specialist 'users'.

The utility factor makes post-academic science answerable for its operations to people and institutions outside the scientific community [7.4–7.10]. This is more than a matter of limiting the freedom of scientists to pursue knowledge 'for its own sake'. It infuses the scientific ethos with Ethics as the world knows it.

Utility is a moral concept. It cannot be determined without reference to more general human goals and values. Scientists cannot be expected to be more on the alert for potential applications of their research whilst closing their minds to ethical considerations. Until recently, academic scientists could dismiss the call for 'social responsibility in science'[55] by claiming that they knew – and cared – nothing about the applications of their work, and therefore need not be concerned whether it might be linked with war-making, political and economic oppression, environmental degradation or other shameful activities. Post-academic science, being much more directly connected into society at large, has to share its larger values and concerns[56].

4.9 Science policy

State patronage inevitably brings politics into science – and science into politics[57]. The more generous the patronage, the more political activity it entails. Governments nowadays find themselves allocating very large sums of public money to various bodies for various forms of research. Academic science is just one part of a loosely articulated 'R&D system', whose activities range from basic science to near-market technological development. This system is too large to be left to its own devices. The emergence of *science policy* – more generally, *science and technology policy* – is a major factor in the transition to a new regime for science[58].

One might say that this is merely an official mechanism for making science more useful and accountable. But academic science faces the policymakers with a problem: they don't know enough about it to oversee it directly. In the past, they have relied on the scientific community to produce knowledge for the common good. The custom has been [3.10] for large blocks of government money to be handed over to independent research organizations, such as universities, research councils and national academies, for subdivision and distribution to individual academic scientists and their research units. Since these bodies are governed and largely staffed by scientists[59] the autonomy of the scientfic community is not significantly infringed.

Detailed arrangements are also made to preserve the personal autonomy of individual researchers. In some countries, such as France and Germany, this is achieved by a complex of legal rights and conventions within a bureaucratic institutional framework. But the general trend recently has been towards the 'soft money' system, invented in the United States after the Second World War[60]. Researchers formulate *proposals* for specific *projects* [8.2], which are submitted to funding bodies, where they are evaluated by peer review and awarded grants on the basis of their scientific merits. These grants are for the additional apparatus, assistants and other facilities needed by the principal investigator to carry out the project. In this way tenured academic scientists are enabled to do their own specialized research, subject only to the reasonable proviso that it is judged to be 'good science' by well-qualified experts.

This system works so well that it deserves deeper sociological analysis. In effect, it provides the scientists with the resources they need for their research by a social mechanism that is consistent with the academic ethos. Thus, competition for grants reinforces competition for scientific recognition [3.8], since these are assessed by essentially the same criteria. By giving the scientists freedom to set their own research agendas and publish the results freely in their own names, governments give them a sense of being in control of their own intellectual lives. In return, it provides governments with communities of active, well-esteemed – and grateful – scientists, on whom they can call for independent advice and expertise[61].

The academic culture is sustained by a tacit 'social contract' between the scientific community and society. It is thus very susceptible to apparently innocuous changes in the terms of this contract. For example, as researchers become more dependent on project grants, the 'Matthew Effect'[62] [3.8] is enhanced. Competition for real money takes precedence over competition for scientific credibility[63] as the driving force of science. With so many researchers relying completely on research grants or contracts for their personal livelihood, winning these becomes an end in itself. Research groups are transformed into small business enterprises[64]. The metaphorical forum of scientific opinion is turned into an actual market in research services[65].

By accepting state patronage on such a large scale, scientists have become very vulnerable to the demands of their paymasters[66]. Science policy unveils apparent discrepancies between what science might possibly produce, and what society actually gets. Priorities are laid down to make good these deficiencies. Research organizations and funding bodies are instructed to build these priorities into corporate plans and flesh them out in specific programmes. Project proposals are invited on specific societal problems, and grants awarded on the basis of their potential for solving such problems. In other words, considerable pressure is put on scientists to work on problems favoured by the government, rather than problems of their own choosing [8.7].

This development was, perhaps, inevitable. The social contract that protected academic science from political oversight was never written into any national constitution. Indeed, it was seldom articulated, perhaps for fear of being challenged. What is now happening is consistent with defining academic science as a component of a national R&D system and supporting it in that spirit.

The notion of the academic scientist as a relatively disinterested source of highly specialized knowledge is also affected [7.10]. Many academic researchers are so reliant on government contracts for research on practical problems that they cannot easily dissociate themselves from government policies. This is often revealed by disputes over the publication of reports on social issues commissioned by government departments. In effect, post-academic scientists relaunch themselves as technical consultants, whose skills as expert advisers include their opinions on contentious practical matters as well as their grasp of the scientific issues.

4.10 Industrialization

It is often remarked that science is being *industrialized*[67]. This points to another major factor in the transition to post-academic science. In many ways, 'industrial science' [2.4] is almost the antithesis of 'academic science'. The actual sociological differences between these two modes of knowledge production need to be explored.

Top-down policy takes little notice of these differences. One of the things that governments do to reduce their financial commitments to science is to 'privatize' some of their research establishments. Institutes and laboratories doing applied research in areas such as health, defence, transport or social policy are sold off to large industrial firms, or turned into independent companies.

The justification for such transfers is almost entirely political: 'Another burden has been taken off the back of the taxpayer.' They are part of the general swing away from public ownership towards private enterprise. No doubt, in a few years' time, the pendulum will have swung the other way, and the public welfare element in such research will again be considered paramount.

The scientists themselves have little say in such changes of ownership. If they are lucky, they keep their jobs and pension rights and go on working at much the same problems. But they are aware that they have crossed a frontier into a different scientific realm. In the past, many public sector research institutes were largely devoted to doing 'science' – that is, essentially academic research with distant strategic goals[68]. Over a period of years this tradition has been fading, as more and more of their funding comes from 'contract' research. Now, clearly, they are all industrial scientists.

The same trend can be seen in the crescendo of calls for closer links between academia and industry. This does not mean that industrial firms are being encouraged to do more basic research. On the contrary, many big companies now regard their corporate research laboratories, where such research used to be done, as expensive anachronisms[69]. It means that academic institutions are expected to undertake more research under industrial auspices, and to produce results of more direct commercial value[70].

The thrust of this policy is to close the traditional gap between the academic and industrial modes of knowledge production[71]. To a politician, civil servant or business executive, this gap is difficult to understand. Technically and epistemically, industrial science has always been the twin of academic science. That is to say, they use the same techniques and technologies and draw upon the same databases of fact and theory. Indeed, this is true nowadays of almost all knowledge-producing organizations, whether academic, governmental or commercial, whether in the natural or social sciences, whether basic or applied.

These similarities even hold at the level of everyday laboratory life[72]. Observed for a few days, the research laboratory of, say, a large pharmaceutical firm is scarcely distinguishable from a university laboratory in the same field. People with the same highly specialized training manipulate the same highly sophisticated apparatus. They use the same jargon to argue from much the same premises to much the same conclusions. An expert ear is needed to distinguish between them.

Indeed, the two great systems of science were always closely connected and dependent on one another. Ever since they were professionalized and separated, more than a century ago, the existence of each was always understood to be essential to the continued vitality of the other. But they have seldom lived easily together under the same roof, and have tended to deal with each other at arm's length. In spite of converging technically and managerially, academic science and industrial science still differ very significantly in their social goals. This difference is often exaggerated by academics, but is not entirely ideological. As we have seen, the social organization of academic science can be described in terms of the Mertonian norms. This description is, of course, highly idealized, but not completely unrealistic. Industrial science, by contrast, contravenes these norms at almost every point.

The reason is, of course, that industrial science is not targeted towards the production of knowledge as such. Its goals, being practical, are extremely diverse [2.4]. The various ways of life that have evolved to achieve these goals are not those of a self-conscious community claiming allegiance to an unwritten ethos. Indeed, the characteristic social practices of industrial science are based on principles that effectively deny the existence of any such ethos.

Very schematically, industrial science is *Proprietary, Local, Authoritarian, Commissioned,* and *Expert*[73]. It produces *proprietary* knowledge that is not necessarily made public. It is focussed on *local* technical problems rather than on general understanding. Industrial researchers act under managerial *authority* rather than as individuals. Their research is *commissioned* to achieve practical goals, rather than undertaken in the pursuit of knowledge. They are employed as *expert* problem-solvers, rather than for their personal creativity. It is no accident, moreover, that these attributes spell out 'PLACE'. That, rather than 'CUDOS', is what you get for doing good industrial science.

The contrast with academic science could scarcely be sharper. The notion that academic science is being 'industrialized' thus means something more than that certain institutions are being taken into the private sector of the national economy. It implies the establishment within academic science of a number of practices that are essentially foreign to its culture. These practices arise naturally out of the demand that the goals of research should conform closely to the material goals of society – that post-academic science should be, indeed, an inseparable component of 'technoscience'[74]. In later chapters we shall explore in detail the effects of these changes on the philosophical characteristics of scientific knowledge.

4.11 Bureaucratization

These are some of the social factors in the transition to post-academic science. For many observers this transition are summed up in a single pejorative term: 'bureaucratization'. Science, they say, is not what it used to be. It is being hobbled by regulations about laboratory safety and informed consent, engulfed in a sea of project proposals, financial returns and interim reports, monitored for fraud or misconduct, packaged and repackaged into performance indicators, restructured and downsized by management consultants, and generally treated as if it were just another self-seeking professional group.

Certainly, all the social factors point in that direction. For example, the movement from individual to collective modes of research obviously entails more formal organizational arrangements. A proper management system is needed to orchestrate the construction of a large research facility, and its use by teams of scientists from many different institutions. Indeed, the remarkable feature of the immense 'collaborations' that carry out experimental research in high energy particle physics is how little administrative work they seem to get away with[75].

Again, as academic science becomes more directly dependent on public funding, it inevitably becomes entangled in governmental red tape[76]. Committee meetings and administrative action are required to formulate and implement policies, priorities and programmes. The transition to steady-state funding can only be handled by elaborate and trustworthy procedures for project evaluation and financial accountancy. Improved linkages between the production and use of scientific knowledge have to be systematically organized and managed. Competitive research grant systems necessarily generate a lot of paperwork and consume a great deal of the time of expert researchers.

Industrial research is widely considered to be peculiarly bureaucratic. This image may be a little out of date. The continued growth of academic science after the Second World War was only one aspect of the overall expansion of the national R&D effort, of which the greater part is actually undertaken by big private sector companies in certain industries. In the public mind, the dramatic technological applications of science are attributed to the research and development activities of such enterprises. In some cases, indeed, these firms owe their present pre-eminence to their heavy investments in relatively basic science over long periods, going back to the end of the nineteenth century.

By the 1960s, therefore, a stereotype of industrial science had developed around the way of life of research scientists and technologists in very large commercial firms, or in a small number of large public sector establishments mainly doing military R&D. This way of life is shaped more by its organizational environment than by its involvement in the production of knowledge. Industrial research was seen as typically hierarchical and bureaucratic because big business was typically hierarchical and bureaucratic.

Since that time, however, industry itself has changed. We are said to be entering a 'post-industrial' era[77], characterized by multinational firms which are decentralized managerially into small, specialised service units, devolving much work to subcontractors, coordinated globally by information technology, etc. And as industrial firms change their working methods, they restructure their research activities along similar lines. Their R&D laboratories are devolved into multidisciplinary matrices and global networks of temporary project teams, buying in specialist functions from independent contractors, and so on.

We recognize here many of the characteristic features of post-academic science. In effect, the new 'Mode 2' of knowledge production[78] is practically identical to the way in which up-to-date firms now organize their research activities. What might be called 'post-industrial science' differs from the earlier stereotype of industrial science by substituting 'market' competition for 'command' management. Researchers work in shifting teams, like small firms producing goods for a competitive market. Commercial enterprise and personal mobility replace managerial responsibility and career stability as organizing principles.

As we have seen, this parallels the way that science is now being organized in academia and in many public sector research establishments. Government funding of basic and strategic research is provided through competitive project grants, customer–contractor arrangements and other market concepts. Academic, governmental and industrial research units are treated on an equal footing as independent purveyors of specialist services, competing with one another or combining in their bids for resources.

The displacement of top-down planning by looser, more open, linkages allows for more feedback in the system, and more autonomy to its components[79]. By comparison with its predecessors, post-industrial science thus looks attractively unbureaucratic. But a market system is not free from unnecessary paperwork, ill-conceived constraints, wasted expert time, and other extravagances. Nor is it really as individualistic as it might seem. At heart it remains proprietary, local, authoritarian, commissioned and expert, even if it does not offer such a safe 'PLACE' as it used to.

The fact is that few of the entities that compete or combine in the pursuit of research funds are actually free-standing. The post-academic scientists who network enthusiastically across the world are mostly full-time employees of universities, government laboratories, charitable foundations or industrial firms. They do not have to take personal financial responsibility for the elaborate facilities that they use in their research. The real economic base of their activities is a complex of governmental bodies, large public institutions and private corporations. Post-academic science will surely not be able to duck the central questions of science policy: who will pay the pipers, and what tunes should they be called on to play?

Indeed, 'bureaucratization' may be the correct sociological term for the whole transition. Paradoxically, academic science – 'the engine of modernity'[80] – exhibits many of the features of a 'pre-modern' social structure. It relies heavily on personal leadership and discretion. Its role in society is legitimated primarily by the charisma attached to it as a community[81]. It is sustained by interpersonal networks of reciprocal patronage. It recruits its members by practical apprenticeship and by individual socialization to its ethos[82]. Its practices are hallowed by tradition but actually evolve and adapt to changing circumstances. Its norms are not codified and yet are enforced by informal communal sanctions.

What we are now seeing may be interpreted quite generally as the belated *rationalization* of this cultural form[83]. The transition from academic to post-academic science is signalled by the appearance of words such as management, contract, regulation, accountability, training, employment, etc. which previously had no place in scientific life[84]. This vocabulary did not originate inside science, but was imported from the more 'modern' culture which emerged over several centuries in Western societies – a culture characterized by Weber as essentially 'bureaucratic'.

What is remarkable is that academic science survived so long as a distinctive culture in this social environment. It would be even more remarkable if the historical clock were now to be turned back to the mode of knowledge production that prevailed little more than a generation ago. Post-academic science, like the modern world in which it is so deeply embedded, is here to stay. Thousands of books have been written about the effects of the advent of modern society on political, social and religious philosophies. Indeed, the rise of science is one of these effects[85]. The aim of this book is to consider the effects of the very same transition, within the scientific culture, on the philosophy of science itself.

5

Community and communication

5.1 What sort of knowledge?

The function of science is to produce knowledge. What sort of knowledge does it produce? Until recently, this question was supposed to be essentially 'philosophical'. Unfortunately [ch.1], the philosophers of science have not come up with a convincing answer. In spite of heroic efforts[1], they have simply failed to come up with a satisfactory definition of 'science'.

In practice, however, most educated people agree that certain bodies of knowledge are, indeed, peculiarly 'scientific'. They do this on the basis of a number of characteristic features, many of which lie outside the scope of conventional philosophical analysis. Indeed, as every would-be meta-scientist soon discovers, the 'philosophy' of any serious scientific discipline is so wrapped up in its subject matter that it cannot be accounted for historically or analysed without a good grasp of its technical language[2]. Since nobody really knows all these different languages, people have to rely on what they know of the social origins of the knowledge in question. In default of better criteria, they put their trust in 'scientists', as individuals and as an organized group, to produce genuinely 'scientific' knowledge.

This reasoning is, of course, incorrigibly circular. But remember that we are adopting a naturalistic stance. If the entity that people identify as science has distinctive social characteristics, then it is our job to identify these and analyse them. In previous chapters, therefore, we have portrayed science as a social institution, of which 'academic science' [2.8] is the ideal type. We have focussed on its practices and norms, showing how these fit together into a distinctive culture [ch.3], and how they are changing in response to various internal and external influences [ch.4].

Nevertheless, this does not mean that we are discarding the notion that science is noteworthy as an 'epistemic strategy'[3]. Academic science has developed, and is committed to, a characteristic mode of inquiry which produces a characteristic type of knowledge. The research process clearly involves entities such as 'facts', 'theories', 'concepts', 'beliefs' etc., which operate primarily in the realm of ideas. It is indeed as much a philosophical enterprise as it is sociological, psychological and material[4]. The sociological and psychological dimensions are required to *complement* the traditional philosophical dimension, not to replace it. All that I am saying is that if this 'disputatious community of truth-seekers'[5] has an 'essence' – which I strongly doubt – it cannot be captured in any one of these dimensions alone.

What now follows, therefore, is a systematic analysis of the epistemic practices of science in its ideal, 'academic' form. As we proceed, we shall note the implications of deviations from this ideal, especially where these correspond to changes towards a 'post-academic' mode of knowledge production [4.4]. This analysis will also cover the 'psychological' dimensions of the research process. That is to say, in keeping with our naturalistic approach [1.5], we shall bring in various concepts and findings from cognitive science, social psychology and other such disciplines, wherever these help us to understand what is going on.

The fact that this analysis is organized around the Mertonian norms [3.2] does not imply, however, that the sociological dimension is paramount. It just happens that these form the most convenient scheme for this purpose. One might have thought that the mighty labours of the philosophers of science would at least have produced a 'philosophical' framework for this part of the book. Unfortunately, having failed to find a keystone for the whole structure, they have left it a pile of 'methodologies'[6], 'regulative principles'[7], 'epistemic norms'[8], 'principles of rationality'[9] etc., without any agreed plan of how they all fit together. Indeed, it is now generally accepted that most of these are no more than 'heuristics'[10], 'maxims'[11] or 'rules of thumb'[12]. In spite of their claims to generality, they usually have very limited authority over the immense variety of intellectual practices that they are supposed to cover.

It turns out, moreover, that the Mertonian norms are not really blind to epistemic or cognitive values. They were, after all, originally set out by a scholar[13] who was himself deeply involved in the academic enterprise in all its aspects. It is impossible, for example, to talk about 'criticism' without reference to the nature of the knowledge being criticised, or the way in which individuals give it or take it in practice. Indeed, the very notion that there are epistemic and cognitive *norms* attributes social and personal values to ideas, and thus crosses the frontier from philosophy into the behavioural sciences[14]. This interdependence between the philosophical, psychological and sociological characteristics of science is at the heart of the present inquiry.

5.2 What are the facts?

The norm of communalism requires scientific knowledge to be public property[15]. This norm ensures that the results of research are published promptly, and in full. It also applies to the way in which research claims are reviewed by experts before being accepted for publication. In other words, for an item of information to be acceptable as a potential contribution to science, it has to reach a minimum standard of credibility and relevance. What is more, in accordance with the norm of criticism, it has to be presented in a form capable of undergoing further communal tests before it counts as 'scientific knowledge' in the fullest sense.

As we shall see, every scientific discipline has its own criteria of 'scientificity'. But what they all have in common is this lengthy period of expert scrutiny, much of it conducted in public. Researchers naturally seek results of the kind that are most likely to survive this process relatively unscathed.

The dominance of *empiricism* in scientific epistemology is thus very understandable. In all their dealings with the world, and with other people, most people believe that their most secure knowledge of matters of fact is derived from direct experience. Thus, the ideal type of an empirical 'fact' is that an event of type A occurred at location B at the time C, before one's very eyes or as reported independently by a number of reliable witnesses.

This is the sort of evidence that carries great weight in legal proceedings. Perhaps this is because it is so difficult to contest in broad principle. It stands or falls on the strength of its details. Was the event correctly described or identified? ('Did she fall – or was she pushed?'). Were the witnesses possibly mistaken about the time or place? And so on. An empirical research report can be fortified in advance at these points of potential weakness – for example, by including a photograph of the location, a printout from a computer clock and other items of mundane information. Then, if it does survive such attacks, it can reasonably be treated as indisputable. It is thus well suited to its desired goal – that is, incorporation into a body of 'established scientific knowledge' held in common by a research community.

In practice, scientists describe their research as 'empirical' in circumstances where this account would seem hopelessly naive. But 'empirical facts' are deemed to have an everyday reality that is extraordinarily difficult to define [10.7]. They are so basic to science that there is little to be gained by trying to redefine them formally in more abstract terms or decompose them into more primitive elements. This is an epistemological scandal which has embarrassed philosophers for centuries. But the redefinition of scientific knowledge in terms of atomic 'sense data'[16], for example, merely reproduces the scandal at a lower level.

For a philosophical naturalist, however, the 'Sherlock Holmes' model is more than a homely metaphor. It illustrates concretely most of the general features of empirical knowledge. In effect, this is our intuitive exemplar for all knowledge gained by *observation*. The 'event' reported might be commonplace, as in meteorology, or unprecedented, as in the discovery of a rare plant. It might be located in a test-tube, as in chemistry, or on the far side of the universe, as in astronomy. It might have occurred over a period of many days, as in social anthropology, or within a few seconds, as in seismology. It might have been experienced by millions of people, as in economics, or by a few unfortunate individuals, as so often in neurophysiology. In each case, the same general formula applies.

With the exception of pure mathematics, there is scarcely any branch of science, natural or social, physical, biological or humanistic, that could exist at all without a certain amount of empirical knowledge gained by observation[17]. It is equally hard to think of any type of observational report or scheme of classification [6.3] that is not shaped to some extent by the preconceptions of the observer[18]. Indeed, this is an essential feature of any cognitive process [5.9].

In effect, 'beauty is in the eye of the beholder'. Science, whose watchword is objectivity, is peculiarly susceptible to this ironic maxim. The meteorologist is looking out for clouds of particular types, the plant hunter has learnt to ignore the common species, the chemist is expecting a change of colour, the astronomer is interested in the spectra of certain classes of galaxy – and so on. Much of the power of science comes from the specialized training of observers. Like a musical conductor or a wine taster, a palaeontologist or ethologist learns to be sensitive to very small differences within a very narrow perceptual frame[19].

Scientists are thus notoriously blinkered to features of experience that lie outside the frame of their specialized interests. Nevertheless, they have many opportunities to observe unusual objects or events, and occasionally have the wit – we call it 'curiosity' – to focus on them. *Serendipity* [8.9] is a characteristic phenomenon of scientific life[20], and plays a very important part in breaking the mould of received knowledge. But the dictionary definition, 'looking for one thing and finding another'[21], indicates that it is not an entirely random process. It occurs in the course of a directed activity[22], and is fruitful only for 'the prepared mind'[23]. In effect, serendipitous observation is an improvised descant to the major tunes of empirical research.

5.3 Eradicating subjectivity

By its very nature, an observation is performed by an individual. But to make it truly communal, it must lose this individuality. To become a *scientific* observation, it must not only be reported to somebody else[24], it must also be stripped of elements peculiar to the particular observer[25]. The apparent solidity and permanence of matters of fact resides in the absence of overt human agency in their coming to be[26]. Much of the 'method' of science is concerned with the eradication of *subjective* influences on research findings.

As we have seen, such influences are diverse and pervasive. It is a logical impossibility to eliminate them completely. But the success of science depends on minimizing them in practice, and then fudging the logic. In every branch of science, therefore, there are elaborate procedures for reducing the effects of subjectivity in empirical research.

Needless to say, this is a particularly significant issue in the human sciences[27]. Observers are trained to take a neutral stance on any matters of value or opinion [7.7]. Questionnaires and interview protocols are carefully standardized. Results that can only be obtained by 'participant observers' in the course of 'action research' are treated very cautiously. And so on.

But these elementary precautions against 'observer bias' scarcely touch the main problems, which lie much deeper. In many humanistic disciplines, for example, controversy rages over whether research is best conducted by 'insiders' or 'outsiders' to the culture being studied[28]. Does the mere fact of being a 'social scientist' necessarily imply a certain attitude – elitist, would-be emancipatory [7.10], or whatever – to the people being studied[29]? These are only entry points to a vast arena of serious scholarly debate.

The Legend has no way of dealing with this issue except by entirely excluding the human sciences from 'science'. Some natural scientists assert that psychology, for example, is not 'a science', except perhaps where it overlaps with physiology. Others limit the behavioural sciences to accounts of purely material phenomena, deliberately rejecting the information obtainable from the subjects of their observations.

But a philosophy of systematic *positivism* that brackets out 'thoughts' – i.e., the meanings that the actors themselves assign to their actions [5.11] – gravely misrepresents the realities that it claims to be describing[30]. It also fails to resolve the epistemological scandal. Indeed, carried to the conclusions of its own logic, it eliminates all forms of empirical knowledge. Even in the physical sciences, making an observation is an active process. As feminist critics of science have pointed out[31], it is always undertaken and interpreted by human beings with diverse personal interests and opinions[32]. This distinction between the natural and human sciences is not therefore as compelling as is usually supposed[33].

5.4 Quantification

From a naturalistic perspective, however, the difficulties with positivism are not fatal to the scientific enterprise. We have to ask what practices have evolved to shape the results of individual observation for collective ownership. What epistemic strategies do scientists find useful, and how far do they actually succeed, in reducing subjective influences to acceptable proportions?

One such strategy is *quantification*. The number of items in a countable assembly of distinct entities is an empirical invariant. Only a grossly incompetent or dishonest observer can be so influenced by 'subjective' factors as to fail to report this number correctly. *Quantitative data* – especially integers representing the results of actually counting heads, vehicles, organisms, stars, electrons, etc. – are thus much prized as scientific 'facts', although immense investments are made in increasingly elaborate systems for processing and analysing them, decoupling them from the act of observation[34] and making them ready for assimilation into the stock of public knowledge.

More generally, vast quantities of numerical data are produced by *measurement*. Indeed, the whole science of physics is built upon what can be said about natural phenomena in terms of observations of distance, duration and other precisely measurable quantities. Much of the philosophy of physics is concerned with measurement processes[35]. That is why twentieth-century physics is so paradoxical. Relativity theory, for example, is about the mutable properties of clocks and measuring rods at very high velocities, and quantum theory denies the possibility of perfectly precise simultaneous measurements of position and momentum.

By definition, a 'measure' is 'a basis or standard for comparison'. Measurement can never be purely subjective, since it involves communal agreement on what entities and procedures should be regarded as 'standard'. Physicists often argue that scientific observation is synonymous with measurement[36]. What they really mean is that by making measurements they can produce numerical data that are tailor-made for communal acceptability.

Standardization – qualitative as well as quantitative – is clearly a very effective epistemic strategy[37]. Over the past century, quantitative measurement has been extended into all the natural sciences, often with impressive results. Empirical quantitative data are vital to chemistry, geology, biology and other disciplines. Increasing emphasis on numerical measurement explains the emergence of hybrid disciplines such as 'chemical physics', 'geophysics' and 'biophysics'. But it goes further. What scientists regard as measurable characteristics of chemical compounds, rocks, organisms, etc. is not confined to their fundamental 'physical properties'. Take, for example, the mineralogists' scale of hardness. This is essentially empirical, since it is based upon whether specimen A scratches specimen B under prescribed conditions. It works because practical experience has shown that certain standardized observational procedures produce consistent results that can be associated with an arithmetical scale.

Nevertheless, there are many scientifically interesting aspects of the natural world that are not amenable to this treatment. In important observational sciences such as biological systematics [6.3] and animal behaviour, all the empirical data have not been successfully reduced to tables of numbers.

The effort to extend the methodology of numerical measurement into the human sciences has met with even less success. It is true that certain observable features of human behaviour can be defined in a way that satisfies the axioms of an arithmetical scale[38], but there is no proof that the quantities thus defined – for example, 'Intelligence Quotients' – are of operational significance beyond the process in which they are 'measured'.

Even the vast arrays of numbers generated every day by financial activity are not to be taken at face value. As every accountant knows, economic entities such as 'profits' and 'assets' are valued in practice according to arbitrary conventions that leave much latitude for individual discretion[39]. The figures that appear on balance sheets are important empirical material for economic analysis, but they are not scientifically equivalent to quantitative data obtained by 'objective' measurement.

This caveat applies even to numbers obtained by counting heads. Social statistics play an important role as rational elements in the modern social order[40], but they are based on very arbitrary and subjective categories [6.2]. Consider, for example, the official figures for 'unemployment', or 'violent crime'. The fact that the formal definitions of such data are frequently changed is not due solely to the reflexive interaction between public concern and political cynicism[41]. The categories on which such figures are based are typically vague, uncertain and strongly affected by the social procedures by which they are 'observed'[42]. To give them scientific status – that is, to establish them as communal property amongst social scientists – would require general agreement on how they should be defined and counted.

Indeed, even the basic demographic data obtained periodically at great trouble and enormous expense by a national census are not as reliable as they seem. The dilemmas encountered in the apparently simple process of counting people can only be resolved by arbitrary rules and standardized practices which ignore or over-ride more subtle features of the situation[43]. In other words, whatever the other virtues of quantification, it cannot succeed in eliminating social or personal factors from the human sciences.

5.5 Instruments

Modern science, like modern society, is highly technological. *Apparatus* is an essential dimension of research[44]. And yet it is largely taken for granted by desk-bound metascientists. A new research technology, such as the telescope, the microscope, the cyclotron or the DNA sequencer, can enormously amplify every aspect of a scientific field[45]. Such inventions are widely celebrated, yet they are seldom analysed for their deeper epistemic significance.

In particular, almost all serious scientific observation is actually carried out with the aid of *instruments*. Since an instrument is free of personal idiosyncrasies, its output is not loaded with subjectivity, and is therefore very suitable for communal acceptance [6.3]. In other words, scientists favour instrumental observation partly because it is an effective strategy for producing empirical scientific 'facts'[46]. Indeed, social and behavioural scientists often describe their questionnaires and interview protocols as 'research instruments', just to make this point.

In practice, a wide range of technical devices are lumped together as observational instruments. At one end of the range there are those that simply enhance ordinary human perception. These would seem to present no problem[47]. It would take a very subtle (and perverse) philosopher to distinguish in principle between a pair of powerful binoculars and a pair of contact lenses at the sharp eyes of the observant naturalist.

At the other extreme, however, there are instruments such as voltmeters and Geiger counters that detect and measure phenomena that are not consciously experienced by human beings. The interpretation of their 'readings' obviously involves a great deal of scientific theory. To describe them as 'observations' stretches empiricism a long way from its home base.

Nevertheless, there is no discontinuity in scientific practice across this whole range. What is the difference in principle between watching a nesting bird through a pocket telescope and recording it with a hidden video camera or observing it by night with infra-red binoculars? Is more 'theory' involved in making sense of the pointer readings of a Geiger counter than of the vivid images, as of sections through a living brain, produced by positron emission tomography? Is the jobbing electrician's ten-dollar voltmeter less 'scientific' than a billion-dollar satellite recording and mapping the earth's whole magnetic field?

It is a philosophical fantasy to suppose that a scientific 'fact' can be freed from the context in which it was observed. That context always contains both 'theoretical' and 'subjective' features, usually closely intertwined[48]. A sophisticated instrument embodies many theoretical concepts. But these are only elaborations and extensions of the theories needed by a trained observer to 'see' what is scientifically significant in her personal experience of the world. A Geiger counter does not detect or measure a magnetic field because that is not its design concept. Similarly, a geologist in an underwater vehicle does not notice novel deep-sea organisms because they are not in his domain of understanding[49].

Astonishing visual immediacy is now achieved by computerized processing and presentation of observational data[50]. But this does not really change the situation. The apparently realistic images are instrumental artefacts that have to be interpreted expertly – that is, subjectively. Indeed, scientific specialists quickly learn to see and think 'through' their instruments to the phenomena in which they are interested[51]. This is a professional skill that cannot be taken over by a lay person or an 'expert system'. The high energy particle physicist points to a blip on a graph of scattering data from the ultimatron and says, 'There! Do you see it? It's the gluino particle we've been looking for!' You have to believe her. Once she is assured by communal consensus that her instrument can be used routinely for this purpose, it becomes as 'transparent' to her as her contact lenses[52].

It is quite wrong to suppose that the perfect scientific observer looks at nature with the innocent eye of a child. On the contrary, scientific observations are necessarily sophisticated and contrived. Modern scientific instruments, whether as immense as high energy particle detectors or as delicate as scanning tunnelling electron microscopes, combine threads of artistry from many distinct scientific and technological traditions[53]. Their results are strongly shaped by structural, theoretical and personal factors. How is it possible, then, for them to coalesce into a body of commonly agreed knowledge?

The answer is that the factors that shape scientific observations and instruments are themselves communal. The observers are members of the research community to which their observations are to be communicated. They are experts in their field. That means that they have learnt, by their professional training and experience, to see the world in the same peculiar way as their mentors and colleagues. The instruments they use and the theoretical concepts that regulate their perceptions are precisely those prevalent in this community.

What scientists count as 'empirical facts' may seem to a lay person extraordinarily artificial, conceptually heterogeneous, and entirely divorced from normal human experience [8.12, 10.7]. But the members of a particular research community are concerned with only a very limited selection of such 'facts'. They are presented with research claims that seem to *them* quite natural, rational and ordinary – claims whose communal acceptability is largely a matter of detailed scrutiny rather than general plausibility.

5.6. Experiment

The scientific norm of disinterestedness [3.5, ch.7] is designed to combat the subjective factor in observation. Elaborate practices have evolved, especially in the social sciences, to neutralize the psychological quirks, and political, moral, economic and career interests, of individual researchers. The effectiveness of these practices may be debatable, but they are integral to the scientific culture.

Where the scientific ethos permits scientists to differ is in their individual *scientific* interests. As we have seen, what they 'observe' is strongly influenced by the body of concepts and methods they share as members of a research community. But this aggregate of 'received ideas' and 'established facts' is always ill-defined, uncertain, incomplete – and open to question. Active researchers offer very varied lists of what they think is agreed on in their field[54]. Indeed, the norms of originality and scepticism positively encourage diversity and idiosyncrasy in the theories that scientists bring to their research [8.10, 9.5].

This diversity cannot be eliminated from the collection of empirical 'facts'. Even the most rigorous observational protocols and the most sophisticated instruments are products of the minds of their users. The only way to deal with such factors is to make them explicit. The scientific community accepts a research result only if it is accompanied by an account of how and why it was obtained. This account normally includes details of its supposed theoretical significance. Allowance can thus be made for this highly variable subjective element in every reported scientific 'fact'.

The many-sided role of 'theories' in the production of scientific knowledge will be discussed at length in later chapters. But the most elementary philosophies of science all emphasize the complementarity of theory and *experiment*. Experiment, also, is a protean scientific practice, which will appear in various forms in later chapters. It plays a very important part, for example, in sustaining the norms of originality and criticism.

For the moment, however, an experiment can be thought of as an act of observation designed to yield a particular type of empirical knowledge. The familiar notion of an experiment is of a deliberate interference with the natural run of things. But this is not of its essence. Thus, scientists often refer to any complex observation with an elaborate instrument

– for example, obtaining a detailed ultra-violet video image of the surface of the sun from a space vehicle poised a million miles above the earth – as an 'experiment', whether or not it involves intentional intervention in the course of nature[55].

The elementary dictionary definition gets to the heart of it: '**experiment**: 1. a test or investigation, esp. one planned to provide evidence for or against a hypothesis . . .'[56]. In other words the whole meaning and purpose of an experimental observation derives from the theoretical context in which it is carried out. Not only is this context made explicit. The observation is designed to produce empirical information that is significant primarily in that context.

Once again, we postpone discussion of how scientific theories actually become established as the collective intellectual property of research communities. The primary role of the individual scientist in this process [9.1] is to make as strong a case as possible for their particular point of view. The experimental researcher is required, therefore, to present her results in relation to a stated theoretical attitude – even if only a generally sceptical, agnostic or exploratory attitude – towards the matter under investigation. She must claim to have observed empirical 'facts' that are favourable or unfavourable to this attitude. Otherwise – who wants to know what she thinks she has seen?

Needless to say, this requirement is the basis of many familiar scientific practices. It goes much further than the use of quantification and instrumentation to reduce subjective factors. The problem is to control the *contingent* elements in observational research. For example, research scientists learn the importance of standardizing their experimental materials, calibrating their apparatus, recording their observations systematically, and so on. In other words, they try to work under conditions where the explanation for an observed 'effect' has to be something more interesting than, say, an impure chemical reagent, a malfunctioning recorder, or sloppy note-keeping. This is not a trivial point. As the sagas of 'anomalous water' and 'cold fusion' show, it is surprising how much wild theorizing can be triggered off by 'effects' due to faulty experimental technique.

As I have indicated, however, the most effective way of projecting an experiment on to a particular theoretical dimension is to undertake it in carefully contrived circumstances where all other potential disturbing factors are eliminated[57]. Every branch of science has its strategies for doing this – studying the behaviour of bacteria on clean petri dishes, rather than in mucky soil, scattering electrons from homogeneous beams of particles rather than in hot plasmas, testing a therapeutic drug against a placebo by a double-blind trial rather than just administering it to all apparently needy patients – and so on.

When people complain that science is 'unnatural'[58] this is presumably what they are really getting at. Many of the situations set up by experimenters are indeed 'unnatural' in that they do not arise in the flow of events outside of science, and may have never occurred before in the history of the universe. Science takes natural objects away from their normal life course. In practice, scientific knowledge is largely designed to account for what happens to carefully constructed artefacts in artificial circumstances[59]. This 'alienation of nature' is not merely a matter of some moral concern. It also raises major epistemological issues, to which we shall return [8.12, 10.7].

As I have remarked, experimentation is a multiform scientific practice, with a complex life if its own[60]. Elaborate experimental systems become like space ships, self-contained, man-made worlds whose polyglot inhabitants have been trained to perform together their carefully assigned tasks and which communicate only occasionally with earth along coded channels. Though such systems may be planned precisely for specific missions, their enormous social and epistemic momentum often carries them serendipitously into regions of knowledge beyond the scope of current theory. For the moment, all I am saying is that experimentation is more than a way of solving problems, or of exploring the hidden potentialities of nature. It is a strategy for generating empirical 'facts' that are relevant to the theories current in a research community – that is, 'facts' that are strong candidates for acceptance as communal knowledge.

Active experimentation is given such prominence by some philosophers that it is sometimes thought to be an essential component of science. But there are whole disciplines, such as palaeontology and archaeology, where this mode of investigation is logically impossible. Nevertheless, empirical evidence about past events can be obtained from their traces in the present – fossils, geological formations, segments of DNA, stone tools, the foundations of buildings, and so on. These are often material objects whose existence can easily be made the shared knowledge of a research community, even if their theoretical significance may not be immediately clear. Such disciplines are therefore perfectly scientific, even if sometimes very controversial.

Indeed the case of all the human sciences is somewhat similar.

Historical research is often strongly guided by theoretical inferences about what might be found, say, in particular types of archive, but cannot be actively experimental. In the other social and behavioural sciences, it is usually very difficult to intervene in the flow of events under controlled conditions[61]. 'Action research' is not necessarily an invalid method of investigation but its results are very risky. The ideology of the participant observer is impossible to neutralize [7.7], and his mere presence may perturb the events being studied.

On the other hand, empirical information about human events can be obtained from the spoken and written accounts of their participants. As we shall see, there is no fundamental reason why such accounts should not be acceptable as sources of communal knowledge. It is as sound a 'scientific fact' as any pointer reading that Napoleon once ruled Europe and was finally defeated at Waterloo, even though it depends overwhelmingly on the written records of contemporaries.

The point is that there are no absolute rules on what constitutes a scientific fact. Formal criteria and classification schemes are soon made obsolete by scientific progress. What counts originally as an 'effect' becomes the basis for a novel method of observation which finishes up as a routine instrumental technology[62]. At any given moment, a discipline or subdiscipline has its recognized research methodologies, its observational instruments, its experimental protocols, its standard scales of measurement and other conventions. But these are maxims, not rigorous regulations. They are not mandatory in principle, and are often ignored in practice.

The common function of these maxims is to ensure that empirical research produces results of the right kind to be incorporated into the communal stock. In particular, these results should have theoretical implications, as judged by the members of a research community [ch.8]. This is not a trivial requirement. It motivates an enormous amount of personal skill, professional craftmanship and social organization. It is also much more difficult to satisfy than many philosophers seem to reckon[63].

5.7 Trust

Scientific knowledge is a collective good, and a collective accomplishment[64]. It is not just the aggregate of the 'contributions' of individual researchers[65]. However aptly presented these may be, they still have to pass the various tests imposed by a sceptical research community [9.1]. Indeed, to be considered at all – for example, by journal editors and referees – they have to satisfy basic standards of prima facie credibility [7.3].

In effect, the observations and experiments reported by each researcher have to be worthy of belief – if only provisionally – by others [8.10]. This is absolutely crucial to the scientific enterprise. It is quite impossible for a single scientist to conduct all the experiments and develop all the theories needed for his own research[66]. But as a member of a research community he shares in a stock of communal knowledge on which he can reasonably rely – at least in the first instance.

In ordinary life, the touchstone of credibility is the testimony of independent eye-witnesses of the same event. We surely require at least the same standard for distinguishing a scientific 'fact' from a hallucination[67]. Many of the practices of academic science can be traced back to attempts to achieve a situation of collective witness[68]. Significant experiments were carried out at actual meetings of the research community, whose members could thus observe their results and be personally convinced of their validity. Indeed, being eminently respectable citizens, they could attest to this to a wider public.

This direct procedure for turning research findings into communal knowledge can sometimes be very effective – for example, in geological fieldwork carried out jointly by independent observers[69]. In general, however, it is impossibly cumbersome. There is no real alternative to the academic practice of relying heavily on the findings reported by individual scientists – or small groups of scientists – working in private. The empirical 'facts' of science are products of, and belong to, research communities. Nevertheless, they originate in the experience of isolated individuals – including our ancestors – on whose say-so they are mainly accepted[70].

Positivists find this distressing. They want absolute criteria for the reliability of scientific testimony. But as legal theorists have discovered over the centuries, this inquiry regresses into a wilderness where every path ends with 'he has always proved completely trustworthy', or with 'that's what I was told'. In other words, the acceptance of consistent, coherent and plausible testimony from knowledgeable, disinterested witnesses is an essential feature of all practical reasoning. Knowing how to evaluate testimony is a basic requirement for membership of any culture[71].

This is why *trust* is an even more important factor in science[72] than it surely is in society at large[73]. A reputation for personal trustworthiness on scientific matters is part of the professional stock-in-trade of all scientists [3.5]. This is fostered by their education, and by a variety of social sanctions and rewards. To some degree it extends to impersonal institutions and research systems – even to 'science' as a whole.

Amongst working scientists, this trustworthiness is part of the informal moral order of each research community[74]. The complex interplay of originality and scepticism that operates in such groups requires absolute interpersonal trust on matters of empirical 'fact'. One of the effects of specialization is the appearance of 'invisible colleges' where 'everybody who's anybody knows everybody else'[75]. As we all know, familiarity with the quirks of our colleagues does not necessarily breed confidence in their personal sincerity on all fronts. But elementary prudence indicates that honesty is the only policy on matters where deceit would quickly be detected.

5.8 Verification

Mention of the possibility of deceit reminds us of psychological and social realities. Trust amongst scientists may be unusually high, but it cannot rest entirely on cultural convention. A reputation for integrity cannot be sustained for centuries on mere esteem. The streetwise punter knows better than to take the word of any such self-regarding group at face value. And yet, academic science is not systematically policed against deliberate fraud [9.4].

Science operates socially as a trustworthy institution because it has, and is known to have, a hard epistemic core. The norm of communalism requires scientists to specialize in producing the type of knowledge that can be taken on board by a research community. This means, for example, that research claims should not deviate too far from reports of events that might conceivably be witnessed collectively. In philosophical language, we say that empirical 'facts' have to be capable, in principle, of *verification*.

This is, of course, one of the major regulatory principles of science. Philosophers have discussed at length the conditions and criteria by which the results of scientific observations and experiments can be made perfectly secure. But the verification of a novel observation necessarily involves prior empirical knowledge of 'the world', so that all such inquiries either close into a circle or regress to infinity[76]. At best, they rationalize commonplace maxims [10.6] such as 'seeing is believing'[77], or 'if you don't believe it, try it for yourself'.

To a naturalistic eye, however, the need for verification explains a number of very interesting scientific practices[78]. Indeed, it is a 'triple point', where psychological, social and epistemic considerations meet and overlap. Take, for example, the customary form of a scientific paper [9.3]. By tradition, this is written as if from the standpoint of an anonymous observer, supposedly reporting in detail on certain events that took place in the author's laboratory. Although clearly fictitious[79], the story told by a 'virtual witness' has the psychological force of personal testimony[80]. Notice, also, that no information is given as to its actual setting. The location of the experiment is moved to a vaguely public space, as if to indicate that it is a social rather than a private experience[81].

Above all, the events reported should be *reproducible*. An account of an experiment or observation should give the reader all the information required to carry out exactly the same procedure and observe the same outcome[82]. This is clearly a severe constraint on the type of knowledge considered acceptable to science. As every scientist knows, this regulative principle both limits and empowers the concept of an empirical scientific 'fact'.

Let there be no doubt about the power of this principle in the service of communalism. Major scientific journals such as *Nature* and *Science* frequently carry reports of the failure of researchers to *replicate* the discoveries claimed by their peers. The rise and fall of research on phenomena such as 'anomalous water' and 'cold fusion' is a chronicle of experiment and counter-experiment by independent research groups, seeking for regularly reproducible results. Once it became clear that these phenomena could not be replicated under the stated conditions, they were excluded from the body of 'established knowledge'.

This power is effective, moreover, without being frequently exercised. In practice, only a very small proportion of the experiments and observations reported in the scientific literature are actually replicated by other researchers. After all, there is little CUDOS to be gained from repeating an experiment whose result is not seriously in doubt – unless perhaps it jeopardizes one's own cherished research claim[83]! Nevertheless, this possibility is sufficiently real to serve as a backstop to all empirical research. Research methods are specifically designed to guard against it. Research claims are muted, to avoid challenge on uncertain points. Indeed, researchers often throw away unexpected results, for fear that they might later be shown to be accidental artefacts of their experimental setup[84].

In general, however, it is seldom as easy as it might seem to reproduce an empirical observation. Indeed, logically speaking, it is impossible.

The events that occurred under particular circumstances at a particular time can never exactly recur. The water in the river now is not what it was then; the earth has moved another million miles around the sun; a thousand babies have been born and nine hundred and ninety people have died; and so on. For similar reasons, simultaneous discoveries at different locations are never identical[85]. Most of the known changes can be discounted, but that is a matter of judgement. A reported observation can only be simulated. What counts as a sufficiently perfect replica depends very much on what theoretical issues are thought to be at stake[86].

This is a very practical problem in biomedical research. Apparently identical organisms belonging to the same species are always individually unique[87], and can differ significantly in their responses to the same conditions. Laboratory animals, for example, are never perfectly standardized or homogenous. To obtain reproducible results, it is usually necessary to collect information over a representative sample. But most of this information is discarded as it is incorporated into a research report[88]. In the end, therefore, even empirical scientific 'facts' seldom record precisely what actually happens to real organisms. They relate to artificial entities which are deemed to be identical in certain selected characteristics[89].

In the human sciences, of course, this problem becomes acute. Even when experiments are feasible, they are difficult and expensive to replicate[90]. Survey methods can produce reproducible statistical data, but often at the expense of conceptual significance. The media joke about the typical family with 1.9 children pokes fun at a genuine difficulty in the theoretical interpretation of such 'facts'.

On the other hand, the human sciences have a strong empirical base in meaningful texts. For all practical purposes, legal instruments, manuscript letters, printed books, transcripts of interviews, etc. are perfectly replicable as *documents*. Palaeographers, editors, archivists and other professional experts separate the messages from the material media on which they happen to be inscribed, and make them available to research communities. In other words, academic science rightly includes disciplines whose members rely for their research on historical, literary, political, autobiographical and other documents. The messages in these documents need to be interpreted and may not be credible. Nevertheless, they are accepted and shared communally as empirical 'facts', along with fossils, *objets d'art*, and other material entities.

In the physical sciences, by contrast, most experiments are almost perfectly replicable. Precisely the same instrument can be set up to make the same types of readings on the same, carefully prepared objects. It is a law of nature that all electrons are absolutely identical, and so are all protons, all nuclei of the same isotope, all molecules of aspirin, and so on. Indeed, if that were not so as an empirical fact, we should proclaim it as a working principle until forced by other evidence to reconsider it.

Nevertheless, scientific experiments are usually performed at the very edge of technical feasibility. An elaborate piece of apparatus may have been specially constructed, or an unusual configuration of sensitive instruments set up, to detect and measure an apparently inconspicuous effect. The apparently unproblematic procedures described in research reports are seldom really perfectly straightforward[91]. The published accounts conceal or play down the particular skills, knowhow, instrumental facilities and other local factors that have actually gone into the research[92]. The physical sciences depend for their meaning on the invariance of laboratory phenomena under changes of time and place[93]. Research specialists share a great deal of 'tacit knowledge'[94] which would not be apparent to the lay reader of their papers. But the particular skills that may be required to 'make an experiment work' in one laboratory may not be easily transported to another[95].

This is one of the arguments against the 'mere replication' of experiments[96]. It is much more original to put the signature of one's laboratory[97] on to a new experiment, better designed, with more sensitive instruments, etc. Indeed, there might be more chance of success using apparatus based on a somewhat different principle. But that would imply prior acceptance of the whole body of theory that guarantees the equivalence of these two techniques for making the 'same' observation.

Consider, for example, the thirty-year-old challenge of Joseph Weber's claim to have observed gravity waves[98]. The essence of this observation is to detect tiny variations in the distance between two massive objects. Weber tried to do this by measuring the longitudinal vibrations of a solid metal bar. Nowadays, physicists are spending millions of pounds trying to measure the same effect – they believe with much higher sensitivity – by looking at interference phenomena in a laser beam reflected back and forth between the two objects. There is general agreement that these are just two different ways of performing essentially the same experiment. And yet almost the whole of our contemporary understanding of gravitational theory, materials science, quantum optics, etc. is implicated in this equivalence.

This argument would apply in principle even if the changes in the experimental setup were apparently trivial – for example, by the substitution of a different type of memory chip into the instrumental control circuits, or a different brand of colour film in one of the cameras. We are back once more to the point that all empirical 'facts' are embedded in theory.

Instead of being 'verified by reproduction', scientific observations are usually *validated by triangulation*[99]. The geodetic metaphor seems peculiarly apt. Scientific theories are analogous to maps [6.4] . Surveyors are trained to dissect a landscape into triangles, and to make more measurements of angles and distances than are geometrically necessary to locate their vertices. The closure of these triangles is a practical sign of the consistency of the survey and the validity of the map. So if the same empirical phenomenon can be observed by two different methods it must surely be genuine – and so also, incidentally, must be the background theory shared by the researchers and their peers[100].

Or so it is generally argued. But this shows that the notion of an empirical scientific 'fact' cannot be separated from the whole question of the epistemological status of scientific theories, which is central to this book.

5.9 The personal element

In this chapter we have seen that many familiar characteristics of scientific knowledge are closely linked with the academic norm of communalism. But there is no *direct* connection between social practices and epistemic principles. All such linkages pass through the minds of the people who carry out the practices and think the thoughts associated with them [1.4]. An empirical scientific fact originates in an observation – an act of human *perception*. To become communal property it has to be accepted by the members of a research community – that is, through acts of human *cognition*. What can be known to science can only be what can be known to scientists [8.14].

For this reason, scientific knowledge is shaped, perhaps fundamentally limited, by the powers of the human mind. The research process cannot be cleansed of personal elements [5.3]. It cannot be entirely mechanized [5.5] or perfectly regulated socially. It is true that a research community is not unlike a complex instrument, whose components are sensitized, standardized and calibrated by their scientific education and experience[101]. But it is energized by individual variations in their capacity for reasoning, feeling, perceiving and moving. The possibility that these variations might eventually be eliminated by automation is beside the point. We are concerned with academic science and its present-day successors, not with the hypothetical knowledge-producing machines of a conjectural future.

Traditional philosophies of science keep all such 'irrational' considerations at arm's length. 'Psychologism' is classed with 'sociologism' as a subversive doctrine, undermining the rule of sovereign reason. But in a naturalistic account of science, both these dimensions obviously have to be taken into account[102]. Indeed, in spite of some notion that they might be in competition for the epistemological high ground[103], they are quite clearly complementary[104]. *Public* knowledge[105] is both the content and context of *personal* knowledge[106] – and *vice versa*[107].

The real problem is that psychology impinges on the scientific enterprise at so many different points. For example, many sociological characteristics of academic science [ch.3] – socialization to research[108], career progression[109], conformity to the ethos[110] and so on – can only be understood in psychological terms. In the present context, it is convenient to limit this discussion to the conventional subject matter of *cognitive science*[111], even though this excludes many mental activities, such as emotion and motivation, which undoubtedly play a part in science[112].

Epistemology is as central to cognitive science as cognition is to epistemology. That does not mean, however, that we must accept in principle the idea that the brain operates like a very elaborate computer. This idea is widely held amongst cognitive scientists, but is also widely disputed[113]. To accept it here would fatally compromise our whole enquiry, for it would effectively prejudge the question whether scientific knowledge, as such, is actually computable.

In any case, we are committed to a naturalistic model. In other words, we start with the assumption that scientific cognition is not distinguishable in principle from the normal thought processes, such as perception, cogitation and belief, in which people engage in everyday life [10.2–10.6]. These may be complex, confused and mysterious, but we really know much more about them than about the brain mechanisms that presumably underlie them.

We do know, however [10.3], that human cognition is a product of biological and cultural *evolution*[114]. Our elaborate cognitive capabilities, including science itself, have apparently evolved by continuous organic descent from the most primitive modes of 'knowing'. This, in brief, is the central plank of *evolutionary epistemology*[115], which can be extended meta-phorically[116] to explain many features of the scientific enterprise [9.7]. An evolutionary approach suggests much more than that human cognition is tuned naturally to the 'mesocosm' – the world of medium-sized objects[117] – since it bears on the fundamental question [10.3] whether other beings might see things so differently, and reason according to such different rules, that they would produce a very different type of science[118].

For the moment, however, we are concerned with *perception* – i.e. the way in which information is gathered in from the external world. This is often differentiated from other cognitive processes, but as even my simple dictionary[119] recognizes, it includes the *interpretation* of this infor-mation. That, surely, is consistent with its evolutionary origins. Perception is not a passive 'I am a camera' procedure. It is normally an active part of a unitary process, involving focussed attention, remem-bered images, conscious or unconscious cogitation, guided motor action, etc.[120].

Paradoxically, perception requires pre-existing knowledge of what is to be perceived[121]. If, like every sane person, we believe in the existence of an external world [10.7], we must still accept that even the most realistic visual images are personal versions of that world, constructed individu-ally in the mind of each beholder[122]. It took a boy's eye to undress the king[123]. What we perceive in a given situation also changes with time. Part of our evolutionary heritage is the capacity to learn from experience. We instinctively interpret sensory information in terms of all that we 'know' at the moment we receive it. This includes a great deal of tacit knowledge[124] acquired from trying to operate for ourselves on the world about us[125].

Indeed, it is very likely that our elementary conceptions of time and space [10.2] derive as much from infantile experience of what we can achieve by bodily action as from what we can see or hear[126]. The blind person 'seeing' his way with a stick is a classic example of this general human capability for active, constructive perception[127]. This is evidently an evolved capability, which forms a natural basis for the way in which scientists often learn to 'see' instrumentally[128]. The world that they manipulate with their apparatus becomes quite real to them [5.5], even when its 'objects' are magnetic fields, quarks, subducted continental plates, and such-like other-worldly entities.

Cognitive science thus fully supports the contention that scientific observation, like historical scholarship, is not a simple matter of record-ing events 'as they really happened'. Human observers do indeed trans-form and interpret sensory data according to their own personal interests, expectations and experience. Visual perception is sometimes likened to the development of scientific understanding[129] – or even taken as the activity on which all scientific practice is metaphorically modelled[130]. Each act of observation has to start with an initial hypoth-esis, which is subsequently tested and modified to accord with experi-ence. Perception, like research, is not the passive reception of signals that convey their own meaning. It involves a dynamical, often highly idiosyncratic interaction between mental representations and natural events[131].

But a realistic understanding of the nature of individual perception does not debunk scientific knowledge as a social product. Considered piece by piece, the results of research are fragmentary and insecure. Nevertheless, ruled by communalism and other norms, the scientific enterprise spins and weaves remarkably robust knowledge out of just such fragile and variegated scraps of information.

5.10 We are not alone

An essential characteristic of any item of scientific knowledge is that it can be shared with other people. What do we mean by 'other people'? For the epistemological naturalist, relying on 'folk psychology', this is not a problem[132]. But the philosophers of science discovered long ago that it was almost impossible to demonstrate *logically* that such entities exist[133]. For this reason, they often try to reduce science to a one-person game, played against 'nature' by an autonomous observer/thinker/knower[134].

Needless to say, this sort of *methodological solipsism* is entirely inconsis-tent with the communal norm. The notion of a community implies active interaction – i.e. *communication* – between its members. Reliance on testi-mony [5.7], for example, depends on the existence of two or more observ-ers with similar cognitive interests, but distinct viewpoints on the same field of events[135].

Intersubjectivity thus plays a fundamental role in science[136]. This insight can be found in various earlier writings[137], especially in the works of George Mead[138], Edmund Husserl[139] and Alfred Schütz[140], but is only now coming to be generally accepted[141]. Perhaps it was just too obvious. Scientific activity would clearly be impossible unless scientists could communicate with one another and often find that they were getting very similar results.

In everyday life, intersubjectivity is equated with realism [10.7]. The similarity of the experiences reported by different persons is almost invariably taken to be due to – or evidence for – their living in the same 'external world'. Indeed, this inference is so basic to the 'Natural Ontological Attitude'[142] that it is treated as a primary fact of life.

Nevertheless, there are some situations in science where that conclusion, even if valid, is not helpful. Sometimes, as in quantum physics, it is not clear what the concept of an 'external world' really means, or whether even such elementary observations as the invariance of the order in which events are seen to occur are a property of that world or of ourselves as conscious observers[143]. This applies also in the human sciences, where the notion that individual members of a culture all inhabit the same 'social world' [8.13, 10.4] has to be interpreted very cautiously[144]. Many affective phenomena, moreover, such as pain and empathy, are reliably interpersonal, but are not made more intelligible by that type of reductive explanation. When, for example, I say to a friend "I know just how you feel", our shared knowledge of toothache is not advanced by the retort: "That's because we both live in the same world."

Wider reflection suggests that intersubjectivity is as basic to the human condition as subjectivity itself[145]. Our self-awareness as individuals – typically summed up in Descartes' dictum, 'I think, therefore I am' – is complemented by our awareness of other beings in the same plight. Our higher-order perceptions of ourselves are silhouetted against our perceptions of others and their perceptions of us[146]. Our thoughts are stabilized by 'conviviality'[147] – that is, opportunities for comparing them with the thoughts of others[148].

Again, from a sociological point of view it is not the existence and activity of individuals that make a culture, but their *co-existence* and *interaction*. An analysis in terms of 'institutions' [10.2] must include the bonds between their members. 'Networking' is now the vogue word in science studies. But the elements of a net fall into two complementary classes. Most accounts of academic science focus on the *nodes* – individual researchers. Intersubjectivity is the essential quality of the *links* – the various modes of communication that combine the activities of individuals into a science[149].

In a naturalistic approach, these considerations arise almost automatically. The question is, should we burrow more deeply into their philosophical foundations? In the case of individual cognition, we were quite ready to explore its characteristics scientifically, even though it could only be considered philosophically as an inscrutable precondition of existence. Similarly, although intersubjectivity is a natural phenomenon capable of scientific study, it also seems completely inscrutable existentially. We can understand how it evolved biologically and emerged eventually as a major function of human consciousness; but – so what?

In other words, we are treating both personal cognition and interpersonal communication as *primary* components in our model of science. Together, they form as firm a foundation raft for epistemology as we are ever likely to find by traditional philosophical analysis. This is not to debar or debunk the effort to give them more profound meanings, or to probe their roots in neurophysiology, sociobiology and other scientific disciplines. It is simply a statement of the boundary we are putting on our own investigations in this book.

5.11 Empathy

For most natural scientists, intersubjectivity is merely a fancy name for observational consistency. It is a fundamental feature of human experience that this consistency is very often possible. Scientific knowledge, then, is the common element in all such observations. In consequence, research findings are accepted by research communities and transformed into scientific 'facts' on the basis of the testimony of several independent witnesses. Caution suggests that science should be limited to what can be discovered and shared in that way.

This multiple witnessing [5.7] is usually purely notional. In principle, it can be re-enacted, theoretically inferred from other observations, or imaginatively reconstructed from material evidence such as fossil remains. In practice, however, these operations always introduce subjective and/or social factors [5.3]. Observational consistency is only obtained by discounting the influence of tacit knowledge, local instrumental quirks, theoretical preconceptions, or even material interests[150]. Purely observational knowledge is not well defined [5.2]. Even loosely interpreted, its range is very restricted. Science, surely, has to have a much wider scope.

The key point is that intersubjectivity includes *empathy*. The power of 'understanding and imaginatively entering into another person's feelings'[151] is essential for participation in social life[152], and for shared understanding of it[153]. Up to now, anyway, it is a scientistic fantasy [10.8] to suppose that human actions can be accounted for without reference to the meanings that the actors give to their behaviour, including psychic states such as 'joy' or 'shame'. These states can be understood only by an observer who has had some personal experience of them and is aware that other people also have them[154].

Scientists, being normal human beings, have all had this experience. They all exercise empathic understanding in everyday life. Indeed, this is one of the ways in which humans differ fundamentally from computers, and would presumably differ from intelligent non-humans if such there were[155]. Its failure to develop in childhood – perhaps due to a genetic defect – is characteristic of autism, a crippling mental handicap[156].

Empathy with the actors is thus an essential feature of observation in the human sciences. This is an old idea, going back to the eighteenth century[157]. Ethnographers employ empathy constantly[158], as when they use interviews to enter people's lives[159] and elicit their personal value systems[160]. In spite of insisting on their objectivity[161], historians have always had to infer the thoughts and motives of the human agents in their stories by re-enacting them in their own minds[162]. How then could a historian or sociologist make sense of, say, a religious movement without some personal experience of religious feeling[163]?

Empathic understanding not only operates between the human observer and the human actor. It also operates between different observers of the same behaviour [6.2, 6.10, 9.2, 10.4]. When, for example, a historian writes that the prospect of a harsh winter made Napoleon decide to retreat from Moscow she is not just relying upon her personal knowledge of how people think and act: she is also taking for granted that her readers have essentially the same knowledge, and will therefore understand and agree with her empathic account of events. This is what makes history a science. Research reports ascribing intentions, motives and emotions to human beings can be accepted intersubjectively by other researchers, and thus become part of the shared knowledge of a research community[164].

The human sciences depend fundamentally on *verstehen* – that is, understanding 'as if from the inside'[165]. They are steeped in *hermeneutics* – that is, the traditional scholarly discipline of 'interpretation'[166]. Nevertheless, hermeneutics is not a solitary, subjective, art form. It too relies on intersubjectivity. It would be pointless unless it were sometimes possible to arrive at agreement between independent commentators[167].

In principle, therefore, the human sciences are not debarred from complying with the communal norm. They have the means of establishing a body of shared empirical knowledge[168]. If they find this difficult to do in practice[169], it is not because the 'facts' are in doubt as such. It is mostly because these 'facts' are so complex and diverse that their theoretical significance is very seldom clear. Consider the immense scholarly literature concerning a major historical event such as the outbreak of the First World War. Which of the millions of well-attested documents in the archives are relevant to which of the innumerable theories about its origins?

But compare this with the situation of the first biologists, faced similarly with the complexities and diversities of the organic world. The scientific 'facts' of ecology and taxonomy only became clear through an elaborate social process of observation and interpretation. The identification of different species of organism is a sensitive skill that is shared by most human beings [6.2], even though it cannot (or cannot yet) be simulated by a computer. The same applies to much of the tacit knowledge that is used in the physical sciences[170].

In other words, a naturalistic approach via intersubjectivity breaches the traditional demarcation line between the 'sciences' and the 'humanities'. Let us not doubt that they differ enormously in their subject matter, their intellectual objectives, their practical capabilities, and their social and psychic functions [7.7–7.10]. Nevertheless, they belong to the same culture, and operate institutionally under the same ethos. As a consequence, the knowledge produced by the natural sciences is no more 'objective', and no less 'hermeneutic', than the knowledge produced by the social, behavioural and other human sciences[171]. In the last analysis, they are all of equal epistemological weight.

5.12 Modes of communication

When we say that a piece of scientific knowledge is the property of a research community, we mean much more than that most members of that community are familiar with it, and believe in it. That would apply to a great many other types of 'common knowledge' [10.6]. The members of a traditional farming community, for example, share a great deal of well-founded knowledge about crops, animals, soils, weather patterns, etc., gained by individual observation and passed on informally in conversation, skilled work, religious observances and a variety of other cultural practices. Indeed, one of the most important recent developments in science studies is the recognition of the diversity, extent and vigour of 'folk science'[172]. In the broadest meaning of the word, all systematic human activity requires 'science'.

Academic science, however, differs fundamentally from 'ethnoscience' in that it produces *codified* knowledge [9.3]. The norm of communalism activates an elaborate *communication* system [3.3]. Research results do not count as scientific unless they are reported, disseminated, shared, and eventually transformed into communal property, by being formally *published*. Scientific knowledge must not only satisfy the norms of universality, disinterestedness, originality and scepticism: it must be 'fit to print'.

This is often treated as a significant epistemological condition. Indeed, most philosophers of science have assumed that scientific knowledge is limited to what can be formulated and reproduced accurately using moveable alphanumeric type. That is to say, they have traditionally concentrated on forms of knowledge that can easily be expressed as linear sequences of verbal or symbolic propositions, such as '[all?] swans are white' or '$E=mc^2$', backed up by tables of numerical data. This has had the effect of grossly exaggerating the role of formal logic, mathematics and other modes of exact reasoning, even in the natural sciences [6.7].

Since the fifteenth century, however, print technology has also enabled the precise reproduction and dissemination of *images* of all kinds[173]. Despite the philosophers, scientists have never denied themselves this potent mode of communication[174]. The most cursory inspection of the scientific literature on almost any subject reveals innumerable diagrams, pictures, maps etc. that go far beyond 'illustrating' the written text[175]. This is associated with the human capacity for intersubjective *pattern recognition*, which plays a vital role in scientific practice and theory [6.2]. It is quite obvious, for example, that many major scientific disciplines, such as botany, geology and anatomy[176], could never have developed at all without some means of replicating and communicating observational information that can only be represented in visual form.

In recent years, moreover, new communication media have emerged alongside the 'Gutenberg technology' of printing. Some of these media, such as 'electronic journals', have features that challenge the very notion of a formal scientific archive [5.13, 9.3]. In general, however, the scope of science is enlarged by technologies such as photography, cine-photography, audio- and video-recording, electronic data-transfer etc. For example, when members of a research community view a film record of human or animal behaviour they approximate to the independent witnesses [5.7] whose testimony is needed – in principle – to establish empirical scientific 'facts'. 'When I made these signs for "banana" and "box", the chimpanzee did this. Do you agree that he understood them well enough to know where to look?' A video recording can, of course, be as much a construct [8.13] as a printed text. But because of its immediacy it clearly facilitates the communication, critique and sharing of intersubjective interpretations of 'what is going on'. In effect, 'multi-media' publication reveals the hermeneutic elements [5.11] in all the sciences – especially those dealing with human behaviour – and subjects them to the empirical test of multiple witnessing[177].

Thus, what is distinctive about a formal scientific communication is neither the medium nor the message: it is that it is *published*. The communal norm requires scientific knowledge to be made 'public'. Unlike an 'informal' communication such as a private conversation or a project proposal [4.9, 8.2], a contribution to the scientific literature is not transmitted along a closed channel from a particular source to a particular recipient. On the contrary, it is deliberately depersonalized and to some extent fictionalized[178], to indicate that it has been released from the control of its originator[179]. Moreover, since it is not addressed to any specific person or persons, it is transmitted as 'information', capable of being captured by an unknown person with a suitably tuned receiver[180]. Ideally, it is fully and freely available for open criticism and constructive use.

Academic science celebrates and fosters this ideal. Unfortunately, it is unattainable. This is not just because scientists have ordinary human failings, such as a tendency to be possessive or secretive about their findings. It is because the notion of complete, unconstrained 'communication' is vacuous. There can be no such thing as a 'message' without a sender and a receiver. They must obviously share the intention to cooperate and a symbolic code such as a natural language[181] [8.14]. This in turn implies that they already have common criteria of rationality [6.8], and agreement on many apparently plain matters of fact, such as the existence of the entities signified by certain symbols[182]. When scientists report their research results a great deal of tacit knowledge [5.8] about apparatus, techniques, specimens, etc. is left unsaid. A contribution to knowledge is never completely detached from its source[183].

Meaningful communication is an exercise in looking at things from the point of view of another person[184]. But we do not have direct access to the 'brain talk' of others[185]. The best that we can ever do is to express our thoughts in the very restricted medium of a suitable 'language' –

including, of course, mathematical formulae and diagrams – and interpret what others try to tell us in the same spirit. Much of the intellectual
action in the sciences is devoted to the development of specialized
'languages' to deal with particular aspects of the world, as revealed
by research.

Nevertheless, the most elaborate symbolic system can never express in
full the richness of the experiences we would like to share with our fellow
humans. All that languages can convey are simplified *representations* of
our percepts and concepts, never the real McCoy. Whether spoken or
written, they are no more than interpretations whose intersubjectivity
cannot be checked in a deeper sense[186]. The communal norm limits scientific knowledge to what can, in fact, be represented unambiguously in
mutually intelligible words, gestures, symbols, pictures, etc.[187]. This is a
fundamental epistemological point, to which we shall return again and
again [6.2–6.7].

But unambiguous communication is scarcely possible except between
people with similar mental models [6.4, 8.14] of the world and of each
other in it[188]. Such models do not construct themselves purely out of
individual bodily experience. They come to each of us in the course of
meaningful participation in social life [10.2], typically through the
medium of a natural language[189]. Indeed, much of the social activity in a
research community is associated with the development and refinement
of just such models – that is, scientific theories and the 'languages' used
to express them. And for reasons that will become clearer as we proceed,
research communities and scientific disciplines *speciate* [8.3]. They divide
and diversify into *specialties*, each associated with a theoretical *paradigm* –
in effect, a model representing some particular aspect of experience.

Even the most empirical research findings are saturated with theoretical notions and targeted on specific theoretical issues[190]. In practice,
therefore, they are only intelligible to a person who is reasonably familiar
with these particular theories and the 'language' in which they are presented [6.6]. Thus, although a formal scientific communication is not
explicitly addressed to named persons, it is really directed towards the
relatively small number of research scientists who share the same paradigm [8.4]. It is transmitted and received along a narrow channel that is
largely confined to this little community.

For the moment we defer discussion of whether these disparate communications in a babel of technical languages are essentially 'incommensurable', or whether they merely portray different aspects of a unified

scientific world picture [*8.14, 9.6, 10.4, 10.8*]. Nevertheless, the fragmentation of the scientific literature into a myriad of highly specialized publications is characteristic of academic science. It is reinforced by many academic practices, such as peer review [*3.7*]. But it is not the outcome of a conspiracy of obfuscation by an entrenched elite. It is simply another aspect of the speciation of academic life and thought into innumerable disciplines, subdisciplines and research specialties[191].

By publishing research results in fragments, academic science clearly offends against the norm of universalism [*3.4, ch.6*]. Nevertheless it has no established machinery for sharing its findings in larger communities. This happens by default, 'informally', rather than by systematic design. Indeed, communications addressed to wider publics are excluded from the formal literature [*9.3*], and may even harm the scientific standing of their authors. We shall return later to some other features of the epistemic gap between what is known to science and what is 'public knowledge' in society at large[192].

5.13 Networking intellectual property

Just as 'community' is the keynote of academic science, so 'network' is the sociological theme of post-academic science [*4.5*]. It emphasizes the concept of science as a communication system[193], where information obtained at certain nodes is transmitted to other nodes, whether these be individual researchers, research groups, specialist communities, corporate bodies or the general public.

Networking clearly favours communalism. 'ICT' – the new technology of information and communication – stimulates and expedites the traditional practices of academic science[194]. Electronic publishing, for example, can significantly cut the time it takes for a 'contribution to the literature' to be disseminated to a research community[195]. This acceleration of communication has significant cultural and cognitive effects. In particular, the increasing density, multiple-connectivity and immediacy of electronic communication draws individual researchers together into collective action.

This is not just a matter of facilitating active teamwork by geographically dispersed researchers. It makes it feasible for novel observations and theories to be discussed in detail with distant colleagues – or even sceptical rivals – as they emerge. Databases and archives can be searched on-line for relevant ideas and information. The continuous informal debate [*9.1*] so typical of scientific life[196] overflows on to the Internet, which becomes a devolved building site for knowledge construction[197]. The research laboratory extends across the world[198].

The material that does eventually get into the official scientific literature [9.3] may thus already have been shared amongst a much wider group than the authors to whom it is officially attributed. It should therefore be less tentative, more convincing, sounder in fact and logic, than was normal in traditional academic science. The trouble is, however, that an electronic text can be amended so easily that there never seems a moment when it ought to be brought to a firm conclusion. The various phases of the research cycle [8.10] – discovery, justification, criticism and revision – merge together in a never-ending, off-the-record process involving a whole cluster of informal contributions. The epistemic quality of academic science depends very much on the existence of a public archive of historically dated and explicit research reports by named authors, who must take personal responsibility for their claims. This function cannot be performed satisfactorily by a collection of texts that are continually being updated and revised.

It is a nice point, moreover, whether launching a research report into cyber space is tantamount to publication. An out-of-print academic book, or an ancient number of the proceedings of an obscure learned society, can usually be tracked down in a copyright library. But membership of a post-academic research network may be limited to recognized specialists on the subject. Even an electronic system for the exchange of 'preprints' – that is, research reports that have not yet been published – can be more like an exclusive club, or even a secret society, than an open-sided 'invisible college' [3.9, 5.7]. Indeed, membership of such a network may be a mixed blessing. It gives early access to new research results, but may be an unconscious barrier to communication with outsiders.

It is true that such networks are often very heterogeneous[199]. Their nodes are widely distributed, by discipline, institution, sector and country. But that means that academic scientists are regularly teamed up with researchers who are not bound by the norm of communalism, and are not professionally dependent on their contributions to 'public knowledge'. The knowledge they are employed to produce is *proprietary* – that is, it is 'owned' by the body that employs them, and may be kept private, for longer or shorter periods, for commercial or political reasons.

Academic science does, of course, have its own notion of 'intellectual property'. A novel discovery 'belongs' to the researcher who first laid claim to it in public – or, as in some disputed cases, can prove they were its original begetter [3.6]. But this right is paradoxical: it can only be turned to advantage by giving it up! The exchange value of new knowledge is realised by disclosing it to the scientific community. In academic science, intellectual property rights are strictly personal, and are limited in value to the esteem they earn as contributions to science[200].

Nowadays, many scientists working in academia are employed on contracts that prevent them from disclosing all their results immediately. In effect, they are forced to trade the intangible benefit of another publication for the material benefit of a job, or a share in the (potential) profit from a patentable invention. The transition to post-academic science has undoubtedly weakened the traditional mechanisms motivating prompt and full disclosure of research findings[201], thus creating serious personal and institutional dilemmas in the scientific world.

But what is the epistemological significance of this deviation from the communal norm? Strictly speaking, none at all. Academic science has never laid claim to all the knowledge that might possibly contribute to its advancement. Good science can often be done simply by exploring the vast hoards of observations and ideas that accumulate in technological, medical and social practice, and selecting some of them for transformation into scientific facts and theories. But that does not mean that the original observations and ideas 'ought' to have been published, or that they did not constitute items of human knowledge before they were made 'scientific' [10.6].

The undisclosed results of research come under the same heading. It makes no difference that they may have been obtained by people with scientific qualifications, using concepts and techniques that are well established in a recognized scientific discipline. Until they are formally 'made known' to science they are not part of scientific knowledge. Our concern in this book is with the epistemology of *science*, not with much more general questions such as how it is that we know anything at all, or what it means to say that anybody 'knows' anything, scientific or otherwise [1.6].

Nevertheless, it is often very difficult to make this type of distinction between 'science' and 'non-science'. In a technically sophisticated context, exactly the same methods might be used to make exactly the same discovery, regardless of whether or not it is intended for publication. The situation is confused by the fact that post-academic scientists often perform dual roles [4.10]. On Mondays, Wednesdays and Fridays, so to speak, they are producing public knowledge under traditional academic rules, on Tuesdays and Thursdays they are employed to produce private knowledge under commercial conditions. At weekends, no doubt, they think about the numerous connections between what they learn in their two ways of life. Formal and informal items of knowledge [3.3] cannot be separated by Chinese walls in the mind of each researcher.

In any case, the knowledge appearing in public out of post-academic science is bound to be incomplete. It is not, perhaps, so serious if a knowledge claim temporarily lacks significant items which are only known to a privileged group, such as the employees of a particular industrial firm. But it does matter if it appears to rely heavily on 'unpublished research data' and 'private communications' which cannot be checked by independent replication. In effect, we are being asked to accept such statements solely on the word of their authors, rather than as the common knowledge of a critical community [ch.9].

Secrecy in science is a form of 'epistemic pollution' to which post-academic science would seem all too open. It is not only a sign of a major change in the social organization of science. It also signifies increasing subordination to corporate and political interests [7.6] that do not put a high value on the production of knowledge for the benefit of society at large[202]. And yet, in the long run, it is precisely the openness of academic science, its respect for the communal norm, and its grounding in reproducible empirical observation, that are the best guarantees of its practical reliability – for good or for ill.

6

Universalism and unification

6.1 Generalization and abstraction

The academic ethos lays down that scientific knowledge has to be the common property of a *universal* community [3.4]. Social realities obviously limit the scope of this norm. Nevertheless, it firmly shapes the type of knowledge that is admitted into the scientific archive. In principle, science deals only with what could be communicated to and accepted by *anybody*, regardless of their other personal beliefs or special circumstances.

The result is that science is primarily concerned with *generalities*. Particular facts are not specifically excluded from the scientific archive. Many serendipitous discoveries – especially in the biomedical sciences – have been triggered by published reports of apparently singular events. But such particularities are very, very seldom of 'universal' interest. It is quite impracticable to 'share' with other scientists the immense quantities of factual information that accumulate in the course of research. If this information is to become communal property, it must be encoded and 'compressed' into a much more compact form[1]. In other words, the detailed facts must be interpreted and presented as specific elements of more general patterns[2] – typically as entities governed by *theories*.

Trying to give a basic, comprehensive account of the concept of a 'theory' is an invigorating but fruitless walkabout in metaphysics[3]. In all that follows, we shall treat 'theories', like 'facts', as primitive epistemic entities – 'natural kinds' as some philosophers call them – with which we are better acquainted from experience than from philosophical analysis. Indeed, from this naturalistic point of view, facts and theories are closely interwoven. It is impossible to talk about knowing something as a scientific fact without reference to a theory [5.5, 5.6]. Scientific facts are not just 'brute facts': they only have meaning in relation to the ideas of those who determine, communicate and receive them[4]. Conversely, scientific theories are not mere 'ideas'. They cannot be completely dissociated from the empirical facts to which they refer.

In this chapter, however, we treat scientific knowledge as a collection of theories rather than as an assembly of facts. This would seem to put us back into the hands of the philosophers. But theories are social institutions[5]. They must conform to the standards of the society in which they are established. In principle, the scientific community is totally meritocratic and multicultural [3.4]. Contributions to knowledge are accepted on their merits, and are not to be excluded because of irrelevant social criteria, such as the religious beliefs of the contributor. By labelling such beliefs as 'irrelevant', the norm of universalism locates them outside science. It thus bars the incorporation into science of ideas that are not acceptable to most human beings.

Universalism is a very demanding regulative principle. As everybody knows, it is almost impossible to formulate theories in the human sciences – to some extent, indeed, in all the sciences – that do not touch on highly contentious political or religious issues [7.7]. On some issues, the gender of the researcher is far from irrelevant[6]. This is the fact of social life underlying most relativist critiques of science[7][10.4]. In systematically guarding against becoming an 'ideology' in the narrow sense of reflecting particular social interests[8], science cannot avoid being seen as an ideology in the broad sense – that is, as a set of beliefs by which a society orders reality[9] [10.7]. A naturalistic account of science must include the way in which it performs, or claims to perform, this role.

Even amongst natural scientists, universalism is not easy to achieve. Indeed, intellectual tension between universalism and specialization [3.9, 8.3] is one of the characteristic features of academic science. Research specialists become committed to particular theoretical systems – 'paradigms'[10] – which dominate their thoughts and perceptions. Such systems are seldom strictly 'incommensurable'[11][8.9, 9.6, 10.4] but their relative incongruities are not always decisively resolved. Despite the Legend of a comprehensive 'scientific world view', scientific knowledge is really no more coherent or systematically ordered than the community that produces it[12] [10.8].

6.2 Classifying the 'facts'

Theories are *schematic*. They introduce *order* into representations of experience at the price of obliterating specific facts[13]. Philosophers have largely concentrated on the *explanatory*, *prescriptive* or *predictive* theories typical of an *analytical* science such as physics, where it seems that an infinity of facts can be derived from a few simple 'laws'[14]. They have largely neglected the problem of systematizing the specific and detailed observations obtained in a *descriptive* science such as anthropology[15], where if there are any general 'laws' they must be nebulous and very complex[16]. Even this conventional dichotomy between 'rationalism' and 'humanism'[17] holds only at the extremes[18]. Many important scientific disciplines, such as brain science and ecology, have emerged in the wide open spaces between hard-core physics and soft-centred sociology.

Indeed, an explanatory theory [10.1] cannot be set up without reference to a *taxonomy* – that is, a *classification* of the 'facts' it explains. For example, Newton's analysis of planetary motion rested on ancient scientific distinctions between fixed stars and planets. Conversely, a well-founded taxonomy often functions as a predictive theory[19], as Mendeleev showed by foreseeing the discovery of new elements to fill the gaps in his Periodic Table. Ernest Rutherford quipped that science was 'either physics or stamp-collecting': he could never have won his Nobel Prize (in chemistry!) if generations of chemists had not previously classified the 'elements' whose radioactive transformations he charted.

Quite simply, unambiguous communication between researchers [5.12] is impossible unless they can identify to one another the entities that they are talking about. They must share a minimal 'theory' about the meanings of words. The very notion of an 'entity' may well be prior to language, both in childhood development and in the evolution of the human species[20]. But once we are into a language, we are into a classification scheme. This scheme may seem so natural that we are normally quite unaware of it[21]. Nevertheless, in talking about, say, 'cats', as distinct from 'dogs' or 'chairs', we are clearly assuming that we are addressing a conscious being who is capable of carving up the world into 'things' that can be classified in this way[22].

As lexicographers know to their despair, the classification schemes implicit in natural languages are infuriatingly unsystematic and multivalent. But like all theoretical systems, they *simplify* the world they are applied to. By lumping together entities that are as near alike as makes no matter in the circumstances, they enable generalizations[23]. In appropriate contexts, all members of the same class can then be deemed to be interchangeable. For legal purposes, your pet poodle is just another dog requiring to be licensed. For the politician, you are just another constituent with a vote to be solicited. For social science purposes, your family has

1.9 children[24]. For the clinical record, you are just like every other case of schizophrenia[25].

It is not the possibility of abstract analysis that makes a taxonomy scientific. It is that it is sufficiently *universal* to be used consistently by a scientific community. For example, something like the Linnaean system became necessary in the seventeenth century to cope with the world-wide diversity of local names for similar – often the same – biological species[26]. Although designed around descriptive 'family resemblances' rather than specific measurable properties[27], this system enabled naturalists to identify the subjects of their observations to one another, and thus brought order into biology.

What do we mean by 'family resemblances'? The honest answer is that we don't exactly know. This corner stone of the scientific enterprise is not built on philosophical bedrock. The best we can say – and this was acknowledged long ago by Immanuel Kant[28] – is that *pattern recognition* is an innate human capability. More to the point, it is an *intersubjective* capability [5.10]. Independent observers can come to unforced agreement that two objects closely resemble one another, or that they perceive the same significant configuration of features in a given object, scene or data set.

A glance at the scientific literature shows that the human capability for pattern recognition is deeply embedded in scientific practice[29]. It is obviously an indispensable factor in the classification of biological organisms, living or long dead. Sciences as diverse as geology, anatomy and ethology would be reduced to trivialities without it. Extended instrumentally [5.5] by photography, telescopy, microscopy, radar, X-ray tomography, auto-radiography, satellite imagery and numerous other sophisticated technologies, it permeates the whole scientific enterprise[30]. Extended cognitively by the use of diagrams and maps [6.5], it underlies much scientific theorizing, even in the physical sciences[31].

By its very nature, pattern recognition cannot be analysed verbally[32]. It can only be simulated computationally over a finite, pre-arranged set of possibilities, as in the electronic detection, analysis and reconstruction of the tracks of high energy particles[33] or the recognition of written characters or spoken words. In spite of much experimentation and theoretical discussion[34], it is still poorly understood as a general mental process. Nevertheless, it does seem to be a universal capability, shared by all healthy human beings down to quite young children[35].

There is a subtle trade-off here between the particular and the general. On the one hand, individual perception does depend markedly on personal experience [5.9]. People can be trained to recognize particular types of pattern – for example, entomologists learn the subtle variations that distinguish one species of beetle from another. It is argued that certain languages make their speakers sensitive to specific features of their natural and social environment – typically, that Bedouin are strong on the differences between camels, whilst Inuit develop a special feeling for snow[36]. In other words, the classification schemes that people find 'natural' are often social constructs that have evolved along with other aspects of their culture[37].

On the other hand, there is good evidence that this personal and cultural diversity is built around a core of biological uniformity[38]. The neurophysiological structures of human beings are as standardized as other features of our anatomy. Almost everybody is born with two legs to walk and run, even if few of us learn to walk a tightrope or dance like a Dervish. Similarly, everybody is born with a capacity for learning a mother tongue, even if few of us become professional interpreters. There is a substantial consensus, for example, on the basic colours in which people say – in a babel of languages – that they see the world[39]. People who have never before seen black and white photographs have no difficulty in learning to 'read' them as three-dimensional spaces[40]. All languages differentiate between noun-type and verb-type words[41]. And so on.

This facility, surely, underlies empathy [5.11] . Patterns of human behaviour are recognized as outward signs of inward thoughts and feelings. Human mental and emotional mechanisms have evolved along with their biological correlates, and are sufficiently similar the world over to make social life possible[42]. One of the ironic features of 'the global village'[43] is the popularity of American TV soap operas in remote Third World villages. The flickering image of 'JR' Ewing is seen not only as a human creature in a material world but also as the sly villain in an all-too-human drama.

TV melodrama may seem a far cry from sober-sided science. But the entomologist observing the courtship rituals of beetles is no more conscious of 'recognizing patterns' than is the devotee of *Dallas* watching the latest episode. She reports the behaviour of two specimens of *dynastes hercules*, just as he sees and hears the exchanges between 'JR' and 'Pam'. In other words, the interpretation of familiar patterns as distinct entities is an integral part of the act of observation. Scientists will automatically interchange 'facts' in terms of the classification schemes and other theories that they share. This is as much a natural condition of science as it is of other aspects of life. We defer to a later chapter [ch.8] the all-important question of how *new* patterns – i.e. novel scientific theories – are recognized and made familiar.

6.3 Systematics

Thus, the human mental linkages between observation and classification are unavoidable. This metascientific fact subverts the Legend of science as a strictly rational enterprise. Any supposed demonstration of the 'truth' of scientific knowledge would necessarily involve such linkages. But human cognition cannot, as yet, be reduced to a computational algorithm [6.6]. We are nowhere near the utopia of artificial intelligence. Indeed, it is questionable whether this utopia is a meaningful concept, let alone an approachable goal[44]. For the present, certainly, no argument that includes human mental operations can be completely proven by formal logic. Despite all the endeavours of generations of philosophers, it is impossible to prove rigorously that science is, or could eventually be made, perfectly 'true'[45].

In each particular branch of science, this situation is seen as scandalous. A great deal of effort goes, therefore, into trying to bypass or replace the human mental links with non-human instruments [5.5] or formulae. For example, the communal procedures of biological *systematics* are carefully designed to reduce inconsistencies and instabilities of nomenclature[46]. Taxonomists are required to give standardized verbal descriptions of the distinguishing features of species, genera and higher taxa. Thus, the 'Death Cap' mushroom, *Amanita phalloides*, is not only recognizable from photographs[47]. It can also be differentiated from other, edible species by characteristics such as '*Cap 4–12 cm across, convex then flattened, smooth with faint radiating fibres often giving it a streaked appearance, slightly shiny when wet, variable in colour, but usually greenish or yellowish with an olivaceous flush…*' etc. In fact, this is a loose English interpretation of an archival formula inscribed in Latin, the universal language of this branch of science. This procedure is genuinely scientific. It assumes that 'anyone' who has been properly trained as a mycologist would behave as a standardized instrument in making such observations. But almost every word in such a formula invokes a visual pattern. The elaborate verbal protocol does not eliminate the human eye.

Biological taxonomy – the grandest and most sophisticated classification exercise in science – exemplifies other metascientific principles. It depends, of course, on two observable regularities. In the first place, individual organisms are found to fall into distinct *species*, each of which can usually be identified by a unique combination of characteristics – not just the *size* of the 'death cap', but its *shape*, *texture*, *colour* and so on. In the second place, there are remarkable similarities between different species in some of these characteristics – the twenty-four British species of *Amanita* all have a cup-like bag at the base of the stem. Thus, the criteria by which species are distinguished from one another can be used to arrange them into larger classes, such as *genera* and *families*.

From the earliest times, people have used these basic regularities to distinguish the 'natural kinds'[48] of creatures in their environment, and to group them into various categories [10.2]. Indeed, such local *folk taxonomies* are surprisingly similar across cultures, and are not really so very different from our scientific taxonomies, either in general principle[49] or in the species they actually identify and differentiate[50]. But they do differ greatly in the overall patterns that they see in nature – that is, in their larger systems of classification. Since the eighteenth century, therefore, the grand project of scientific taxonomy has been to establish a *universal* classification scheme covering the whole of biology.

The question here, as with all taxonomic enterprises, is whether such a scheme can be derived by direct inspection of the entities being classified. Does the living world divide itself into natural 'taxa' at a higher level of generality? Is there a well-defined pattern in the similarities and other regularities in the characteristics of biological species? Can this pattern be established by a rational, analytic procedure, such as a mathematical computation [6.7]? These questions suggest a research programme of great metascientific significance.

The first step in this programme is obviously to determine each characteristic unambiguously – preferably as a number measured by an automatic instrument [5.5]. Needless to say, descriptive biology has made enormous scientific progress through the standardization and instrumentation of its methods and techniques. Physical measurements and chemical analyses of the characteristic features of organisms are essential tools in the great enterprise of classifying the living world.

Yet even this first step involves human judgements[51]. The initial choice of the characteristics to be measured depends both on their apparent relationship with each other in the whole organism[52] and on our previous experience of the organisms actually encountered in the natural world[53]. You propose, say, to differentiate 'elephants' from 'horses' by the length of their trunks: pray how would you differentiate the 'trunk' of a horse from its muzzle: how do you even know that there *are* any creatures with such an unnatural organ?

Nevertheless, computers now make it feasible to analyse vast quantities of such data, looking for patterns of similarity and difference. For example, the data for a large population of different organisms might be expected to fall statistically into distinct 'clusters', corresponding, perhaps, to different natural species. One would then hope to find analytical criteria by which such clusters could be arranged into distinct sets – even into a hierarchy of sets – which could then be taken to define a truly objective scheme of classification.

The fact is, however, that 'numerical taxonomy' has not succeeded so far in its programme of eliminating human judgement from biological systematics[54]. Computational techniques for finding patterns in large data sets will surely continue to improve, but there is no certainty that such a programme must eventually achieve its goal. Indeed, quite apart from its practical difficulties, *phenetics*[55] – the automatic classification of organisms by similarity of features – is very questionable in principle.

Every natural object has an infinity of 'features'. The way that it is classified depends on which finite set of such features is included in the computation, and on the relative weights to be attached to them. Why, for example, should the similarities of shape and mobility between dolphins and sharks be subordinated to the differences in their metabolic and reproductive characteristics? It is easy to see that such decisions cannot be automated without enlarging the scope of the analysis until it eventually includes the whole ecosystem – perhaps every object in the universe! Anyway, whatever we humans may think of dolphins and sharks, the poor fish on which they prey would surely put them in the same class.

In metascientific terms, classification, like observation [5.2], is a 'theory-laden' activity. It cannot be done entirely without reference to its intellectual and social environment[56]. The resulting scheme always reflects conscious or unconscious influences [7.5, 10.4], such as socially potent metaphors[57], formal mathematical patterns[58], the supposed functions of component elements[59], relationships to unobservable structures[60] or the need to reconcile conflicting conceptual or practical paradigms[61]. Some of these influences are quite clearly incompatible with the norm of universality. A scientific taxonomy of animals, for example, obviously has to transcend the diversity of taboos and totems attached

locally to different species. Nevertheless, however hard we try to eliminate them, general human beliefs about the nature of things [10.5] cannot
be taken out of the loop.

Even our most modern scientific classification of the living world rests
upon just such a belief. Biological systematics nowadays is completely
dominated by Charles Darwin's theory of evolution[62]. According to the
principle of 'common descent', it is possible to classify organisms as if
they were members of a single enormous family [9.8]. Taxa at various
levels of generality should correspond to groups that are more or less
closely 'related' – that is, to historical branches of varying degrees of
antiquity. Every branching episode, however, preserves many common
features, so that closely 'related' organisms show many similarities – and
so on.

This kinship structure is delightfully reminiscent of the folk taxonomies of many clans of native Australians[63]. Indeed, like the 'myth'
behind such taxonomies, it is not directly observable. But it can be
inferred from the observed similarities between living organisms and/or
fossils. Indeed, the further evidence accumulated in more than a century
of research has fully confirmed the overall consistency of Darwin's interpretation of the 'pattern' of natural kinds. It takes a socially potent metaphor, such as a fundamental religious doctrine, to stand in the way of
accepting it.

Biological systematics and evolutionary theory are now inextricably
interwoven. Let us be clear, however, that the relationship between them
is not one of logical implication. Take, for example, the remarkable suc-
cesses of molecular genetics in confirming or revising the supposed
family relationships between biological taxa. At first sight, this looks like
the dream of numerical taxonomy come true. Rigorous computational
algorithms apparently deduce *cladograms*[64] – in effect, 'family trees' –
solely from the similarities and differences between the DNA sequences
of a group of related organisms. But note that this analysis seldom produces a unique answer. And even when it does, our satisfaction depends
upon a prior belief that this particular type of similarity should take precedence over all other classifying criteria.

Along with the vast majority of scientifically educated people, I take
Darwin's concept of 'evolution by natural selection' to be a convincing
and reliable interpretation of many features of the living world. Indeed,
as a metascientist I note that a coherent *descriptive* theory of all life forms

could probably not have been developed without an *explanatory* theory of this power [*10.1*]. For my money, the triumphs of 'molecular phylogenetics' are a marvellous application of this concept. These triumphs add to the widely shared conviction that this is a thoroughly reliable domain of scientific thought and action. But they do not, single-handed, *generate* the taxonomic system they support. Nor do they prove *logically* that Darwin's 'dangerous idea'[65] is necessarily, absolutely, uniquely true.

6.4 Theories as maps

Inside every taxonomy, there is an explanation [*10.1*] struggling to get out. Conversely, every scientific explanation of 'facts' is a mode of classifying them. What is common to all theories, weak or strong, descriptive or analytical, is that they represent the world as *structured*. Linnaeus' system of biological classification is not just a list of all known species. It represents them as related to each other in a branching hierarchy, a 'tree'. Newton's theory of gravitation is not just a mathematical formula explaining the motion of the planets. It represents them as interacting to form a planetary 'system'.

As philosophers and other metascientists are coming to realize[66], theories are very like *maps*. Almost every general statement one can make about scientific theories is equally applicable to maps. They are representations of a supposed 'reality' [*10.7*]. They are social institutions [*6.1*]. They abstract, classify and simplify numerous 'facts' [*6.2*]. They are functional [*7.2*]. They require skilled interpretation [*5.9*]. And so on. The analogy is evidently much more than a vivid metaphor.

In effect, every map *is* a theory [*6.5*]. An analysis of the most commonplace map explores almost all the metascientific features of the most recondite scientific theory. From a naturalistic point of view, the London Underground map exemplifies these features just as well as, say, the 'Standard Model' of particle physics. Of course we feel much more at home with the former than we do with the latter. But by looking at an ordinary map as a very familar, indubitable theory, we learn a great deal about very unfamiliar theories whose validity may be much more doubtful.

For the moment, let us set aside such important questions as how theories – scientific or cartographic – come into being, how they are tested and how they relate to other aspects of our lives, such as personal beliefs and social interests. Thus, we leave for later consideration what it might mean to say that a theory is 'objective', 'novel' or 'well-founded'. In other words, in keeping with the overall plan of this book, we focus here on the implications of 'communalism' and 'universality' for scientific theories, postponing discussion of epistemic features primarily associated with other norms, such as 'disinterestedness', 'originality' and 'organized scepticism'.

I am not suggesting that all scientific theories can be 'mapped' in the everyday sense – that is, laid out concretely, in two or three dimensions, as a *material map*, for direct visual inspection and interpretation. *Iconic* representations can be very effective media for intersubjective communication [5.12]. A whole new science, such as geology, may evolve out of a new way of mapping observational data[67]. The recognition of 'patterns' in such representations is one of the major sources of scientific theories [6.1]. Scientific concepts are often grasped and manipulated cognitively as 'mental models' of this type [8.14, 10.1].

But there are many scientific concepts – for example, Einstein's concept of gravitation as a 'warping' of the four-dimensional space-time continuum – which cannot be completely represented on a material map in the real, three-dimensional world [8.14]. Indeed, even in the social sciences, theoretical structures often have be so complex and convoluted that they are very difficult to map out for visual verification[68]. One of the major scientific developments of recent years has been the extension of the notion of a material map to cover, say, computer algorithms capable of extracting and presenting various 'sections' from many-dimensional sets of data.

Nevertheless, our everyday conception of a map as having more than one dimension is a vital element in the analogy. It reminds us that a scientific theory is normally much more than a *list* of data or a *chain* of propositions. To be of any value, it must have sufficient structure to suggest unanticipated patterns of fact or inference[69]. There must be ways of, so to speak, 'getting around' blockages, observing objects from 'several different points of view'[70] or 'making connections' between previously unrelated entities or ideas[71]. Scientific discourse is paved with the graves of one-dimensional metaphors.

This is why scientists are no longer satisfied with the tradition of presenting scientific knowledge in terms of 'laws', 'formulae' and such like 'one-dimensional' relationships[72]. Thus, to state that a certain 'effect' has a certain 'cause' [10.1, 10.5] merely corresponds to pointing to an 'itinerary' on a more general theoretical map[73]. Such a statement is meaningless unless combined with other information about the scientific context in which it is to be interpreted – that is to say, how it is connected with other statements of fact or theory about the entities involved[74].

Indeed, scientific theories can often be mapped as abstract *networks*, where nodes of fact and/or concept are cross-linked in many dimensions by laws, formulae, family resemblances or other functional relationships[75]. Strictly speaking, the 'network model' of science is only a special case of the map metaphor, emphasizing – often exaggerating[76] – its overall connectivity. Nevertheless, it has the same significant implications for the production, testing, revision and acceptance of scientific hypotheses.

6.5 Maps as theories

A map is not just a *picture*[77]. In everyday language, a picture is a representation of some segment of the world, as seen from a particular viewpoint. For scientific purposes, the ideal form of a picture is a *photograph*, pinning down a specific configuration of events and differentiating it from others[78]. By contrast, the essence of a map is that it does not have a particular perspective[79]. In principle, every location is equally unprivileged as a possible viewpoint. In other words, as we have already remarked, a scientific theory, like a scientific 'fact', must be sufficiently free of subjective elements to be equally acceptable to all members of a scientific community.

But the notion of a 'view from nowhere'[80] is not meaningful. A map has to represent a particular territory, and cannot be interpreted or used without *some* idea of its relative viewpoint[81]. The standard cartographic convention is that a map should be read as if it were a picture taken through a telescope, looking straight down from a great height – e.g., like a photograph from a space satellite[82]. This convention is so familiar that it is often forgotten. Nevertheless, it reminds us that a scientific theory is not a pure abstraction like a mathematical theorem. It is only meaningful as a representation of a particular aspect of reality as it might appear in principle to *some* human intelligence.

In practice, most maps are not derived from single space photographs. Even apparently lifelike satellite images are often mosaics of pictures and survey data taken from a number of points, reconciled and harmonized into one big picture[83]. As cartographers know, the unity of this picture is an illusion. The curve of the earth is artificially flattened and the play of light, shade and colour varies unnaturally over the whole scene. In other words, scientists have to do a good deal of work on the empirical 'facts' to fit them together into coherent theories – and these are usually more contrived and less consistent than we are often led to believe[84].

This choice of a common viewpoint is particularly difficult in the human sciences [7.7, 8.13, 10.4] where people have their own individual 'pictures' of the social world and of themselves in it[85]. These overlap and can often be fitted together, through empathy, into a generally agreed 'map'. But as every novelist knows, it is not easy to establish a common standpoint outside any particular human mind from which to look at this map as a whole[86]. Thus, the social scientist often has to choose between adopting the stance of an all-knowing anthropoid god, or theorizing about the social world in non-human terms, as if it were a biological organism or a machine.

Maps are drawn to different *scales*. A world atlas fills a page with Europe, another page with England, and another page with London. To plan a car journey from Oxford to Edinburgh, I use a motorway map of 1 centimetre to 50 kilometres. To plan a cross-country ramble to the next village, I use a map of 5 centimetres to 1 kilometre. Each map turns out to be reliable for the use I make of it. Each map is entirely truthful on its own scale. Yet even where they include the same geographical feature, they present it differently. On the motorway map, the villages are not shown and Oxford is a mere dot: on the rambler's map, Oxford is an irregular patch, as large as my hand, crowded with lines, coloured symbols and names.

Indeed, the form of a map depends greatly on its intended *use*[87]. I regularly use maps of London in four different forms, each for a different purpose. For getting around by car, I have a Highway Map, showing the network of main routes – including one-way streets. To find my way to a specific address, I then turn to a Street Directory, which helpfully indicates house numbers. Often, however, I travel by bus, and need to consult a Bus Route Map – typically quite complex, in spite of showing only the roads where buses run. Finally, there is the Underground Map, whose schema of tube lines, stations and interchanges is etched on my memory from frequent use [8.14]. These four maps all cover the same region on much the same scale, and in spite of various simplifications are all essentially 'truthful'[88]. Yet they emphasize such different characteristics that they are not equivalent in use or meaning. Indeed, as we all learn from the experience of exploring a foreign city, it takes quite an effort to identify their common features – the relation of bus stops to underground stations, and so on.

In other words, scientific theories have to be understood as *purposeful* generalizations[89]. Indeed, one of the achievements of the social sciences is to provide people with unsuspected 'meanings' for many cultural features of their lives[90]. The entities that figure in a scientific theory are selected and simplified to suit its scope and function[91]. Theories of different degrees of generality or with different purposes include or emphasize quite different features [9.3]. Factual distinctions that are of significance in one theory are glossed over or lumped together in another. Relationships that are merely incidental for one theory are explanatory concepts for another [10.1]. Even many-sided entities that we quite properly believe to be unitary are not always shown as such. For example, the entity labelled 'DNA' is presented differently and plays a different theoretical role in organic chemistry, biochemistry, physiology, genetics and ecology[92].

The fact that maps are human artefacts does not, of course, imply that they are untrue to nature. Indeed, the whole purpose of making a map is to convey reliable information about the domain it claims to represent. But cartography is never a purely mechanical process. The need for a map emerges out of a sea of other needs and interests[93]. A publicly available map is a social institution. It is designed to serve a specific social function, such as running a railway or avoiding traffic jams.

In practice, the skilful cartographer has considerable freedom in selecting and symbolizing the geographical entities and relationships that best serve the functions of their products. This freedom is very obvious in the marketplace. For example, try to buy *The Motorway Map*. What you will be offered will not be a unique document: it will be one or more out of a whole *family* of variants, all perfectly genuine as maps but differing in format, scale, date of revision, details of junctions and service areas, etc.

The same goes for scientific theories. It cannot be denied that they are 'constructed' to satisfy human purposes [8.13]. This does not automatically imply that the knowledge they claim is untrustworthy. It only means that in evaluating a theory we have to take into account the needs and interests of its makers. Indeed, the whole argument of this book is that theories constructed by academic scientific communities [8.10] should be evaluated in the first instance on this basis – i.e., as primarily serving the needs and interests encapsulated in the academic ethos [ch.3].

Scientific theories, like maps, are *under-determined*. They are products of their time and place. They emerge out of the exercise of originality and scepticism in a disputatious community. Of course there are moments when a novel scientific theory seems precisely right. But its form and substance are neither pre-ordained nor permanent. Even the most compelling theory is usually shaped by unconscious aesthetic and utilitarian criteria[94]. Theoretical entities such as 'DNA' have to be redefined to meet the demands of new fields or other disciplines[95]. As aspiring textbook authors soon discover, scientific theories, like maps, are not uniquely specified. They come in *families*[96] [8.4], where sometimes the differences are of form rather than substance, but sometimes the supposed relationships and resemblances are more notional than real[97].

Let me emphasize, moreover, that we are not talking here about obvious inconsistencies calling for resolution by further research. Even good scientific theories, like good maps, can present the same 'domain' in a great variety of very different forms. But this theoretical pluralism is very disconcerting for the Legend of a unique scientific world picture [10.8]. To avoid further embarrassment, the Legend insists that such apparent pluralism is not genuine: it is just a consequence of our present ignorance. The diverse theories of today are merely provisional: in due course, so it is argued, they will be seen as different approximations to the 'theory of everything' that will eventually be completed.

But any such 'theory of everything' – including, of course, all the contingent features of the world as well as their fundamental equations of motion – is not merely hypothetical: it is not a meaningful concept[98]. Here again, the cartographic example is instructive. A 'map of everything' would have to be drawn on an enormous scale so as to make visible every microscopic detail. As Lewis Carroll pointed out over a century ago[99], the concept of a total 1 inch to 1 inch map is not only impractical – the farmers would object to all their land being covered by it – but also absurd in principle. As formal mathematical theory has now shown, any domain as complex as the real world cannot be fully 'mapped' by anything less extensive than itself[100].

A naive cartographical realist might insist that we should indeed consider the world as its own perfect map. But the whole concept of a map implies that it can be detached from the actual domain of which it is an icon, and used for other purposes. That is to say, a map is not the same as the geography it represents[101]. Nor is it an approximate version of a perfect abstract replica – a hypothetical, super-duper, Ordnance Survey map – of that geography. Like any other scientific theory, it is simply an image built upon factual data obtained by direct exploration of the domain it represents.

6.6 Formalization

For most people, including most philosophers, the characteristic feature of scientific theories is that they are peculiarly *logical*. Indeed, according to the Legend, a proper scientific theory should be stated and argued *formally*, almost like a geometrical theorem[102]. Students of the physical sciences certainly get that impression from their textbooks, where topics such as quantum theory are presented as systems of postulates from which observational data, solutions to practical problems, predictions of novel phenomena etc. can be deduced mathematically. In practice, the textbooks do not always agree on which are the theoretical postulates and which are the empirical predictions[103]. But that is entirely consistent with the map metaphor. It simply suggests that the theory is a tight-knit logical network [6.4] where many of the deductive links can be traversed in either direction.

In effect, scientific theories are widely held to be primarily exercises in *algorithmic compression*. That is to say, the ideal theory is considered to be a symbolic formula – preferably mathematical – that encodes empirical data very compactly. In this spirit, Ockham's razor sagely advises the theorist to be economical with postulates[104], sciences are arranged in a hierarchy of public esteem according to their degree of formal codification[105], imaginative writing about science is resented because it stimulates intuition rather than engaging reason[106] – and so on. This principle operates throughout the length, breadth and depth of all modern conceptions of science.

Without question, *analytical thought* is the mainstay of the scientific culture [8.14]. Indeed, such a culture could not have emerged before the invention of formal modes of exposition and argument – an invention that was made possible by new techniques of symbolic communication such as writing[107]. Nevertheless, scientific disciplines obviously vary enormously in the extent to which they are formalized[108]. In physics and economics, for example, theories are nothing if not mathematical: in ethology and social anthropology, by contrast, theoretical concepts are often difficult to express clearly, let alone formulate rigorously.

The Legend asserts, of course, that unformalized disciplines are either 'immature' sciences, or not yet sciences at all, that the growth of knowledge will inevitably bring them under full logical control, and that in the fulness of time, the 'Newton' of sociology will discover the Principles of Social Philosophy, the laws of communal dynamics and the equations of motion of institutions. But this is a scientistic slogan rather than a well-founded metascientific principle[109] [10.8]. Meanwhile, resuming our naturalistic enquiry, we are prompted to ask what it means to formalize a scientific theory, and how this benefits the scientific enterprise.

The answer is very simple: a formalism is a *universal* mode of representation, ideally suited to the exercise of *criticism* [ch.9]. It thus satisfies perfectly the requirements of a culture conforming to these norms. For example, the dictum [8.11] that a theory is 'falsified' if its observational implications are not confirmed[110] applies only to rigorously formulated theories whose implications can be deduced logically[111]. Again the force of such critical practices depends on the assumption that logic transcends cultures – that logically sound reasoning can be communicated in any human language, so that strictly logical scientific ideas can be made compelling to any rational human being [6.8, 7.5, 10.4].

But formal theories perform a more basic function than facilitating criticism. They enlarge the channels of *unambiguous communication*. Sophisticated technical languages have evolved to enable scientists to assemble their individual observations into a body of common knowledge [5.12]. Every stage in this process is permeated with theory. Theoretical concepts enter into the choice of what to observe, how it is reported and how the report is interpreted. Theoretical concepts are expressed in theoretical terminologies. In scientific practice, therefore, observational languages are always essentially theoretical languages[112] – however faulty those theories may be.

In an active field of research, however, there are always several conflicting theories in vogue. The differences between such theories are associated with significant differences of terminology. As a consequence, uncertainty about the precise meaning of apparently standard terms muddles and obscures all scientific communication, even about empirical 'facts'. An agreed formal language is required to resolve such ambiguities. In practice, communal consensus on such a formalism [9.2] usually signifies the triumph of one of the competing theories over its rivals. It is the final outcome of critical thought, not its basis[113].

But theories are often formalized as they evolve, as part of the struggle to establish themselves eventually as the communal standard. Indeed, diligence in defining scientific knowledge more precisely, in order to make it more uniform and more open to criticism, is taken to indicate strong commitment to the norms of universality and disinterestedness[114]. Formalizing zeal is sometimes little more than a rhetorical stance, designed to lend credibility to a theory on moral grounds.

The ultimate purpose of a theoretical formalism is thus to ensure unambiguous communication. But that requires it to be free of *contradictions*[115]. Otherwise, even the most formal statement could be interpreted in two ways, according to which side of the inconsistency it was approached from. Indeed, it can be proved logically that this one contradiction could, in principle, render uncertain every deduction using the formalism[116].

But even if we did, say, reduce all the laws of physics to a set of abstract axioms, we would not really be that much better off. According to Gödel's Theorem [6.8], we could still not be sure that this set was absolutely complete and self-consistent[117]. In any case, the language of normal scientific discourse is not symbolic logic[118]. It is sometimes instructive to strip a scientific theory down to an arrangement of symbols[119], but these symbols cannot really be detached from what they represent[120].

Theoretical non-contradiction is clearly a very powerful principle of scientific practice. Applied in detail, it imposes severe conditions on certain theoretical structures, such as the concept of spatio-temporal location[121]. In general, a theory that is intrinsically self-contradictory must eventually die or be excluded from science[122]. A theoretical contradiction is a scientific scandal, yet its discovery is often hailed as a significant contribution to knowledge.

Nevertheless, regardless of strict logic, a certain amount of formal inconsistency is seldom a serious obstacle to intelligible scientific communication[123]. Sometimes the inconsistency is only apparent, arising perhaps from an attempt to combine two theories designed for different uses – like the Underground and Bus maps of London[124]. Good theories sometimes involve *antinomies*, such as between free will and determinism in the human sciences, for which there are good arguments on either side[125]. Sometimes scientists simply do not appreciate the lack of coherence in a supposedly well-established body of knowledge[126].

Indeed, as we have seen, the typical situation at the research frontier is that several incompatible theories are competing for acceptance [8.10]. This does not, in practice, make communication impossible between their protagonists. Scientists become adept at giving provisional credibility to incompatible theoretical propositions[127]. Following the norms of universality, originality and criticism, they may accept both propositions as *bona fide* scientific theories, even though they know that they are inconsistent. Then, instead of treating the inconsistency as a logical impasse or as a cause for social conflict, they may redefine it as an 'anomaly' requiring further experimental or theoretical research [8.9].

In practice, scientists seldom deploy formal logical arguments, and are not even peculiarly competent at solving logical puzzles[128]. And yet they have a reputation for being outstandingly 'logical'. This reflects the distinctive form of written discourse, very unlike everyday speech, which has evolved in scientific communities to ensure clear, unambiguous communication.

This distinctiveness arises, however, from the difficult of expressing scientific ideas without the use of technical jargon, taxonomic names, mathematical equations, tables of data, diagrams etc. These cannot just be strung together into a symbolic formula like a computer algorithm. They have to be embedded in a *text* whose basic structure is that of a *natural language*. A scientific paper is written, and read, as English, or Russian, or Chinese, or whatever, and has to conform to the grammar of that language. Scientific discourse is not really distinct from other 'didactic'[129] modes of ordinary speech.

The sloppiness and diversity of natural languages [8.14, 10.2, 10.4] is thus a very serious challenge to the norm of universalism. Scientists think, speak and write in a variety of languages, each with a different general vocabulary and a different grammatical structure. And even the most precise of natural languages contains numerous ambiguities of meaning and usage. How is it possible to formulate and communicate clear-cut scientific theories in such inexact media? Even when meaningful grammatical sentences can be put together out of technical scientific terms, these must previously have been defined in ordinary words – words that may themselves have been translated from, say, a textbook in quite a different tongue.

Yet this challenge is met and overcome daily throughout the scientific world. It has not been necessary to banish natural languages from scientific discourse. It is true that English has become the principal international language of modern science. But English is not the language in which the majority of scientists commonly think or express themselves to their immediate colleagues and has not superseded other languages for the formal communication of research results. Anyway, English is just as inexact, ambiguous, irregular and idiosyncratic as any other language on earth – except in its spelling which is uniquely disorderly.

In practice, the universality of science is enabled by the universality of certain structural features of all human languages. Linguists are now generally agreed that children are born with an innate disposition to acquire competence in any existing natural language, and that deep down in all existing languages there lies the same basic *generative grammar*[130] [8.14] It would take us too far afield to speculate on the biological, cognitive and cultural correlates of this linguistic discovery. The evolution of language is one of the major problem areas of science, where very little is known for sure.

Nevertheless, the empirical fact of the existence of a universal grammar must surely be a major feature of a naturalistic account of science. This is what makes it possible to take a meaningful string of scientific terms – a sentence, say – in one language and translate it into a sentence with essentially the same scientific meaning in another language [10.4]. In effect, it is a necessary precondition for the extension of intersubjectivity to scientists of all nations and cultures. Like intersubjectivity itself [5.10], we cannot explain it, but have to accept it as given. This is where the metascientific buck stops.

Even if science were conducted in only one natural language, it would rely on linguistic principles of which we are seldom conscious. Thus, it would have the means for expressing logical operations such as negation, conjunction and the uses of number[131]. Its usages would embody the elementary properties of space and time that are codified in geometries[132] [10.3] and the separateness of the material objects, living organisms and other entities classified in taxonomies[133] [6.3]. Above all, generative grammar would empower the effortless transformations of verbal messages quasi-logically – e.g. from active to passive forms – that occur as freely in scientific discourse as in everyday speech[134].

This is not to say that scientific theories are essentially linguistic entities[135], any more than they are logical, pictorial or mathematical. But human languages have evolved out of everyday human experience with the world, and their common deep structure is adapted to the communication of that experience[136]. When we state a scientific concept in words, we shape it accordingly. In effect, we build into science the natural logic of the everyday world, where there are 'solid objects' – i.e. entities that mutually exclude one another; 'causes' – i.e. actions giving rise to specific events; 'relationships' – e.g. as between siblings; and so on. One of the intellectual challenges of high energy particle physics is to simulate this natural logic electronically in order to differentiate between various types of particle tracks and events[137]. Indeed, what we take to be the completely impersonal, uncompromising and universal laws of logic evolved as structural principles of language and thought – perhaps of the brain itself[138]. The notion of a completely naturalistic, *evolutionary* epistemology will be discussed more generally in a later chapter [10.3].

For most people, however, formal scientific discourse is a very *unnatural* language[139]. It uses exotic technical terms and sophisticated modes of argument that can only be understood after years of study. In vocabulary and idiom it has already evolved away from any ordinary natural language. Its usages are attuned to the philosophy, institutions and worldview of the scientific culture, differentiating it from other forms of life[140]. Why not go the whole hog and invent a complete new language for the presentation of scientific knowledge?

The project of constructing a perfect scientific language stretches back to the beginnings of modern science in the seventeenth century[141]. Its failure then can easily be understood in the light of present-day difficulties in achieving much the same goal. The fundamental obstacle to the perfection of artificial intelligence, machine translation and other 'computer models of mind'[142] is the impossibility of integrating into a computational algorithm the tacit knowledge derived from personal and social experience – knowledge that is essential for meaningful communication [5.12]. Nevertheless, the project lives on, even up to dreams of discovering a language that could make direct neuronal connections between human minds[143].

6.7 **Mathematics**

A scientific theory is never just a 'formula'. Nevertheless, scientific theories are full of 'formulae' – that is, *mathematical* formalisms. As we have already noted, the results of observation and experiment are often communicated *quantitatively*. To count or measure implies a grasp of the principles of *arithmetic*. Again, to communicate knowledge by a map or diagram is an application of *geometry*. A scientific theory that claims to interpret 'facts' of this kind has to symbolize them and represent abstractly the relationships between them.

A mathematical formula such as '$PV = RT$' not only sums up a potentially infinite quantity of data. As a statement of the Boyle–Marriotte

Law, it is also much more compact to write than 'The Pressure of a fixed mass of a gas is proportional to its absolute Temperature, and inversely proportional to its Volume'. Each of the symbols is precisely defined, so that it can be read and understood unambiguously as part of a scientific text in any language[144]. In effect, mathematics is a universal language [6.6], perfectly adapted for the communication of scientific observations and the presentation of scientific theories[145].

But the real power of mathematics as a scientific language lies in the exactness of its 'generative grammar'. A mathematical relationship can be transformed into other relationships, to each of which it is logically equivalent[146]. Thus, using the elementary rules of algebra, I can rewrite the Boyle–Mariotte Law in the form '$P = RT/V$' (which happens to be the simplest way to state it verbally!), or '$PV/T = R$' (which is an easy formula to test numerically) or '$T = PV/R$' (which would be useful in calibrating a gas thermometer) – and so on. But I could transform it into a much more elaborate formula, such as '$\exp\{\ln P + \ln V - \ln R\} = \cot\{\arctan T \text{ (modulo } \pi) - \pi/2\}$'. Don't worry if you don't understand this bit of mathematical gobbledygook. The point I am trying to make is that an endless variety of mathematically equivalent, potentially applicable quantitative relationships can be spun out of even the simplest formula.

What is more, mathematicians have explored the implications of various types of formal relationship, and discovered amazing connections between them. For example, if I were to plot the values of P and V (at a fixed value of T) as points on a two-dimensional graph, they would lie on a hyperbola, which is the geometrical curve you get when a plane surface cuts a circular cone. This curve, moreover, satisfies the equations of motion of, say, a space probe from a distant galaxy passing through the solar system under the gravitational attraction of the sun. Algebra, geometry, calculus and theoretical mechanics are all firmly bound together by powerful theorems.

Needless to say, these connections are widely used as structural elements in scientific theories. In theoretical physics, for example, mathematical relationships typically bind abstract concepts and hypothetical entities together into extensive networks [6.5] which are far too complicated to describe in ordinary words[147]. Indeed, one of the characteristics of an interesting theorem is that its proof should be 'transparent' in thin slices but 'opaque' – i.e., far from obvious – overall. A successful theoretical prediction derived through such a theorem is thus peculiarly convincing [8.11].

Consider, for example, the General Theory of Relativity. This is accepted by most physicists. But this is not because it is an elegant mathematical system incorporating the physical equivalence of inertial mass and gravitational weight. It is because it has some quite unintuitive consequences. After an elaborate mathematical calculation, using the abstract theorems of the tensor calculus, Einstein was able to explain a long-known anomaly in planetary astronomy – the rotation of the perihelion of the orbit of Mercury – and to predict a previously unobserved optical phenomenon – the bending of starlight passing close to the sun. Although the gist of these arguments can be conveyed to experienced physicists in words and pictures, no amount of 'handwaving' would make them convincing without a full mathematical derivation.

An interesting feature of Relativity Theory is that the tensor calculus had already been developed by pure mathematicians without any premonition of its later use in physics. This is not uncommon in the history of science. Perhaps pure mathematicians are so imaginative in their invention, discovery, construction or exploration of mathematical structures[148] that they already have on hand a portfolio of formal patterns that might be used in scientific theories[149]. Perhaps theory-makers simply use whatever mathematical tools that happen to be available – especially generalizations of formalisms to which they are well accustomed[150]. Or perhaps mathematics should be treated as yet another language[151] which has co-evolved with human cognitive and cultural development[152][9.7, 10.3], and is therefore well adapted to the communication of certain aspects of human experience[153].

What these diverse views suggest is that a mathematical argument has the same epistemological status as the scientific theory in which it occurs. However successful it may be as a means of representing the natural world[154], it is no more characteristic of that world than, say, the German grammar that was also used by Albert Einstein in his scientific papers[155]. Mathematics is often thought to be completely logical, which it cannot be because of Gödel's Theorem [6.5]. Like symbolic logic, it tells us nothing we didn't already know in principle about the symbols and abstract entities whose relationships it depicts and regulates[156].

Mathematical reasoning is such a powerful instrument of thought, and mathematical formalisms are so transcultural, that mathematics is often taken to be *the* universal scientific language. Just as many scientists would like all scientific 'facts' to be reported quantitatively [5.4], so they would like all scientific theories to be formulated mathematically. But as we have seen, not all the scientifically observable features of the world can be measured, and not all the results of scientific measurement can properly be treated as variables in mathematical formulae.

Take, once again, the results of intelligence tests. It may well be that the ratio of the measured 'mental age' of a person to his or her biological age is a quantity that stays fairly constant through childhood, and is not a bad guide to later competence in other rational activities, as compared with other people. But of course the formal arithmetical definition of 'IQ' as a 'quotient' is nonsensical when applied to a middle-aged adult. And what would be the other variables in a theoretical formula for its magnitude? There might well be a statistical correlation with, say, the number of 'intelligence genes' in his/her DNA plus the family income in tens of thousands of real dollars. But that is a meaningless arithmetical construct, like the total number of apples and pears in a fruit bowl. Nothing can be deduced mathematically from the sum of quantities symbolizing such different dimensions of reality, each with its own conventions of measurement and its own calculus for the representation of equivalences and differences[157].

We shall discuss later [6.10] the well-known danger of making unrealistic or over-simplified assumptions about real-world entities in order to set up a tractable mathematical *model* of their behaviour[158]. The objection to mathematics as a universal language of science lies deeper. On the one hand, mathematical formalisms have the advantage that they are *semantically* wide open – that is, the terms that occur in them can be designated to mean whatever we want them to represent theoretically in each case. On the other hand, mathematical formalisms are *syntactically* very restricted – the relationships symbolized by mathematical operations on these terms are highly specialized, and are very often meaningless[159].

For example, let x and y stand for the IQs of two people. Apply the operation of addition. According to the laws of algebra, this operation is 'transitive' – that is, $(x+y)$ ought also to stand for 'an IQ'. But that is clearly nonsense. Even the average, $\frac{1}{2}(x+y)$, does not represent the IQ of a particular person: it is just a number typical of this set of people. Again, the difference $(x-y)$, although perhaps of interest for some theoretical reasons, is not 'an IQ' – it might even be negative, which is nonsense squared. In other words, the elementary arithmetical operation of addition symbolizes a meaningless relationship between these two terms.

It is possible, of course, to give to the symbol '+' a different operational meaning when applied to an IQ. But then we should lose all the deductive power of ordinary algebraic manipulation, and would have to develop a whole new set of theorems – an 'IQ calculus' – to put the formalism to work. As we have noted, pure mathematicians have developed a wonderful variety of alternative formal systems, of which one might prove to fit the circumstances of IQ theory. But there is no proof that this must always be possible, or even that it would produce interesting results. In practice, most of the entities that figure theoretically in the human sciences are not 'arithmomorphic'[160], and are not at all amenable to formal mathematical analysis[161].

6.8 Rationality

In the last two sections, we have seen both the capabilities and the limitations of formal reasoning in the presentation of scientific theories. For reasons explored in this and later chapters, scientific knowledge is never strictly ruled by formal logic[162]. But wherever they can be used, *near-logical* modes of presentation and reasoning, such as mathematical analysis, are enormously powerful. They strengthen scientific knowledge at every stage in its production – in observation, measurement, experiment, inductive generalization, deductive inference, concept formation, hypothesis testing and all the other operations that scientists regularly perform in the practice of research.

Nevertheless, the domain of science extends far outside the scope of formal reasoning. Research communities studying various aspects of nature and society come to agree on vast bodies of 'fact' and theory that do not satisfy the rules of any regular mathematical language. How can such agreement be reached [9.2] without appeal to formal proof? More significantly, how can an established consensus be challenged without the same means?

The usual answer is that science is peculiarly *rational*. But when we ask for a definition of rationality, we are either referred upstairs to the laws of logic, which are too strict, or back to the English word 'reasonable', which is a mundane synonym for the upmarket Latin. The definition is perfectly circular[163]. Indeed, when people nowadays quote scientific reasoning as their ideal of rationality [10.6] they are simply affirming their trust in science as a reliable social institution[164].

But scientific reasoning varies from discipline to discipline. Cosmologists make very bold conjectures about very distant events in very high mathematics; sociologists are suspicious even of headcounts of ordinary people engaged in very ordinary occupations. According to the circumstances, valid scientific reasoning may involve the evaluation of testimony, empathic understanding of human behaviour, pattern recognition[165], category formation, classification, generalization, analogy, unification and, above all, the grammar of a natural language. Most of these modes of reasoning are not deductive[166], and none can be completely validated by a computer algorithm. Yet they are all quite normal, and often fully persuasive, in everyday situations, far beyond the world of science.

In effect, scientific rationality is no more than *practical reasoning*[167], carried out as well as possible in the context of research [10.2, 10.4, 10.6]. When, as in the social sciences, the whole context is cultural [8.7, 10.2], then common-sense terms, meanings and inferences are natural components of rational discourse[168]. The key point is that *all* research is to a significant degree socially situated. Scientific facts and theories have to be communicated to and accepted by a research community. Scientific communities are disputatious[169]. They are bound by the norms of universalism and criticism. They reject communications that are patently unintelligible, ambiguous or contradictory, or that cannot stand up to well-founded criticism. In other words, they impose a communal standard of rationality on the science in their area of research[170].

These standards are embodied, for example, in the specialized languages that are typical of different scientific disciplines [5.12, 6.6]. Apparently ordinary words are enlisted, or novel words coined, to refer to precisely defined entities and concepts. Ordinary notions of place and time are tightened up into spatial coordinates and clock readings. Abstract verbs such as 'homogenize', 'differentiate', 'calibrate' and 'interpolate' are invented to describe operations that scarcely exist outside science. Arguments are strung together with logical connectives and stock phrases such as 'hence we see . . .' and 'it can be shown that. . .'.

These languages are not static. Scientific theories are not written on clean blackboards: they are proposed against the background of the knowledge of their day. Standards of rationality evolve along with each science. Einstein not only knew physical facts and theories that were quite unknown to Newton. He had to use a more sophisticated style of argument than seemed compelling to the readers of Newton's *Principia*. Darwin's wonderfully convincing presentation of his theory of evolution would be regarded as hopelessly sloppy in a modern scientific text on evolutionary biology.

Scientific rationality is quite distinct from *rationalism*. It has long been clear that reason alone, however technically impregnable, cannot guarantee scientific validity[171]. But a passage of scientific reasoning is not easily separated from its knowledge context or its specialized vocabulary. This vocabulary creates, and is created by, a local 'map', a little world of meaning [6.4]. On such a map, a scientist can normally find her way, and point out new ways, by quite elementary modes of reasoning. It is very unusual to encounter a research situation that is not amenable to ordinary ways of thought. That is why quantum theory has proved such a challenge to scientists and philosophers alike.

A naturalistic account of science requires a naturalistic account of rationality. The *social* norm of universalism cannot be put into practice without some degree of *cognitive* universalism. In fact, people the world over share certain basic criteria of rationality that cannot be summed up in formal logic [10.4]. This follows from *linguistic* universalism – the empirical fact that it is possible to establish meaningful communication between members of the most disparate cultural groups [6.6]. Any effective mode of translation between languages requires that their speakers have certain elementary modes of reasoning in common[172]. The notion of total intellectual incommensurability applies only to lunatics and Martians, not between 'rational' human beings.

But this common core of practical rationality [10.6], although much larger than formal logic, is not well defined or commonly agreed[173]. How could it be? By Gödel's Theorem, that would require an appeal to some other universal mode of reasoning – which doesn't exist[174]. Indeed, even a highly literate society such as Japan may not have a concept of formal logic[175]. That does not mean that scientific ideas cannot be communicated clearly in Japanese, or that Japanese scientists are unable to distinguish between sound and unsound scientific hypotheses[176]. It merely reminds us that we are in a world where translation between natural languages is always possible, but never perfect[177].

In effect, the norm of universality limits scientific rationality to *transcultural* modes of practical reasoning[178]. In the physical and biological sciences, these are well defined. Even the obscure tribes of sceptical philosophers and self-reflective sociologists do not practise what they preach about not being so sure that we are all living in the same material world. But in the human sciences, the scope of a *consensus gentium*[179] – if it exists at all – is always highly disputable [10.6].

Take history, for example. This relies on historians being able to convey to one another their empathic interpretations of human actions [5.11]. But agreement on these interpretations depends, in the end, on sharing some understanding of how the actors are motivated – that is, of their personal and social *values*. These values are often deeply hidden, and change over time. They are part of the whole culture of the actors – and of the historians, too – and vary enormously from culture to culture [7.4, 7.7]. On the other hand, if the value systems of different societies were entirely incommensurable, there could be no scientific basis to history[180].

Cultural relativism [10.4, 10.7] is thus a very serious issue, which applies in principle to all sciences at all levels. Scientific knowledge is not only sensitive to major differences in its general social environment: it is also sensitive to differences between scientific sub-cultures, even within the same academic discipline [7.5]. Only a naturalistic stance can disarm this relativist critique, and limit the epistemological damage that it could cause throughout academic science.

For the moment, however, this is a timely reminder that there are vast repositories of reliable human knowledge that are not 'scientific' by the criteria of this book [10.6]. The invention of the intellectual and social procedures that we discuss here was a revolutionary episode in the evolution of the human mind. Nevertheless, it did not render valueless the use of narrative and myth to attribute significance and meaning to life and nature[181]. Systematic, would-be objective, universally intelligible observation and analysis cannot draw out of our personal and cultural experience all that is worth communicating to our fellow humans[182].

In fact [10.5], we learn most of what we know about what makes life worth living, and how to live it well, from non-scientific sources – biography, narrative history, serious journalism, and religious texts, not to mention novels, poetry, drama and the visual arts[183]. For Europeans, at least, there is more insight to be got from a single volume by Jane Austen or Gustave Flaubert than from a whole shelf of treatises on the social psychology of bourgeois love and marriage.

6.9 Systematization

Scientific theories are about aspects of nature on which it is possible to arrive at communal agreement. Scientists have become interested in such an immense variety of such aspects that they have developed innumerable specialized languages and images to convey their observations and concepts to one another[184]. As we have seen, scientific theories are not always precisely formulated or strictly logical. They cannot be entirely mathematical, completely abstract, or perfectly empirical. What counts as serious scientific discourse can be extraordinarily varied and heterogeneous in form and substance.

Nevertheless, scientists from different disciplines recognize each other as committed to the same intellectual enterprise. They not only engage in a standard repertoire of research practices, such as 'observation', 'theory', 'measurement', 'experiment', 'deduction', etc. They also see these practices as having the epistemic goal of 'understanding' or 'explaining' the world. To achieve this goal they draw on a common stock of intellectual strategies, such as 'hypothesis', 'inference', 'prediction', 'confirmation' and 'discovery'.

Thus, although scientific research is not governed by a hard and fast 'method'[185], all scientists do have in common a distinctive methodological culture. Do they also share a distinctive *conceptual* culture? Is there a standard repertoire of theoretical structures typical of scientific thought? Are there *metascientific* features common to all scientific theories?

For example, the urge to *describe* the world and *classify* its contents is characteristic of scientific thought. The primary 'map' of a discipline is usually a *taxonomy* [6.2]. Much of the knowledge produced by science is to be found in enormous *databases* – that is, elaborately ordered archives of information about biological organisms, diseases, human genes, chemical compounds, historically significant people, etc., etc.

What is more, scientific taxonomies are not just collections of 'facts'. They are not only steeped in theory: they are also theoretical structures in their own right [6.3]. Immense scientific resources are devoted to collecting the relevant information, testing it, purifying it, reworking it, presenting it intelligibly, classifying it, indexing it and storing it safely. Taxonomists in different disciplines encounter very similar intellectual problems, and draw upon the same techniques of information science and computer programming to solve them. *Classification* is thus a generic feature of scientific theorizing, right across the board.

But a *description* is not an *explanation* [10.1]. Even in everyday situations, *pictures* [6.5] have to be *interpreted*. The urge to understand the world by *analysing* it is characteristic of scientific thought. But scientific understanding involves much more than observing 'phenomena', discovering their 'causes', and formulating general 'laws of nature'. Indeed, despite all the philosophical attention given to them, these traditional metascientific terms are scarcely to be found nowadays in genuine scientific discourse.

As Immanuel Kant put it, more than 200 years ago, scientific theories are *schemas*[186]: they 'enable the understanding to apply its categories and unify experience[187]'. In other words, the grand strategy of research is not just to trace out itineraries of cause and effect: it is to *represent* 'schematically' those aspects of nature that can be grasped in principle and/or used in practice.

Thus, the conceptual culture of science includes a number of characteristic schemas. Of these, the simplest and most widely used is the notion of a *system*. Astronomers talk about 'the solar system'; physiologists talk about 'the immune system'; economists talk about 'the monetary system' – and so on. In each case, they mean 'a group or combination of interrelated, interdependent, or interacting elements forming a collective entity'[188]. In effect, they are mapping an aspect of nature as a *network*. They schematize the seamless web of the real world by representing it as a more or less closed and coherent set of relationships between potentially separable entities.

It would take us too far away from our main theme to go into the (metascientific!) theory of systems[189]. The key point is that the system as a whole can have properties or functions that are different from those of its individual components[190]. Indeed, it may be an assembly of *subsystems* or *modules* each with an internal structure – and so on. Nevertheless, this does not necessarily mean that the natural world can be represented scientifically as a *hierarchy* of systems. Nor does it imply that a scientific understanding of the properties of a system can always be obtained by *reducing* it to its components [10.8].

The system schema arises naturally in the scientific culture because it can so easily be communicated pictorially[191]. A diagram showing a network of interconnected 'black boxes' is easily drawn and easily understood [5.12]. Given that the components have been clearly defined and classified, it is an unambiguous representation of their relationships. The question then shifts to the nature of these relationships[192]. The physical sciences are accustomed to dealing with 'hard' systems – typically *mechanisms* – made up of perfectly defined, interchangeable components interacting according to precise rules[193]. But as we move through biology to the human sciences, the systems become 'softer' and vaguer. Psychological and social interactions that can only be grasped empathically are always imprecise, and are difficult to communicate unambiguously from scientist to scientist[194].

In effect, we here approach the region where the scientific culture opens towards other modes of knowledge production and belief [10.5]. It is highly disputable, for example, whether the theoretical representation of a social institution – let alone a whole society – as an assembly of interacting individuals can ever do justice to its *holistic* properties, such as collective values, loyalties and goals[195]. These can only be brought to the surface by other tools of thought, such as play-acting, manual manipulation, poetry and other aesthetic activities[196], which are not constrained by the norm of universally unambiguous communication.

6.10 Models and metaphors

An image of an aspect of nature as a 'system' is a descriptive theory. This may be scientifically instructive, for example by drawing attention to inconsistencies[197] or missing elements[198]. But for an explanatory theory one requires a *model*. This can take a variety of forms – scale models, mechanical models, analogical models, ostensive models, toy models, mathematical models, etc.[199] Indeed, despite philosophical objections[200], the word 'model' is so widely used in scientific practice[201] that it has become almost a synonym for a 'theory'. In the human sciences, moreover, 'theory' has come to mean little more than 'theorizing', without reference to a coherent system of concepts[202].

Like other metascientific concepts, the notion of a model defies formal definition. One might say, perhaps, that a theoretical model is an abstract system used to represent a real system, both descriptively and dynamically[203]. The significance of the 'Rutherford–Bohr model', for example, was not simply that it depicted a hydrogen atom as a light, negatively charged electron attracted to a heavy, positively charged nucleus. It prescribed equations of motion from which the actual energy of the electron could be calculated, and eventually – after the imposition of certain quantum conditions – compared with experiment. In effect, it was a demonstration that a number of theoretical physical principles, some explicit and some tacit, some well established, and some hypothetical, could be brought together and made to 'work'.

In practice, although a model of a 'hard' system in the physical sciences or economics may well be easy to formulate mathematically [6.7], the equations are normally too complicated to be solved analytically. But the behaviour of the model under various circumstances can often be computed and displayed visually[204]. From astrophysics to zoometry, *computer simulation* is now a regular feature of 'good science'[205]. Traditional philosophy may have no place for arguments based on the output of incomprehensible computations[206]. But from a naturalistic perspective there is everything to gain from detailed studies of the behaviour of non-linear, irreversible, functionally irreducible models of – the non-linear, irreversible, functionally irreducible systems [10.8] that we find in the real world[207].

Indeed, computer simulations of 'hard' systems, have a much broader epistemological scope than analytical mathematical models. For example, they make it possible to explore the properties of a wide range of *hypothetical* systems [8.10 – 8.12], such as non-crystalline solids with novel constituents, under extreme conditions that could not easily be realized in the laboratory, and hence to determine, say, the quantitative causes of significant qualitative effects[208]. In spite of many caveats[209], computer power is a Tertium Quid[210] that closes the gap between 'theory' and 'experiment' – a gap that has always seemed a fixed feature of scientific practice.

Nevertheless, as we move from the physical sciences to the human sciences, we soon find that computational modelling has irreducible limitations. This is not because biological, psychological and social systems are peculiarly complex, non-linear, irreversible, etc. It is because they are irredeemably *soft*[211]. There is no precise way of classifying their components and formulating the interactions between them [5.4, 6.7]. Their 'equations of motion' – if such could properly be said to exist – are just too vague to form the basis for a realistic model.

It is true that the behaviour of a simplified mathematical model – as in 'catastrophe theory'[212] or 'artificial life'[213] – is often an instructive metaphor in the understanding of biological and social phenomena[214]. Complex human activities such as wars and business deals can even be simulated by 'games'[215], in which people play assigned roles – 'soldier', 'tycoon', 'gambler' or 'thief'. Intersubjective empathy [5.11] is probably more realistic than any formal 'decision algorithm' in modelling such emotion-laden circumstances.

Outside the physical sciences, theoretical models are usually as 'soft' as the systems that they represent[216]. They may be purely schematic, or, as in econometrics, deeply compromised by questionable assumptions, inaccurate data and arbitrary 'adjustments'[217]. That does not mean that they lack scientific validity. There is a wealth of explanatory knowledge in, say, a series of diagrams illustrating the successive stages in the response of the human immune system to a foreign protein. It is not necessary to know the details of the various biomolecular processes to understand how the system works [10.1]. This 'qualitative' model was not only the outcome of a celebrated scientific achievement: it remains a powerful and reliable guide to further research and practice in this field.

A good theoretical model is a *communal* intellectual resource. It is an account of how a system is thought to *function*. Like the more traditional notion of causation, it harnesses the universal human capacity for envisaging *action* [8.10]. The intersubjectivity of pattern recognition [6.2] is extended to include temporal change. From life-long sensorimotor experience of manipulating objects we acquire a feeling for *mechanism*. Anybody who has ever kicked a football can look at a model and see that if component A moves so as to strike component B, then the latter is also set in motion. The dynamics of the system is thus tacitly communicated from scientist to scientist along with the schema of its structure.

This shared understanding about how things 'work' applies even to highly abstract models of human social activities. A sociological model of a legal system, for example, might indicate, verbally or diagrammatically, that action in a court of type A has an 'impact' on a court of type B – just as if it were a boot making contact with a ball. This conveys more than a whole page of differential equations. Thus, noting that the B-type courts are strongly networked to other institutions, we opine that this disturbance will probably spread to them – and so on.

In other words, even the most austerely 'scientific' models operate through *analogy* and *metaphor*. The Rutherford–Bohr model depicts a hydrogen atom as a miniature solar system. Darwin's concept of 'natural selection' is analogous to the 'artificial selection' practised by animal breeders. 'Plate tectonics' is about thin, flat, rigid areas of 'crust' floating on a highly viscous but fluid 'mantle'. Linguists talk of the 'brain mechanism' by which grammatical language is generated. And so on.

Scientific theories are unavoidably metaphorical[218]. Indeed, how could even the most original scientist construct, make sense of, or communicate to others, a theoretical model that did not incorporate a number of familiar components? Sometimes a ready-made model can be taken over from another branch of science – for example Fresnel's model of light as the vibration of an elastic medium. Sometimes the key elements come straight out of everyday life, as in von Neumann's model of economic behaviour as a 'game', or the molecular-biological model of DNA as a genetic 'code'.

This heterogeneity is not a serious defect. The scientific value of a theoretical model, as with all metaphors, does not require it to be literally equivalent to the system it represents[219]. It resides in the variety of phenomena that it makes plain, or suggests [10.1]. This understanding seldom comes through elaborate formal analysis. It arises directly from our knowledge of the typical properties of its components and of the way that they might be expected to interact.

Indeed, analogy and metaphor cannot be driven out of scientific reasoning. Scientific ideas cannot be communicated through the 'literal' medium of formal logic [6.6]. But all natural language is metaphorical[220]. Just look at the previous two paragraphs, for example, and unwrap the life-world corpses mummified in such abstract words as 'unavoidably', 'construct', 'incorporate', 'literally' and 'resides'. Scientific communities devote much effort to the development of complex, well-defined, communal languages [6.6]. But their best endeavours merely ensure that the metaphors embedded in their specialized taxonomies and models are unambiguous and mutually intelligible[221].

Indeed, the history of a scientific discipline can be traced through its changing repertoire of models and metaphors – what Gerald Holton called its *themata*[222]. Modern physics, for example, deals in 'forces' and 'fields', or 'waves' and 'particles', and has no place for pre-modern themata such as 'sympathies' and 'attractions', or 'essences' and 'effluvia'[223]. The choice of themata often has deep cultural significance. Thus, a feminist perspective [7.5, 10.4] on the theory of complex interactive systems might give greater weight to stabilizing themata such as 'glue', 'nexus' or 'linchpin' than to dynamic themata such as 'cause'[224].

Despite their dangers[225], fashionable themata often store the solutions to old problems, and can be recycled to deal with new ones[226]. But they are not all rooted in timeless everyday images. Scientific progress has given us quite new themata, such as 'feedback' and 'gene'[227], as well as novel mathematical metaphors such as 'fractals' and 'chaos'[228]. The transfer of themata between different disciplines – for example, the notion of a 'code' from information theory into molecular genetics – is often an effective recipe for good science[229].

It is clear that scientific maps, models, metaphors, themata and other analogies are not just tools of thought, or figures of speech. They are of the very substance of scientific theory. As sources of meaning and understanding, they stand on an equal footing with explicit verbal and symbolic representations[230].

A theoretical model starts life as a personal or social construct [8.13], designed for a specialized scientific community[231]. But that does not mean that it is necessarily more arbitrary, contrived, or unconstrained by empirical facts, than, say, an axiomatic formulation of a 'law of nature'[232]. The norm of universalism requires scientific theories to have general applicability. But scientific knowledge has to be communicated between people, so it can never be completely detached from its human sources and contexts. What is more, the form in which a scientific theory is presented has little to do with whether it is ultimately accepted as a reliable or valid representation of a particular aspect of nature.

6.11 Scientific domains

To take part in a science one has to know its maps and models. To be a physicist, for example, one has to learn to 'think physically'[233]. A scientific observer perceives, interprets and talks about the world in the light of her special experience and knowledge [5.9]. Observation is laden with theory. New 'facts' only become visible against old expectations.

The theoretical models of an advanced science are notoriously fragmented, uncertain and inconsistent. Nevertheless, they make up a 'big picture'[234] that is as elaborate, detailed, cross-connected and apparently coherent as the street map of a great city. A practising scientist gets to know it until it is as familiar as 'the Knowledge' that a London taxi-driver has to learn by heart to get a licence. It becomes a 'domain' of thought and imagined action [8.14, 10.7] that is very different from the domain of everyday life[235].

In later chapters we shall consider other features of the numerous 'paradigms' that arise so naturally in scientific practice [8.4]. How do they come into being? How sound are they? Where are they to be found? Are they mutually 'incommensurable'? What does it mean to say that they are 'real'? These questions cannot be answered definitively, but they indicate a number of epistemological issues that have to be further explored.

What we have learnt in the present chapter, however, is that the norm of universalism drives the scientific enterprise towards the generation of elaborate general theories, typically embodied in conceptual models spanning abstract scientific domains. The 'map' of such a domain, like a natural language, is both *collective* and *personal*[236]. It is not only a collective *accomplishment*, in that it is generated and sustained by the combined efforts of many individuals [8.13]: it is also a collective *resource*, in that it is designed and shaped to be shared by a research community[237]. It is not only a personal accomplishment, in that individual scientists contribute personally to its construction: it is also a personal resource, in that each of them carries it 'inside their head' as a guide to research, teaching and practical use[238].

The striking thing about scientific knowledge domains is that they really are extraordinarily different from the corresponding domains of the *life-world*[239] – the world of everyday knowledge[240] [10.2]. The accepted scientific map of, say, inorganic chemistry not only classifies all the chemical compounds of a certain general type, and represents the reactions that take place between them; it is a vast assembly of diagrams [5.12] that the trained chemist reads almost as if they were words in a book. Test-tubes, balances, retorts, centrifuges and all the other paraphernalia and physical operations of a real-life laboratory are not even mentioned.

This is not to suggest that the scientific image in terms of abstract structures is in that respect unsound. It is just to remind us that the maps and models in the scientific domain are much more sophisticated, and require much more interpretative effort, than the maps and models we use in everyday life.

And yet they are genuine maps, with all the cognitive power of genuine mental models[241] [8.14]. The chemical researcher looks at two structural diagrams, manipulates them mentally, and concludes that the addition of a certain other compound – represented by another diagram – would transform the one into the other. Her task as a teacher is to make this interpretation a natural mode of thought for her pupils. The chains of meaning between ideal and empirical representations of the world are not deductive[242], are often very lengthy[243], and have to be taken for granted in formal scientific communication[244]. Nevertheless, a great part of formal science teaching is to explain and practise movement between the two domains until the student can go back and forth between them, quite intuitively, in problem situations[245].

The contrast between an everyday object and its scientific model is often so extreme[246] that they are held to be philosophically distinct [10.6]. Indeed, in written Chinese they are always denoted by different characters[247]. But as the Underground/Bus map metaphor makes clear, this distinction is difficult to sustain in principle. Eddington's two tables – the hard, solid object and the buzzing cloud of electrons and atomic nuclei – are one and the same table[248], seen from different points of view.

The incongruity between the vague, indeterminate world of immediate experience and the sharply defined, clear cut elements of scientific theory is only apparent[249], and does not apply in the biological and human sciences. And the notion that scientific domains are typically 'unperceived'[250] is refuted by such sciences as ecology, where the scientific observer perceives real organisms, even whilst locating them mentally as abstract entities on conceptual maps[251].

Indeed, in the human sciences the two domains become intertwined[252]. The empathic understanding of human action, on which all such sciences are based, cannot be abstracted from its life-world context. Computer algorithms for pattern recognition are inadequate scientific models for human perception [6.2]. So far, at least, every attempt to patch them up introduces life-world elements, such as remembered experiences or affective goals.

In the end, there can be no real value in a scholarly work in the human sciences that is completely meaningless to the people that it claims to talk about[253]. This situation becomes even more complicated when the human observer has to be included in the model of his own environment[254]. A truly reflexive sociology may well be impossible, for it would have to incorporate itself in its own representation of the life-world[255], and thus become part of the 'folk sociology' that it scorns[256]. Indeed, although the human sciences are very poor at prediction [8.11], they are rich in theoretical ideas through which people can give meaning to their own personal worlds and social actions[257].

Our knowledge of the life-world is not unproblematic, nor is it entirely different in principle from the way we know about science [10.3, 10.6]. But there is such a gap between the scientific and life-world domains that much attention is now being given to improving 'public understanding of science' [258]. What is more, this lack of understanding also applies between various scientific disciplines, each with its own particular domain of maps and models [8.4]. Scientific progress that fosters the development and consolidation of each of these domains tends to reinforce the distinctive features that keep them apart.

Epistemologically speaking, the notion that science should be 'universal' suggests that the scientific domain can be covered by a single map. In other words, it sets the *unification* of knowledge as a desirable and attainable goal for science. Science thrives on the resolution of contradictions. The removal of frontiers of apparent incommensurability between previously independent disciplines is one of the major achievements of research. Even such 'archipelagos' of scientific activity as the study of the brain and human behaviour are now linked together by fragile bridges, if not yet by solid causeways.

But the notion of a *unified science* has metaphysical features – for example, the licence it gives to *reductionism*, and the implied downgrading of other modes of knowledge – which need to be explored more carefully [10.8]. In practice, moreover, the unifying impulse is inhibited by the mundane necessities of academic *specialization* [3.9, 8.3], which bear even more heavily on research careers and programmes[259].

Indeed, even in sociological terms, science is far from the norm of universality. We began this chapter with the notion that the findings of research should be acceptable to *anybody*, regardless of their cultural background. As we have seen, however, scientific knowledge has become so elaborate and esoteric that new research results are unintelligible except to 'scientists' – that is, the relatively small number of people who are already at home in the scientific domain. Moreover, this domain is highly fragmented [8.8]. In effect, there are many different scientific domains, each served by and serving a different scientific community[260].

Thus, the actual group of people who are involved in accepting and sharing a research result is only a tiny sub-set of the universal, multicultural ideal. Epistemically and socially, this group practises a distinctive version of the scientific culture – almost a culture in its own right. In later chapters we shall consider whether scientific knowledge is subject to *cultural relativism* [10.4]. One might suspect that this fundamental philosophical characteristic is more likely to arise from its own internal divisions than from the diversity of the larger cultures in which academic science is immersed.

7

Disinterestedness and objectivity

7.1 Striving towards objectivity

When Robert Merton pointed out that scientists are constrained to be 'disinterested'[1], he was referring mainly to their professional behaviour. As we have seen [3.5], there are many conventions that severely limit the operation and public display of personal motives in the regular practice of academic science. The direct effect is thus to sever the connections between scientific knowledge and its personal origins.

According to this norm, it might seem that scientific knowledge should always be presented as cognitively *objective* – i.e. as if referring to entities that exist quite independently of what we know individually about them[2]. But the implication that science rests on and requires absolute *realism* calls for further discussion, which we postpone to a later chapter [10.7].

As a *social* norm, however, disinterestedness functions primarily to protect the production of scientific knowledge from personal bias and other 'subjective' influences. Strictly speaking, this is impossible [5.3]. There is no denying that scientific facts and theories are produced by human beings, whose minds cannot be completely cleansed of individual interests. Academic science therefore strives to attain *consensual* objectivity by merging these interests in a collective process.

The norm of disinterestedness thus combines naturally with the norms of communalism and universalism to strip scientific knowledge of its subjective elements and turn it into a genuinely *communal* product. Indeed, the impersonal style of formal scientific discourse is designed to make research claims appear immediately acceptable. They are presented as imperfectly glimpsed fragments from the 'Book of Nature', ready for incorporation into the communal archive[3]. The scientist who uses phrases such as 'hence we see that...', and 'it follows that...', is desperately inviting us to believe that 'we' already includes all her peers, and that they will not question whether 'it' really does 'follow' as clearly as she hopes.

This is a rhetorical device that deceives nobody. After all, one of the conventions of the academic culture is that every publication should proudly proclaim the name(s) of its author(s) [3.8]. But this style of presentation is obviously an essential step in generating a body of knowledge that can eventually be detached from all its human roots, including even the 'thought collective'[4] where it was born and shaped. The norm of disinterestedness, along with the rest of the academic ethos, functions to ensure that this cut can be made cleanly, without leaving loose ends of personal prejudice, injustice or mendacity by which the communal consensus might be unravelled.

7.2 What makes science 'interesting'?

The academic ethos strongly encourages the production of knowledge. But it doesn't seem to care what that knowledge is *for*[5] – except perhaps the production of more knowledge. Academic research is directed towards, and celebrates, the solution of problems[6] – but primarily problems raised by previous research [8.1]. This notion of producing knowledge 'for its own sake' defies logical analysis. Nevertheless, it is a powerful motivator. Academic scientists do indeed clamber up mighty knowledge peaks, as much because, like Mount Everest, they are 'there', as because great CUDOS comes to the first person to reach the top [3.8].

What everybody knows, of course, is that scientific knowledge is often of immense practical use. Society at large fosters science principally for that reason [2.4, 4.8]. To make a pun, science is 'interesting', not only because it arouses *intellectual* interests, but also because it can serve *material* interests. The utility factor can actually be quantized. Overall, research is an immensely profitable investment. Hard-faced economists reckon that every £100 of public money put into academic science yields, on the average, a perpetual return approaching £30 *per annum* in industrial, commercial, medical and other social benefits[7].

The *instrumental* power of science must surely figure in its philosophy[8] – but where? Epistemologically speaking, the *utility* of scientific knowledge is no more questionable than the utility of a map. Science provides us with public, communally shared representations of certain aspects of the world. From them we can read off directly, or infer rationally, information about where we are, and where we might be going. True, the value of this information depends on its being reliable and relevant. True, specialized expertise is needed to make good use of it. True, in the human sciences – even in a formalized science such as economics – the map itself is part of the action[9]. But only a very subtle metaphysician would want to dispute that there are practical circumstances where soundly based scientific knowledge is potentially applicable and materially beneficial.

Indeed, this feature of science is so salient that it is sometimes taken as its defining characteristic [2.4]. But *pragmatism* in this narrow sense does not do justice to the intellectual creativity of academic science. Scientists regularly generate, and convincingly validate, theoretical concepts which are far removed from worldly practice. Technological applications often test such concepts very severely, but no more so than contrived experiments [5.6]. Scientific and technical disciplines operate in extraordinarily diverse practical contexts, each with its own criteria of pragmatic cogency. Predictive success in such contexts is only one of many considerations in the communal assessment of research claims [8.11].

Thus, when philosophers assert that scientific knowledge is validated 'pragmatically'[10] they mean this in the broad sense of being 'coherent with experience'[11]. In effect, they are reaffirming our naturalistic observation [9.2, 10.7] that scientific validity arises out of scientific practice as a whole rather than through a special process of 'justification'[12]. They are also reaffirming the openness of the frontier between 'pure' and 'applied' science [2.7]. As a mode of knowledge production, science conceived as 'mastery over nature' – i.e. research for specific, temporary advantage – merges imperceptibly into science as 'the pursuit of understanding' – i.e. as if for eternal, universal use[13].

7.3 **What makes science reliable?**

Philosophical pragmatism does point, however, to a major epistemic value – *reliability*. The map that we consult, the scientific knowledge we propose to apply, is worse than useless if it leads us astray. The validity that we require of science is not that it should be absolutely 'true'. Indeed, it is often the expression of ignorance or uncertainty that makes science seem such a trustworthy guide to *action* [10.1, 10.5]. That is why the question is often asked – and not always answered convincingly – whether the human sciences have really produced knowledge that goes beyond the common-sense wisdom gained from thoughtful reflection on experience in practical life-world situations[14].

People regard scientific knowledge as peculiarly reliable partly because it fits in well with the world of making and doing. Scientific work is typically *performative* rather than *contemplative*[15]. The dynamism of a good scientific model is valued less for depicting a process correctly than for getting its *outcome* right. When the explanation comes long after the event, as in a historical science such as geology, it is presented as a detached commentary on the workings of nature. Nevertheless, non-scientists can read it as a potential guide to what might be expected to happen if similar circumstances were to recur[16].

Research itself is often a highly technological mode of action [8.2]. Scientific 'interventions'[17], such as experimenting [5.6], test the reliability of the design principles of the apparatus, rather than the validity of these principles in general. Indeed, they sometimes generate bodies of knowledge that can be transferred, almost unchanged, to cognate practical 'interventions' such as inventing. Scientific theorizing, like engineering design, makes full use of existing scientific knowledge. Conceptual model-building [6.10] is similar to the construction of prototypes. Thus, in spite of many differences [2.2, 2.4], scientific and technological domains often contain practices and entities that are recognizably alike in their relevance to action, and that are required to be 'reliable' in the same practical spirit.

Above all, scientific knowledge also has a reputation for *moral* reliability. It is deemed to come from *trustworthy* persons and institutions [5.7]. At first thought, this takes us back to the ideal of 'disinterestedness'. In ordinary life, we are prepared to trust informants who report their findings to us piecemeal, soberly, and with duly modest doubts[18]. Unlike pretentious 'system' builders, they seem not to stand to gain or lose by the effect on our own actions of what they tell us. For this reason, well-established scientific knowledge is extraordinarily compelling. From its vast impersonal archives I obtain undisputed information that could never have been contrived specifically to influence little *me*. It seems natural to trust implicitly a source which has an evident urge to inform, and no obvious motive to deceive.

Needless to say, this is a false and dangerous argument. It would apply to any systematic body of knowledge that had been purged of explicit inconsistencies or obviously self-serving interests. This is how religious and political movements [10.5] gain followers. It often misleads scientists themselves, when they come under the spell of a widely accepted theoretical paradigm [8.4, 9.6]. The fact is, rather, that scientists and scientific communities have a *positive* interest in the reliability of the knowledge they produce.

Nobody imagines, of course, that scientists are bloodless robots, indifferent to the reception of their research claims. They have the strongest possible interest in gaining public recognition for their discoveries. The whole social apparatus of academic science – its norms, its communication system, its rhetorical style, its critical mechanisms – assumes that every scientist thirsts for its psychic and material rewards [3.8].

The interest that every scientist has in having acceptable ideas[19] is perfectly transparent. But this is no different from the transparent interest of every shopkeeper in making a profit, or the interest of every politician in winning power. Tenacity in presenting and defending one's ideas in public is one of the hallmarks of 'good science'[20]. This is what makes science tick.

The trick is to nullify these individual interests by setting them against one another. In effect, the scientific ethos delineates an agonistic arena, where a hidden melodrama of clashing egos[21] is transformed into apparently dispassionate intellectual debate [9.1]. As in a free commercial market[22], the particular bias of each individual is neutralized in the collective outcome[23].

It is true that the formal mask of disinterestedness sometimes slips, revealing power ploys, wheeling and dealing, quasi-military alliances[24] and other sordid social manoeuvres[25]. It is a sad fact of life that academic communities are prone to clannishness, careerism, nepotism and other worldly vices[26] – including downright fraud[27]. Like most self-regulating institutions, academic science is not good at detecting and eliminating petty social prejudice or occasional cases of deliberate deceit [9.4]. In distributing scientific recognition and its material rewards, scientists often favour their own colleagues and fellow-countrymen[28]. In pursuit of material resources they often behave as swinishly as any other political lobby[29]. Amongst themselves, scientists freely acknowledge that they are susceptible to fashionable trends[30] – witness the current vogue for cultural relativism in science studies[31]. And they often demonstrate the characteristic conservatism of comfortable vested interests[32], though seldom to the degree of collective prejudice[33] exemplified by the geologists' long interdict on Wegener's hypothesis of continental drift[34].

The fact is that academic science could never possibly live up to its romantic image as a selfless quest for a Holy Grail of Truth [2.7]. It is a normal social institution inhabited by normal, morally frail people. What else would one expect? A significant cultural event, such as the public recognition of a scientific discovery [8.9], is bound to involve a certain amount of social negotiation, if not extreme *angst*. Contrary to the Legend, such issues can never be settled by appeal to a non-human transcendental authority. Indeed, the intensity of these negotiations is as much evidence of the intrinsic significance of the event[35] as of hidden forces that enlist it in their own mysterious conflicts.

What these revelations do not prove is that academic science is thoroughly corrupt. On the contrary, they show that its formal communication system [3.3, 5.13, 9.3] is very far from an empty ritual. Here, if anywhere, the norms of originality and criticism rule. Entrenched views have to be defended publicly against novel lines of attack [9.1]. When two scientists privately agree to back one another in a scientific controversy, they must still contrive plausible arguments to support their position. They are fully aware that the intellectual bargaining counters in their negotiations must be weighty enough to survive in the open court of scientific opinion. No one will listen to their case unless it respects outwardly the epistemic norms, rhetorical values, metaphysical commitments and other cognitive interests of the scientific culture[36].

What is more, most new scientific arguments are spun out of the published results of earlier research by *other* scientists. Indeed, scientists themselves are the most active 'users' of scientific knowledge, and have the most need for it to be reliable. They are deeply embedded in webs of *trust* [5.7], both in their dependence on others and in the acceptability of their own contributions[37].

These webs are stronger than they seem to an outsider. The reliability of every thread is seldom separately tested. Nevertheless, the webs themselves are associated with theoretical networks [6.4] where faulty elements eventually show up. This possibility is well understood by working scientists, and is much stronger guarantee of reliability than the threat of formal sanctions.

In effect, academic science is a culture where a reputation for reliability – that is, *credibility* [8.10, 9.2] – is the prime personal asset[38]. This asset is so valuable as a long-term source of material support and social esteem[39] that it is not to be risked for short-term gain. This is strongly emphasized in the education of scientists and their apprenticeship to research, and reinforced by a variety of social practices such as peer review

[3.7]. For the mature scientist, reliability on relevant scientific matters ceases to be learned behaviour or a calculated strategy: it becomes, as the saying goes, 'second nature'.

Let me emphasize that I am not trying to present academic science as a uniquely virtuous institution. All I am saying is that the issue of credibility exemplifies the interweaving of epistemic, social and psychological factors in the scientific culture. This culture is remarkably powerful in shaping the behaviour of those who are caught up in it. Through its ethos and its established practices, it transforms their conflicting personal interests and cognitive powers into a shared collective interest in the production of reliable knowledge and in the anonymous, institutionalized credibility of that knowledge[40].

7.4 Interests and values

In the interest of its own credibility, academic science strives to produce knowledge that is not influenced by other interests than its own. It scarcely needs support from sociological theory[41] to demonstrate that complete *social* objectivity is an impossible goal. Science has always been a social institution, embedded in the larger society of its time and place [ch.4]. Scientists have always been citizens, consumers, producers, owners, employees, parents, teachers, believers and at times even warriors. It is nonsense to suppose that such people can be brought together into scientific communities to produce knowledge that is completely untainted by the collective interests and cultural values that drive and shape their non-scientific lives.

Nevertheless, the norm of disinterestedness dictates that scientists should not be influenced by any such external considerations. What this means, in practice, is that all reference to economic, political, religious or other social interests is rigorously excluded from the formal scientific literature [9.3]. This is not merely a public stance of detachment. It is a traditional feature of the 'scientific attitude'[42]. Academic scientists are taught to think of themselves as persons who know nothing, and care less, about social problems[43], who solve intellectual puzzles without reference to their practical significance[44], and who do good automatically by producing valid knowledge that sometimes turns out to be useful[45].

Historically speaking, this tradition enshrines an ideology that emerged on the side of 'experimental' science in the rationality wars of the seventeenth century[46]. It lives on as an unwritten ethical code governing the relationship between science and society at large. Academic science cannot formally regulate the behaviour of its individual practitioners [3.10]. But it gives its collective imprimatur only to knowledge that has supposedly been produced in accordance with this code. Through it, therefore, academic science protects its institutional reputation for credibility, reliability and social objectivity.

Notice, moreover, that this code extends the norm of disinterestedness from the *presentation* of science to its *performance*. The Legend holds that research claims should be assessed solely on their intellectual substance. In real scientific practice, as in ordinary life, the motives of an informant are not considered irrelevant to the credibility of what they offer. In particular, research claims are scrutinized very carefully, and even rejected, if they might possibly be biassed in favour of some external body, such as a commercial firm that has supported the research.

It is not just moral squeamishness, for example, that makes university medical scientists wary about entering into research funding relationships with tobacco companies. They fear that their professional credibility would be compromised by the suspicion that they might have slanted their findings in favour of tobacco use. Such thoughts would seem uncharitable, if it were not for recent revelations that tobacco companies have suppressed the scientific findings of their own in-house researchers when these were adverse to corporate policies and profits. Indeed, to an ordinarily streetwise observer, the secrecy that often shrouds such relationships [5.13] would be *prima facie* evidence of an undeclared bias in the knowledge that is produced.

Epistemology thus spills over into ethics. Scientists are bidden to disclose any potential conflict of interest between their academic research programmes and their financial holdings in high-technology companies[47]. Or they are urged to stand aloof from the policy implications of their findings[48], and especially not to challenge governmental policies, for fear of losing their reputation for objectivity as independent technical advisers[49]. The academic ethos embodies the insight that science cannot be expected to 'speak truth to power'[50] unless power is forbidden to talk back.

Unfortunately, this frontier of the academic tradition is now giving way [7.9]. The ethical code supporting the norm of disinterestedness cannot stand up to the external pressures to exploit the ever-growing instrumental power of science. But before we turn to the effects of such influences on post-academic science, we should consider a much older and perhaps deeper question – the shaping of scientific knowledge by general social forces.

7.5 Social interests in the natural sciences

Since scientific knowledge is a social product, larger social interests may have played an unseen role in shaping it. This has always been a terrifying thought for scientists and philosophers. But they have failed to find a way of putting scientific knowledge outside the reach of any such influences. That does not mean, however, that we must throw in our hand and treat scientific knowledge as if it were of no more weight than a political party manifesto. The real challenge of much recent sociologizing about science is to identify these interests, measure them up and measure up to them[51].

The trouble is that an overall 'social' bias in scientific knowledge cannot be measured without reference to 'non-social' criteria. But there are no such criteria. Even such bastions of rationality as the rules of formal logic [6.6] and the credibility of empirical evidence [5.7] are not absolutely independent of their cultural milieu. We cannot question the epistemic practices and products of any community except by reference to those of another community, real or imagined. Thus, to get some idea of the epistemological influence of some very general feature of the social context of science, we ought properly to consider what science might be like in a society where that feature was absent or very different.

In particular, much political debate is about substantial changes in the distribution of power between social groups. Modern science has evolved in societies where it has effectively been monopolized by a socially domi-nant group – e.g. an economic class, an ethnic group, persons of a partic-ular gender. It is often asserted that it operates unobtrusively to serve the interests of that group. So the real question is whether the production of scientific knowledge would be very different in an advanced society with a different distribution of social power.

Unfortunately, this is a hypothetical question. It is connected with more general political questions such as why such societies have not yet appeared and how they would actually function. Theoretical speculation on such matters is a legitimate research exercise in the social sciences, and is capable of producing valuable insights into many of the issues dis-cussed in this book. But general theory only tells us that this effect should exist in principle. It does not tell us whether it is so significant that it ought to be allowed for explicitly in practice.

The observable *direct* influence of social interests on 'laboratory life' in the natural sciences actually seems to be quite limited[52]. In their everyday research, academic scientists typically put their cognitive, career and institutional interests far ahead of general societal interests – including their own social class, ethnicity or gender[53]. Even in a pathological social environment such as Stalinist Russia, they persevere remarkably in trying to do 'good science', often finding a moral refuge in their commitment to the pursuit of knowledge for its own sake.

These observations are not really out of line with social theory. Consider, for example, the influence of gender. Academic science is a distinctive and potent culture. In the natural sciences, its epistemic practices are firmly based upon widely shared human cognitive traits, such as the recognition of logical inconsistencies [6.8] and the perception of meaningful patterns [6.2]. There is no convincing evidence that men and women differ significantly in this respect. In these sciences, every effort is made to rest the communal acceptability of research claims on the exercise of these universal capabilities. Until very recently, the overwhelming majority of natural scientists have been male. Nevertheless, there is nothing to show that predominantly female research communities would apply significantly different cognitive criteria, and thus arrive at significantly different judgements, in their assessment of research findings.

It can be argued, however, that such judgements can *never* be made entirely according to publicly acknowledged cognitive criteria, and that other considerations, including general social biases, *must* enter into them. For example, as feminists often point out, the very language in which the findings of science are presented is permeated with predominantly masculine metaphors [6.10, 9.7, 10.4]. That may well be true. Indeed, personal commitment to the scientific ethos requires at least a minimum commitment to the specific institutional and social interests in which scientific work is embedded[54]. Values serving these interests are incorporated in the presuppositions and practices of normal research[55]. The bias favouring these values is as difficult for a scientist to perceive as the viscosity of water is for a fish swimming in it.

The difficulty is that the direction of any such bias is usually underdetermined by the circumstances[56]. It is often far from obvious which of two competing scientific theories – for example, heredity versus environment in explaining a particular instance of animal behaviour – is favourable to which particular class, ethnic or gender interest[57]. Indeed, the fact that a particular scientific finding is seized on by one side or another in a social conflict does not prove that it actually serves their interest in the long run.

It takes us beyond the theme of this book to discuss the way in which an inoffensive item of scientific knowledge can be sharpened into an ideological weapon in political or religious strife. But the fact that this has happened does not prove that this piece of knowledge – for example, the 'myth' of ritualistic cannibalism[58] – was originally 'constructed' for just that purpose [8.13]. In due course, no doubt, our successors may be able to detect and interpret such trends as expressions of our own peculiar *zeitgeist* – but even then only dimly, in the fitful light of hindsight[59]. For most practical purposes, one has to assume that the effects of general social interests on knowledge production in the natural sciences are so weak and self-cancelling that they mostly get lost in the noise of other larger uncertainties in the research process.

7.6 **But who sets the research agenda?**

In the previous section I have presented the conventional argument for the *effective* social neutrality and objectivity of the natural sciences. In many ways, it is very convincing. But it has a great hole in it. It deals with the influence of social interests on the *outputs* of science, but ignores their influence on its *inputs*. It justifies the *decisions* of the research community, but does not consider its *agenda*. It establishes that academic natural scientists produce relatively reliable and disinterested solutions to the problems that they tackle, but does not ask how and why those problems are posed.

As we shall see in the next chapter, this is a large question, with many ramifications[60]. It is central, for example, to the norm of originality and to the detailed specialization that is typical of academic science. But it takes little formal philosophy or sociology to see that this is where external interests really exert their influence over science. Indeed, this is no more than elementary folk psychology. Every streetwise politician or crook knows that once an investigation is under way, its outcome is largely determined by discoveries that cannot easily be altered to favour an interested party, but that social power can often be exercised to prevent the initiation of an investigation whose results might turn out to be acutely embarrassing.

Any discussion of the objectivity of scientific knowledge would thus be incomplete without an analysis of the influence of social interests on the research agenda. This theme will be taken up later in the present chapter. Meanwhile, however, we must not forget that academic science is a much broader church than the traditional natural sciences. The scientific ethos stretches without a natural break right across the academic spectrum. In many respects, the professional norms and epistemic practices of the human sciences and the natural sciences are essentially the same, and merge into one another in intermediate disciplines such as psychology and geography. But the disciplines at the two ends of the spectrum do differ profoundly in their ability to comply with the norm of disinterestedness, and hence in the 'objective reliability' of their findings.

7.7 Disinterestedness in the human sciences

We are entering here, of course, one of the most vexed areas in science studies. The central issue is very simple. The whole concept of a human science depends on the intersubjectivity of empathic interpretations of human action [5.11]. A science of history, for example, is feasible because two or more historians, having read the same documents, can come to a useful minimum of agreement on what the authors of those documents and other contemporary actors most probably had in mind when they behaved as they said they did.

But any such agreement necessarily involves more than an appreciation of the social interests apparently operating on those actors. It also requires that present-day historical scientists should share a sufficient range of social values to come to a rational consensus on the matter [9.1]. In other words, historians find it notoriously difficult to keep their own social interests out of their accounts of the past.

This critique is very cogent. It applies with real force to all the human sciences. To a sceptical eye it seems to unmask them as yet another means of doing politics[61]. To a conservative eye, it casts social theorists as utopian thinkers or social critics whose research results are so subjective that they have no practical implications[62]. To many natural scientists it robs the human sciences of the right to call themselves 'sciences' at all.

Let us be clear, however, that the same objection applies in principle to the *whole* scientific enterprise. *All* scientific knowledge contains subjective elements [5.3, 5.9]. Even the 'hardest' of the physical sciences is not necessarily free of social interests. Serious historians of science have argued, for example, that Newtonian dynamics was designed to serve the interests of sea-borne capitalism[63], and that quantum mechanics was shaped by the social uncertainties of inter-war Germany[64]. I am not saying that these particular arguments are very convincing. If they have any substance at all, it is surely very subtle[65]. But they cannot be controverted by reference to some superior 'method' that puts physics above any such considerations.

For researchers in the human sciences, however, this is not just an academic issue: it is a very practical, day-to-day challenge to their art. Some respond by systematically eliminating obviously subjective elements from their epistemic practices[66]. They accept the physical scientists' definition of a science, and try to produce knowledge that looks as 'scientific' as possible by those criteria [7.8]. Thus, they collect and classify precisely defined quantitative data, and articulate formal concepts into elaborate theoretical models whose behaviour they simulate on high-powered computers [6.10]. But even economics cannot be made to look just like physics [6.6] without cutting it adrift from its empirical foundations in suffering humanity. Indeed, many perfectly sound natural sciences, such as evolutionary biology or ethology, would not survive such mutilation.

Another, more timid response is to undertake social and psychological research entirely within an established 'life-world' frame [10.4]. For example, practitioners of 'professional social inquiry' [67] use 'ordinary knowledge' to interpret their observations of people engaged in activities such as voting or shopping. They take a certain set of conflicting material interests to be a natural feature of the social environment in which they and their observands operate, and present their research findings and theories on that basis. In effect, they ignore the norm of universalism. The scientific domain that they explore [6.11] is thus limited to concepts and behavioural regularities that are shared by members of a particular culture – typically, a modern pluralistic democracy – without reference to alternative viewpoints sited in other very different cultures[68].

As will be clear from the whole theme of this book, the most constructive response to this challenge to the human sciences is to accept the reality of social interests and produce knowledge that acknowledges their existence. This is not such a grave defect as it is sometimes presented. But it certainly reduces the practical 'reliability' of research results [7.3], especially as a foundation for theoretical prediction [8.11].

A social institution is not a closed system. How could we use even the best information about it to work out what will happen in it without knowing from some other unspecified external source the relative strengths of the more general interest groups that have influenced our information? Is it feasible to allow for *reflexivity* [6.11] – that is, the understanding that such groups might have about the future effects of their own actions – without introducing an infinite hierarchy of ever-more extensive frames of motivation[69]? And so on.

But there are several other major reasons why human action is scientifically unpredictable – including the unexplained yet familiar fact that people often behave in regular, stereotyped, 'predictable' ways[70]. The categories and concepts of the human sciences are intrinsically vague, many-sided and heterogeneous [5.4, 6.7, 6.10]. Many of the empirical facts on which a forecast might be based – such as the stereotyped responses of particular types of individuals in routine situations – are essentially hermeneutic [5.11]. They involve interpretation that goes beyond direct observation, and are therefore strongly value-dependent[71].

In addition, the functioning of any model of a social institution depends on the rational behaviour of its human members, which will surely be based on similarly limited and imperfect information. Indeed, the actual influence of even the cleverest interest group is soon confounded by the very unpredictability to which it partially contributes. The most striking feature of social action is that its consequences are mostly unintended![72]

This lack of predictive power is not fatal to the human sciences. Although successful prediction is often sufficient evidence of the validity of a scientific theory, it is not a *necessary* condition for its communal acceptance [8.11]. If that were made an essential criterion of 'scientificity' then even the proudest faculty of natural science would be decimated.

Nor does it mean that the human sciences are purely contemplative[73], and have nothing to offer in the world of practice [2.4, 7.2]. Their direct instrumental role is clearly limited[74]. The notion of using them to 'engineer' people and their societies is totally misconceived – perhaps malignant[75]. Nevertheless, action in the modern world continually calls on social science knowledge[76]. Indeed, even the technical problems supposedly soluble by direct application of knowledge derived from the physical sciences always involve social experience, values, interests and goals[77].

Perhaps the simplest way of putting it is to use the 'map' metaphor. The human sciences do not produce triangulated surveys or naval charts on which we can locate a building site or plot a safe course. But they do greatly enlighten and give meaning to social life by presenting it intersubjectively through sketch maps, outline drawings, explanatory diagrams, impressionistic pictures and other shared representations [6.4, 6.11].

These shared representations serve vital practical purposes [10.5, 10.6]. They reveal hidden dangers and the ways around them[78]. They tell us where we are, as *persons*, in the social life-world[79]. By presenting the human scene from many different viewpoints, they remind us of the diversity of our values and goals, and of the plurality of the paths towards them[80]. They provide coherent schemas of argument – and also, for those who need them, ideologies[81] – legitimizing individual and collective action in the civil sphere[82].

John Maynard Keynes quipped that what the practical man called economic common sense was usually the ideas of 'some long dead economist'[83]. People in advanced societies now largely fashion their cultural maps out of the findings of academic research in the human sciences [10.6]. Since such societies are typically pluralistic, the corresponding sciences must be equally pluralistic. The acceptance of a research claim thus depends much less than in the natural sciences on making a convincing contribution to a supposedly coherent body of 'established' knowledge.

In disciplines such as political science, for example, a state of 'pluralistic objectivity' is achieved, where a research community provides policymakers with a conceptual language revealing the variety of conflicting interests and values involved in a practical decision, without striving to resolve their contradictions[84] – or even establishing that they are really influential[85]. In other disciplines, such as education, academic scientists often collaborate with practitioners to do 'action research', which not only helps to uncover the interests latent in a practical social problem but also reveals to them their own unsuspected value orientations in such matters[86]. Again, in multidisciplinary areas such as cognitive science, pluralistic research in the human sciences monitors and moderates the monomaniac technologies proffered by the natural sciences.

It is a commonplace that the unprecedented material development of modern civilization since the seventeenth century could not have occurred without the 'academic' knowledge produced by the physical and biological sciences over the same period. It is not always realized that the human sciences were equally essential to this process. An unprecedented organizational elaboration of society was required for the invention, production and distribution of sophisticated technological products and services[87]. This elaboration occurred in a culture permeated with the economic, historical, political and sociological ideas that we associate with academic developments in the human sciences since 'The Enlightenment'[88]. The economic theories of that Scottish professor, Adam Smith, were surely as instrumental as the physical theories of that other Scottish professor, James Clerk Maxwell, in the creation of the modern world.

7.8   Free from interests – or free to be interested?

The academic ethos [ch.3] defines an idealized culture shielded from external interests. In reality, the norm of 'disinterestedness' was always difficult to sustain. Even university teachers engaged in 'pure' research have strong professional interests, and are not completely detached from the economic and political pressures of the life-world. Nevertheless, academic science has for long lived remarkably close to its ideal of an autonomous social institution.

This autonomy renounces external influences, but cannot positively exclude them. In their public rhetoric, academic scientists code it as 'academic freedom'[89]. This was traditionally maintained by liberal financial patronage of institutions and by tenured employment of individuals [3.10]. Just how far this freedom should extend has, of course, always been disputed. Freedom to speak the supposed truth is prescribed by the norm of communalism; but does that include freedom *not* to speak out against evident *untruths*?

The norm of originality clearly requires freedom to undertake research of one's own choosing [3.6, 8.7]. But this presumably permits one to break one's contract of employment by choosing not to do any research at all[90]. What is more, academic scientists cannot be forbidden to do research that directly furthers their personal material or ideological interests in other ways than success in the rat race for communal esteem [3.8]. Indeed, university-based researchers in practitioner sciences such as medicine and law have never been expected to make any such distinction [3.11, 4.8]. As practitioners, they are committed to the simultaneous advancement of pure knowledge, of their personal careers, and of the communal interests of their profession – commitments that do not always pull in the same direction.

This is the weak point in the defence of academic science against external influences. Ideally, academic scientists programme their research around 'basic' problems – that is, problems whose solutions would be a significant contribution to knowledge, regardless of their practical implications. This usually amounts to choosing from a communal port-folio of more or less difficult research questions that have not yet been solved [8.1]. But even the assertion that the chosen question is 'basic' [2.5] need not mean that it has been selected without any concern for its even-tual usefulness[91]. In practitioner sciences this is seldom the case, and even in the purest of the human sciences many researchers are very con-scious of the larger social values that motivate their research interests.

Indeed, even the communal academic research agendas of the natural sciences are often slanted unobtrusively towards the exploration of knowledge domains that are closely connected with material societal interests [7.5]. Could one really deny, for example, the general influence of naval and mercantile interests on the scientific study of celestial mechan-ics in the seventeenth and eighteenth centuries, or the influence of the electronics industry on problem choice in academic solid state physics in the mid-twentieth century? These influences flow in both directions. New knowledge produces new practices, and *vice versa*. As a consequence, academic science may almost unwittingly produce large blocks of new knowledge that might seem to have been deliberately constructed to serve the commercial or industrial interests with which they co-evolved.

What is more, in exercising this freedom, academic scientists typically neglect socially 'uninteresting' problems – that is, problems which relate to the welfare of relatively inarticulate social groups or whose answers might turn out to be embarrassing to strong vested interests in society at large. Thus, until recently, research seldom got started on social issues relating to gender, ethnicity and class, where most academic scientists shared attitudes that were deeply entrenched in the society about them.

It might not have been that they were peculiarly prejudiced on such matters. They just could not 'see' them as problems, or recognize them as worthy of research. For example, academic medical science was biassed against research on the health problems of women and of the poor, not because solutions to these problems were particularly undesirable or dis-reputable but because they did not arouse the scientific curiosity of pre-dominantly male doctors with mainly middle class patients.

There can be no doubt, therefore, that academic science is affected by external interests[92]. The real question is whether these interests are so influential and systematic that they turn science into their unwitting tool. This is often postulated, but is very difficult to demonstrate empiri-cally. Indeed, research on such effects typically interprets them as devia-tions from a relatively well-honoured norm, indicating that they would only be visible against such a background. In other words, even radical metascientists who profess to the contrary tacitly assume that academic science – especially their own discipline[93] – is usually free enough to live up to its ideals of disinterestedness and institutional autonomy.

And isn't this just what politicians and other public figures mean when they accuse academic scientists of being uninterested in 'exploiting' their discoveries [4.8]? They see academic science as a self-regarding community devoted to producing knowledge 'for its own sake', and are highly critical of its lack of incentives and opportunities to transform this knowledge into profitable industrial products. Such criticism is based more on anecdotes, such as the notorious failure of British molecular biologists to patent the invention of monoclonal antibodies, than on systematic investigation. Nevertheless, the charge of socio-economic irresponsibility attests to the reality of the institutional and epistemic autonomy that has long been enjoyed and celebrated by academic science.

7.9 Problem solving in the context of application

The academic ethos uses the norm of disinterestedness to define the frontier between academic and industrial science. But this frontier is now regarded as a serious fault line in the national R&D monolith [4.9]. Indeed, the development of much closer relationships between academia and industry is one of the major features of the transition from academic to *post-academic* science.

More generally, knowledge production in 'Mode 2' arises directly or indirectly 'in the context of application'[94]. This is an elastic concept, which can be extended 'strategically' [2.5] to cover almost any scientific activity except cosmology. It does not mean that all post-academic science must be obviously useful, as if to invert the tradition of academic purity. Indeed, post-academic science gains overall coherence and fitness for application[95] by drawing on, and generating, problems, techniques and research results from all parts of the conventional 'R&D spectrum' [8.7]. But manifest utility then ceases to be a demarcation criterion. Basic research and technological development already interpenetrate one another: in the long run, they become indistinguishable.

'The context of application' is a misty, unknown – even hostile[96] – territory for academic science [4.10]. But it is certainly not an empty land, waiting to be colonized by enterprising post-academic researchers. On the contrary, it is already heavily populated with professional and technical communities. Immigrants from academia soon discover that it is closely subdivided into regions governed by other major social institutions – industrial enterprises, commercial firms, government departments, health-care organizations, practitioner associations, and so on. In other words, it is a place where scientific communities interact with a variety of organizations strongly linked to the major interest groups of society at large.

In 'Mode 2', therefore, knowledge is produced by teams of researchers networked [5.13] across a wide range of organizations . The members of these teams often have quite different epistemic responsibilities, depending on their terms of employment. Those who work for governmental organizations or industrial firms cannot disregard the political or commercial potentialities of their research. University employees doing commissioned research on short-term contracts are not in a position to take an independent line. The barriers against external influences are thus much weaker than in traditional academic research.

These influences are not hypothetical. Post-academic science is organized on market principles[97]. Research is performed by semi-autonomous research entities that earn their living by undertaking specific projects supported by a variety of funding bodies, including private sector firms and government departments [2.8, 3.10]. Some of these bodies make elaborate efforts to foster scientific originality and integrity. But even quasi-academic public agencies such as research councils are instructed to favour projects with manifest 'wealth-creating' prospects, or with practical medical, environmental or social applications.

In effect, a post-academic research project [8.2], however remote from actual application, is tagged with its potential for use. This potentiality may be quite naive or opportunistically speculative[98]. Nevertheless, it assigns the project to the sphere of influence of bodies with the corresponding material interests. Thus, for example, basic research findings in molecular genetics have potential applications in plant breeding. Agrochemical firms and farmers are therefore deemed to have a legitimate right to influence the course of this research, from the formulation of projects to the interpretation of their outcomes.

Post-academic science is thus deeply entangled in networks of practice. If, as I have suggested[99], 'Mode 2' is essentially a post-industrial hybrid of the academic and industrial modes of research [4.10], then it does not just 'produce' knowledge. Indeed, some metascientists[100] would see *all* science as an instrument for the *construction* of knowledge in accord with the commercial, political or other social interests of the bodies that underwrite its production [8.13]. Although 'Mode 2' may also incorporate traditional scientific values – including, of course, the sheer obduracy of physical reality – it is clearly an activity where socio-economic power is the final authority.

This interpretation of the social role of science invokes disputable theories of the nature of modern society. What cannot be denied is that the academic norm of disinterestedness no longer operates. Even the genteel pages of the official scientific literature, where feigned humility is still the rule [7.1], are being bypassed by self-promoting press releases[101]. In any case, scientific authors indicate, by the 'affiliations' and 'acknowledgements' in their papers, that other interests than their own personal advancement have had a hand in the research.

At first sight, the abandonment of this norm might be expected to have a devastating epistemological effect. So much of the status of scientific knowledge depends on its reputation for 'objectivity', which is usually attributed to the detached, impartial, unbiassed, dispassionate (etc., etc.) attitude with which scientists undertake research. As we have seen, however, this epistemic virtue is a *collective* product. It is the outcome of the whole process that transforms a mass of individual research claims into a communally accepted body of knowledge.

The production of objective knowledge thus depends less on genuine personal 'disinterestedness' than on the effective operation of the other norms, especially the norms of communalism, universalism and scepticism [ch.3]. So long as post-academic science abides by these norms, its long-term cognitive objectivity is not in serious doubt. Both communalism and universalism are under some pressure from outside interests, and are far from perfectly maintained within the scientific community. But provided that 'organized scepticism'[102] [3.7, ch.9] continues to be practised conscientiously, we need not revise our belief – or otherwise – in the 'objective reality' of the 'scientific world view' [10.7].

Indeed, post-academic pragmatism reinforces cognitive objectivity. Research findings in sciences such as physics, chemistry and biology are automatically put to the test along with the technologies where they are applied. If a new aircraft crashes, then aerofoil theory naturally comes up for criticism – even though it is not necessarily at fault[103]. 'Back to the drawing board!' is a harsh lesson in validity. Such practical tests are less systematic and less easy to interpret than the results of contrived experiments, but they often bring high-flying theories down to the ground of empirical 'facts'. Technologies survive because they have been found to be *reliable*. This reliability is naturally attributed also to the scientific knowledge that has gone into their making [10.7].

What does go out of the window with 'disinterestedness' is a reputation for short-term *social* objectivity. Professional *credibility* is the hallmark of the academic scientist [5.7, 7.4]. Members of a research community have to be able to trust one another in the presentation of scientific facts and theories. This trustworthiness is seriously compromised by any suggestion that other interests might be involved. But in post-academic science every research claim is labelled with the names of interested parties outside the research community. Can the traditional web of mutual trust be maintained under such circumstances?

This is a very serious issue, which extends far outside the scientific community. People believe in the power of scientific knowledge, but are seldom in a position to judge for themselves the cogency of the grounds on which that knowledge is based. In general, the word of an accredited expert is all they have to go by. But experts often disagree, especially when substantial material interests are at stake.

In the eyes of the public, the major virtue of academic scientists and their institutions is that, even when they do disagree, they can be trusted to present what they know 'without fear or favour'. Whether or not this high level of credibility is really justified, it is what gives science its authority in society at large. Without it, not only would the scientific enterprise lose much of its public support: many of the established conventions of a pluralistic, democratic society would be seriously threatened[104].

Nevertheless, it is an open question whether public trust in science is on the wane. Many highly articulate critics insist that it is, or ought to be, but the results of opinion surveys are far from decisive. It is a bit like global warming. We are talking here of a major cultural change, which will surely take a long time to show up in aggregate data. The transition to post-academic science began less than a generation ago, and is still far from complete. Research scientists in many institutions and some disciplines still follow traditionally 'disinterested' practices, especially in their formal communications. Indeed, their attachment to the convention of mutual trust often infects the other researchers with whom they collaborate. This is still the stereotype of 'a scientist' [2.8]. Many people try to live up to this stereotype, however they may actually be employed.

In the natural sciences, moreover, the type of knowledge that is produced is too easily checked to be fudged cavalierly. Ideally, it is based on quantitative measurements [5.4] by standardized instruments [5.5], analysed mathematically [6.7] according to precise and uncontestable theoretical principles. In practice, even physics is seldom quite like that. But even in biology, where research results are often purely qualitative, they have to be capable of replication and independent confirmation [5.8].

This brings us back, of course, to the central dogma of the Legend. This is the doctrine that scientific knowledge is, by its very nature, perfectly objective, and that if it were not objective it would not be perfectly scientific. We have scrutinized the actualities of academic science too closely to accept this dogma. But interpreted naturalistically it points to research fields where it has proved much easier to achieve consensus than it has in others [9.1].

Thus, the relative 'objectivity' of physics is not due to the fact that it describes the world 'as it really is'. It is because, as we have seen, physics has evolved as an epistemic culture devoted to the measurable aspects of the world. These are, of course, just those aspects on which an open community of independent observers can be brought most firmly into agreement[105] – and where, therefore, self-interested bias is most certain of being detected.

The other natural sciences cannot be expected to reach the same level of quantitative consensus as physics. Nevertheless, their research communities often come close to complete agreement on various qualitative features of the particular objects they study. In ordinary social practice, therefore, natural scientists are still regarded as highly credible expert witnesses [5.7]. It is considered perverse to challenge research results and theoretical inferences that apparently conform to their well-tried methods of observation and argument.

In general, therefore, post-academic natural scientists can usually be trusted to tell 'nothing but the truth', on matters about which they are knowledgeable. But unlike academic scientists, they are not bound to tell 'the whole truth'. They are often prevented, in the interests of their employers, clients or patrons, from revealing discoveries or expressing doubts that would put a very different complexion on their testimony. The meaning of what *is* said is secretly undermined by what is *not* said.

This proprietorial attitude to the results of research [5.13] has become so familiar that we have forgotten how damaging it is to the credibility of scientists and their institutions. For example, a government minister may announce, quite truthfully, that 'his' experts – scientists employed by his ministry – have reported that there is no convincing evidence of the transmission of 'mad cow' disease to humans. The public remain sceptical, for he has refused to make public the minutes of meetings where those same experts have probably mulled over masses of fragmentary information that might have led them to a different conclusion. The credibility of the scientists is thus tainted by the equivocations of their political or commercial masters.

Let me emphasize that I am not opposing on principle the legal right of a political authority or commercial firm to control the publication of the findings of research that it has funded. Such rights are perfectly proper in law, and are covered explicitly in most research contracts. Indeed, their precise scope in particular cases is often a matter for serious negotiation across the interface between academia and industry.

All I am saying is that 'the context of application' is largely defined by the material interests of bodies outside science. Post-academic scientists rely heavily on these bodies for their research funds [3.10, 4.9]. But they cannot claim professional credit for the knowledge they produce without taking some responsibility for its fate. If it is obviously misrepresented or improperly held back, then the loss of credibility is reflected back on the scientific enterprise[106]. In an age when most citizens distrust the public pronouncements of government and industry, this means that science becomes tarred with the same brush.

7.10 Objectivity or emancipation?

The transition to 'Mode 2' has clearly damaged the street-cred of the human sciences much more seriously than of the natural sciences. As we have seen, research findings in the human sciences are inherently uncertain, and cannot be dissociated from human values and social interests. They offer little scope for experimental replication, predictive confirmation or long successful use – the traditional indicators of 'reliability'. There is much less communal consensus, a much less definite background of 'established scientific knowledge', against which to challenge a suspected bias of fact or theory.

Academic specialists in the human sciences are always presumed to have an expert knowledge of the research findings of past and present colleagues. But their own scientific statements have never been expected to be completely 'credible' – that is, offering entirely trustworthy advice on what can safely be taken as absolutely sound and dependable in practice[107]. Economists, for example, are notoriously at loggerheads on the explanation of how our modern commercial culture actually ticks. Even when they are not uttering the wishful thoughts of a political party or sector of industry, they do not produce knowledge which a sensible person ought to trust unreservedly in deciding to build a factory or buy a block of shares.

In post-academic science, a variety of other moneyed interests are encouraged to supplement or replace general state patronage [4.9]. The traditional notion of the human scientist as a relatively aloof source of highly specialized technical knowledge then begins to look very naive. For example, many university-based researchers are now so reliant on government contracts for research on practical problems that they cannot easily dissociate themselves from government policies. This is often revealed by disputes over the publication of reports on social issues commissioned by government departments.

In effect, to obtain resources for research, many post-academic human scientists are forced to relaunch themselves as technical consultants. They compete with one another for research projects on political, social or industrial problems commissioned by public or private organizations. Very often, however, their research findings and expert advice are desired as much to rationalize a particular policy on a contentious practical matter as to present an independent analysis of the situation[108]. Social scientists in socially 'relevant' disciplines such as criminology or educational studies cannot always avoid being 'labelled' according to their presumed socio-political sympathies and interests, and their results discounted accordingly. In consequence, when their research findings apparently undermine the vested interests or entrenched opinions of a supposedly hostile party they are often dismissed cynically with 'they would say that, wouldn't they', rather than contested directly.

In a pluralistic society it is certainly highly desirable that contending interest groups should have scientifically well-informed advocates. But that leaves open the question whether the party that can afford the services of the most authoritative scientific experts necessarily has the strongest scientific case. And there are interest groups that are so weak and dispersed that their situation has scarcely aroused any scientific attention or relevant research [7.8].

In any case, the issues in which such experts are engaged usually turn more on moral values than on reliable facts. Indeed, a genuine scientific controversy [9.1] is not like a law suit that can be decided by the presentation of the case for each side to a neutral jury. The very notion that there are only two sides in such matters is not consistent with the scientific culture.

There is growing public appreciation that knowledge as such is not necessarily entirely beneficial[109] and seldom really solves the problems for which it is sought. Nevertheless, the urge to improve human understanding [10.1] in order to realize specific human interests and values is insatiable[110]. A genuinely pluralistic society cannot be expected to be tranquil. It is likely to make demands on the human sciences for knowledge and insights far beyond their established themes and resources[111]. Post-academic science has evolved to foster this enlarged research agenda[112] by taking it out of the 'invisible hands' of research communities [3.1] and putting it under the thumbs of policy and profit.

But a pluralistic post-academic mode of knowledge production is not all that is required of the human sciences in a pluralistic post-industrial polity [4.10]. In particular, the traditional academic stance of 'objectivity' still has a part to play. Post-modern sociologists scoff at this tradition, which they interpret as the self-serving ideology of a professional elite, smugly aloof from democratic values whilst profiting behind the scenes from phoney claims to expertise[113]. But theoretical deconstruction does not completely devalue the practical role of this stance in social life.

As we have seen, the notion of 'objective knowledge' is linked with the academic norms of communalism, universalism and disinterestedness. The culture built around these norms in the natural sciences has produced a solid core of *established* knowledge. There is no *absolute* basis for our belief in the cognitive objectivity of scientific entities such as gravitating planets, atomic nuclei, tectonic plates or the genetic code [10.7]. Nevertheless, this is a well-founded belief, as natural as our attitude to the rest of the shared furniture of all human lives. What is more, these are entities that exist independently of our individual thoughts about them, and operate even-handedly, for everybody. This consensus is so overwhelming that nobody can seriously deny that they are simultaneously 'socially' and 'cognitively' objective in full measure.

This double-barrelled objectivity is one of the features that make academic science so valuable in society[114]. As a source of well-established, disinterested knowledge, science plays a unique role in settling factual disputes. This is not because it is particularly rational or because it necessarily embodies 'the truth': it is because it has a well-deserved reputation for impartiality on the material aspects of what is at stake.

A pluralistic society has to be a *society*. The complex fabric of an advanced, *democratic* society is held together by trust in the claimed objectivity of scientific experts[115]. Without science as an independent arbiter, many social conflicts could only be resolved by reference to political authority or by a direct appeal to force. That was the historical experience out of which scientific institutions such as the Royal Society emerged, and its wisdom remains our cornerstone.

The post-modern critics of science insist that even its claims to cognitive objectivity are false, and actually conceal powerful entrenched interests. It is noteworthy, however, that anti-establishment groups also appeal to scientific knowledge to advance their causes – for example in environmental disputes. Rebels against intellectual authority often insist that they are particularly free from self-serving bias[116]. But even metascientific anarchists should realise that by unmasking the ideology of 'objectivity' they are breaking their own swords in the struggle against their most feared opponents – the corporate and governmental enterprises that drive post-industrial society[117].

Any loss of confidence in the *cognitive* objectivity of science thus has serious *societal* implications. This illustrates perfectly the main theme of this book. In academic science, the communal norm of disinterestedness [3.5] is linked to this epistemic principle, each reinforcing the other. This linkage is not a metascientific abstraction, but it is seldom explicitly voiced. Apprentice researchers do not learn it by precept. It is a tacit element of the scientific culture[118], absorbed through participation in such 'socio-epistemic' practices as double-blind trials of new drugs or elaborate statistical analyses of experimental data in elementary particle physics.

But post-industrial research has no place for disinterested practices, and post-modern thought has no place for objective ideals. 'Mode 2' is concerned with the solution of problems arising in the context of application [4.8], not with curiosity-driven, blue-skies research in pursuit of truth [2.7]. It is no good telling post-academic scientists that they should adopt a 'scientific attitude' in their work. They have neither examples of disinterested behaviour to emulate, nor formal standards of social objectivity to live up to. Constructive reinforcement thus gives way to deconstructive decay.

As we have seen, the general public still have confidence in the hard core of established knowledge at the heart of the natural sciences. The increasing tendency to question knowledge claims outside this core is very healthy, especially when the public welfare is at stake. There are often good grounds for scepticism [ch.9] about research findings that are not yet well established, even when they are obtained by reputable scientists using reliable techniques. This applies particularly to the *trans-*

scientific or *trans-epistemic* questions[119] that usually arise in 'the context of application' – that is, questions deeply interwoven with social interests and values.

The real trouble comes as one moves from the natural sciences to the much 'softer' domains of the human sciences. Subjective values and external interests begin to play a more obvious part in the knowledge production process – and healthy scepticism tends to give way to cynicism. Even though society contains very real structures that are independent of our individual images of them [10.7], the very notion of cognitive objectivity becomes suspect.

This development bodes ill for the democratic social order. The human sciences complement the natural sciences in providing legitimacy and a rationale for our modern pluralistic society[120]. This is the form of society in which they both evolved [9.7], and to which they are functionally adapted.

Indeed, the 'two cultures' [2.9] are not really distinct. They merge into one another in many disciplines, such as cognitive science, geography, archaeology – and science and technology studies [8.8]. Even though they are divided into many sub-cultures practised by innumerable specialized research communities, they share an over-arching ethos [3.2] and paradigm of rationality [6.8]. Recent history has not been kind to political regimes that try to foster the natural sciences whilst suppressing their human counterparts.

By insisting on complete ethical independence [7.4], purely academic science produces knowledge that lacks social legitimacy [121]. Few would now doubt that post-academic science, being much more directly connected into society at large, has to share its larger values and concerns. But by completely compromising the relative societal independence of the human sciences, it could call into question the legitimacy of *all* the knowledge it claims to produce[122].

The norm of disinterestedness has always seemed paradoxical. People who commit themselves to the epistemic austerities of the scientific culture often have a strong interest in realizing its emancipatory ideals. Somehow or other, these ideals have to be worked into the notion of 'good science', not necessarily as an alternative epistemology[123], or new type of epistemic institution[124], but as a way of producing knowledge to which people can turn with confidence [10.7] in dealing with the problems of everyday life[125].

8

Originality and novelty

8.1 Problems

Academic science is energized by the norm of *originality* [3.6]. This norm requires scientists to produce *new* knowledge – that is, communally acceptable information that was not previously known. To do this, they engage in *research*. But they can be credited with CUDOS [3.8] only for what they discover through research that they have themselves decided to undertake[1]. Academic science – and, in the end, cognitive change – is thus steered by innumerable independent decisions of this kind[2].

The significance of this process is often under-estimated. Philosophers constantly insist that scientific knowledge is provisional, but they seldom remind us that, even though it is continually expanding, it is very patchy in its coverage. This is not just a regrettable weakness, which can be forgiven because it will in due course be made good. It is a fundamental *epistemic* characteristic of academic science, closely connected with its social structure and cultural practices.

To put it simply: at any given moment, what we know and how well we know it depends on what our predecessors decided to study in the past. What we shall know in the foreseeable future will depend on what research we undertake now. For example, we would not nowadays be studying the medical uses of genetic information if Francis Crick and James Watson – not to mention many others – had not individually decided, nearly 50 years ago, to investigate the structure of DNA. And for many future decades, research findings in this field will have been shaped as much by the collective decision to map the human genome as by the material configurations of base pairs that they will report. The way in which such decisions are made must surely figure in a naturalistic epistemology. This is the theme of the present chapter.

Another way of putting it is that scientific change can be interpreted as an *evolutionary* process[3], where a great diversity of research claims are subjected to critical selection and communal retention [9.7]. But that does not mean that these claims are as random and 'blind' as genetic mutations. They arise, it is true, out of activities whose outcomes are necessarily uncertain and unpredictable. 'Research' is not just a folk term for the cultural practice of 'doing science'. As the word indicates, it is a characteristically intentional, often highly methodical, mode of *action*. Indeed, there are few human activities that require such elaborate and careful planning as some scientific experiments[4]. The hands that guide scientific progress are not, surely, entirely invisible.

But how can one 'search' for one knows not what[5]? This is the paradox at the heart of the scientific endeavour. It is true that scientists often undertake *exploratory* investigations, where an obscure aspect of the world is systematically scrutinized and 'mapped' for the first time [6.4]. But the paradox is too sharply posed. In reality very few scientists are privileged to live out the heroic metaphor of embarking over an uncharted ocean to undiscovered lands. In spite of their occasional protestations to the contrary, they almost always set out – as indeed did Christopher Columbus and Vasco da Gama – with presuppositions as to what they might find.

Indeed, most scientists would say that their research is directed into a specific area of ignorance opened out and framed by such presuppositions[6]. They tell us that they are actually seeking the answer to a certain 'question', or the solution to a certain 'problem'. This question may be little more than a vague rationale for an enterprise that began with quite a different purpose. The so-called 'problem' may be a tiny bump on a graph, which only another researcher in the same field could be persuaded to see as a significant barrier to understanding . Nevertheless, by using words such as 'questions' and 'answers', 'problems' and 'solutions', they clearly locate the research in a specific cognitive frame and show that it is directed towards specific cognitive goals.

These particular words seem to point to a 'challenge–response' mode of knowledge production. But they should not be taken too literally. By describing their work as essentially 'problem-solving'[7], scientists let it be understood that research is not just idle curiosity, and that there are indeed 'solutions' whose discovery will justify their endeavours. But these are really code words for any of the many different epistemic goals of scientific activity[8].

Nevertheless, personal commitment to such a goal is clearly a critical career decision. Shrewd choice of a problem or problem area is often the key to success in research, and is widely regarded as one of the most creative of all the scientific talents[9]. It is easy enough – especially in the human sciences[10] – to think of 'important' scientific problems: the real question is whether they have discoverable solutions[11]. As Peter Medawar famously remarked[12]: 'If politics is the art of the possible, research is surely the art of the soluble'.

In any case, research problems are not there for the 'choosing', or even for 'finding'[13]. They have to be *formulated*[14]. The world is not laid out for study in pre-ordained categories [6.2]. Dame Nature does not present the scientist with a list of carefully crafted questions like a well-made examination paper. In principle, each researcher has to make up for herself the problems on which her performance will be assessed. The highest peaks of scientific genius are not for sheer virtuosity in 'solving problems'. They are reserved for the immortals who have posed these problems in the first place – by asking the right questions about an age-old mystery[15], by perceiving a genuine conundrum inside a mundane enigma[16], or by inventing tools that open up quite novel fields of inquiry[17].

In practice, the typical academic scientist is not expected to demonstrate such genius. She works on artificial 'objects of inquiry' [5.6], within a specialized research tradition[18] which makes a certain type of question particularly meaningful at a particular moment[19]. Nevertheless, even in the narrowest scientific specialty, the 'search space' [8.14] always turns out to be much wider and more open than it seemed to the eminent scientist invited to 'survey the field' at the latest research conference. In the words of Immanuel Kant[20], 'Every answer given in principle by experience begets a fresh question which likewise requires its answer'. In effect, there is usually infinite scope for *formal* originality in the formulation of research problems, even if most of the problems that come immediately to mind are either technically insoluble or intellectually trivial.

At whatever level of virtuosity it is practised, skilful exercise of 'problem choice' is an idiosyncratic factor in the art of scientific research. Scientific historians and biographers try to reconstruct and analyse its successive steps, probing its roots in the psychological, social and intellectual trajectories of individual star performers. What were their prior theoretical commitments[21]? What were their career expectations, intimate desires, interpersonal relationships, political interests and other non-scientific circumstances[22]? How did they establish and stabilize the goals of their research in the shifting hierarchy of political, economic, technical and cognitive considerations which might have influenced it[23]? How did they recognise, isolate and reconstruct a badly posed problem in a complex, fragmented, research setting and turn it into 'good science'[24]? But like most attempts to explicate individual instances of personal creativity[25], these studies merely confirm that it cannot be reduced to a formula.

But that does not put a stop to further discussion of how academic scientists decide what research they should do. These decisions are made in *local* contexts. As I have already indicated, they are strongly influenced by the material, social and epistemic environment in which they have to be made. Taken overall, these influences may well over-ride the uncoordinated actions of individual researchers. We must now consider the way in which scientific activity is orientated and scientific knowledge shaped by its human environment.

8.2 Projects

In scientific mythology, research problems are not publicly formulated until they have effectively been solved. They are supposed to germinate in the subconscious of the researcher, begin to take shape in private notebooks or personal letters to trusted colleagues, and only be fully revealed in an epoch-making publication such as *Philosophiae Naturalis Principia Mathematica*, *The Origin of Species*, or *Die Grundlage der Allgemeinen Relativitätstheorie*. But this extreme individualism is only feasible for a researcher who already commands all the cognitive and material resources that will be needed along the way. Some academic scientists are still in that ideal situation [3.10]. For example, they may hold tenured posts in theoretical disciplines such as pure mathematics or neo-classical economics, with unlimited access to computers and libraries.

Generally speaking, however, scientific problems cannot be solved without instruments, laboratory facilities, technical services, expert colleagues and so on. Many would-be scientists have to bid for a living wage or time off to undertake the research they want to do. This personal need for material resources obliges academic scientists to mobilize social support for their intended research[26] by disclosing its epistemic goals and explaining how they are to be achieved.

This obligation is at the heart of the concept of a *project* [4.9]. In principle, this is a formal statement of a perceived scientific problem, together with a plan of action designed to lead to its solution. It may be as sketchy as 'one side of A4', or as detailed as a thousand pages of technical specifications[27]. It may merely indicate a scholarly area to be explored over a certain period, or it may set out a step-by-step timetable of specific operations. It may involve a single individual, a small research team or a whole scientific institution[28]. It may refer to a self-contained single experiment [5.6], or – in the guise of a *programme* – to a long series of vaguely connected investigations.

The presentation of projects as formal *proposals* has become a standard feature of academic science in many countries [4.9]. Busy researchers regard this as a wasteful administrative chore – especially when a high proportion of their bids fail to get funding. And yet they know that their scientific career depends as much on writing plausible project proposals as on making convincing scientific discoveries. In effect, project proposals emerge at the nodal points of the modern scientific culture[29], where its personal, material, social and epistemic dimensions intersect[30].

Project proposals thus provide much food for thought about science. They show, for example, that research is typically *opportunistic*[31] – that is, that the importance ascribed to a scientific problem depends as much on the instruments, facilities and technical skills available for attacking it [4.6, 5.5] as on its location on the frontier of knowledge[32]. They also show, through their elaborate literature reviews and emphasis on the 'track record' of the proposers[33], how deeply most research is rooted in its past [4.8]. Thus, as the medium through which projects are defined before selection by peer review[34], they discourage risk-taking by favouring problems that can easily be recognized as such in the research community[35] [8.7]. And they are often the means by which vested interests, national priorities and other 'trans-epistemic'[36] influences quietly shape the research process[37] [7.5, 7.9].

From an epistemological point of view, one of the striking things about research *proposals* is how closely they have come to resemble research *reports*[38] [3.3, 9.3]. One finds all the same features – the impersonal stance, the technical jargon, the deference to established authorities, the introductory survey of the theoretical situation, the discussion of the practical considerations governing the design of the proposed experiment, the acknowledgement of institutional support, collaborative contributions and access to facilities, right down to the analysis of the inferences to be drawn from the likely results. In many cases, the insertion of a statement that 'the results of the investigation were as expected' is all that would be needed to produce a paper ready for publication!

To some extent this is just a stylistic convention. Scientists have had to work hard in learning this special way of presenting their ideas, and usually know no other. After all, a project proposal is addressed to the same readership as a research paper – that is, the members of a specialized research community – and should therefore be written in the same style.

But project proposals and research reports are rhetorical documents. They are designed to win favour from highly critical colleagues and/or potential competitors[39]. Thus, they try to present themselves as if they were almost unquestionable, and were already part of the archive [9.3]. From an epistemological point of view, this is a bizarre development. What is the status of the conjectural outcome of an experiment that has not yet been performed? What has been contributed to scientific knowledge by a plan to undertake such an experiment? Are non-experts so impressed by the precision of the plan that they lose sight of the uncertainty of its outcome?[40]

Nevertheless, there is a trend towards treating project proposals as if they were, to some small degree, contributions to science. Summaries of projects that have recently won grants are regularly published in professional journals and could at least be cited as 'informal communications'. One of the difficulties of peer review is that a 'peer' has occasionally been known to steal a novel scientific idea from a proposal he is reviewing for a funding body, and to present it as his own. In other words, a confidential project proposal is now considered as much a target for plagiarism [3.6] as a manuscript of a research paper.

The whole system of allocating resources to academic science on the basis of project proposals obviously puts a great strain on the norm of originality. To protect this norm [3.10], funding bodies often operate in the *responsive mode*, where projects are judged entirely on their scientific merits, regardless of how they accord with other considerations such as social benefits or national priorities[41]. This is not as easy as it sounds [7.6, 8.7], but at least it leaves individual researchers at liberty to present novel projects for serious, expert consideration and possible implementation.

But freedom to write imaginative research proposals is a far cry from freedom to carry out imaginative research – especially if few of one's projects are actually funded[42]. The originality of the researcher is respected in principle but deeply compromised in practice. This loss of autonomy seems unavoidable. Viewed in the large, it is a typical example of the inexorable 'collectivization' of modern science[43] [4.6]. The outright individualism of the academic ethos is no longer compatible with the need to mobilize large resources to do meaningful research. As a result, the ultimate authority over 'problem choice' has been transferred to the research community, through its specialized review panels and other representative organs.

But is individual freedom of 'problem choice' really so essential to academic science? As we have seen, any such freedom is necessarily very restricted. The options facing the would-be researcher are complex and tightly constrained[44]. The problems that seem worth tackling are defined by the intellectual environment. The means of tackling them are limited by practical considerations. Many scientists find that the need to demonstrate originality under such circumstances makes unbearable demands on the psyche[45]. They are quite ready to give up *strategic autonomy* in setting the goals of their research, provided that they have sufficient *technical autonomy* in framing the precise questions to be posed and answered[46].

Indeed, the concept of a project straddles the boundary between research as the pursuit of knowledge and research as a technical accomplishment. The skilled *technicians* and elaborate *apparatus* that are deployed in research can no longer be treated as the invisible instruments of the probing scientific intellect. In the project proposal, as distinct from the research report, they have to be listed, their use justified, and their services accounted for in hard cash. The traditional class distinction between workers by hand and brain does not stand out clearly in what is said and done in modern laboratory life. It belongs more to sociology than to epistemology.

Many highly trained 'scientists' off-load the burden of intellectual originality by perfecting their practical skills. They thus become, in substance if not in name, professional technicians. In the teamwork required by 'Big Science', they make significant contributions to the advancement of knowledge by allowing their skills to be deployed at the service of other scientific workers who are presumed to have a better grasp of the longer-term intellectual objectives of the research[47]. And yet those advances may lie as much in the technological practices that they originate as in the theoretical 'problems' supposedly being solved[48].

Thus, many aspects of the research culture tell of a tension between the austere ideal of individual autonomy and the human realities of intellectual and psychological interdependence[49]. For example, much research in the natural sciences has always been done by groups of doctoral students, post-doctoral assistants or junior colleagues, working on suites of inter-related problems set or suggested by a more senior scientist[50]. On the face of it, this hierarchical arrangement strongly favours the 'principal investigator', who exploits their scientific labour through his control of research funds and job patronage[51]. But it also benefits inexperienced researchers by providing them with a supportive intellectual and social environment, and shielding them from the risk of embarking on projects that are beyond their powers[52]. In the human sciences, where research projects are typically less risky and less tightly linked, graduate students are supposed to be much more autonomous in their choice of research topics – and often lose their way for lack of guidance[53].

For the moment, we postpone discussion of the epistemic effects of the 'collectivization of problem choice'. In general, this reduces unevenness in the technical quality of the research that gets started. Projects have to be formulated more systematically, to standards consistent with the communal concept of 'the state of the art'. Even very experienced researchers have to show that they can still do 'good science'[54]. By the elimination of projects that do not satisfy orthodox criteria of rationality or practicality, science thus loses a lot of useless junk. But it also loses the occasional benefit [9.8] of quite unexpected results [8.9] from idiosyncratic projects that were strangled before they could be brought to birth.

8.3 Specialties

For most non-scientists, one of the major characteristics of science is that it is intensely *specialized*. This is an inevitable consequence of the demand for originality [3.6, 3.9]. In principle, no two academic scientists should be doing exactly the same research. Their competition for CUDOS and resources [3.8, 4.9, 9.1] drives them to differentiate their work into distinct *problem areas*, or *fields*. They specialize their research interests and capabilities in order to exploit distinct 'niches' in the ecology of the scientific world[55] [9.7].

We have already noted some of the sociological and psychological aspects of specialization. But let us not forget that research specialties are defined as *problem areas*. They are the local intellectual arenas where *knowledge claims* are disputed[56]. One of the principal responsibilities of a recognized research scientist is to be sufficiently familiar with the literature of her specialty to assess the novelty and/or plausibility of a research report or project proposal in that field [8.2].

This requirement can be quite a burden, especially when the field is changing rapidly. One can even argue[57] that this is what keeps an active research specialty from growing too large. A hard-working researcher cannot 'keep up with her subject' if this is accreting more than, say, a thousand papers annually. But the international archival literature [9.3] of a science such as physics may well be expanding by several hundred thousand papers annually [9.8]. Even the most learned 'authority' cannot be expected to be completely knowledgeable about more than a few per cent of the subject matter of a typical academic discipline[58].

This narrow peak of expertise is often set in a broad plain of almost complete ignorance. Scientists are not just the archetypical specialists 'who get to know more and more about less and less, until they know everything about nothing': by contrast, they are made to feel – by the specialists in other fields – that they know nothing about everything else.

Thus, scientific specialties often seem to be shut off from one another by walls of mutual ignorance. Even the experimentalists and theoreticians in the same field of Big Science may have their offices on different floors of the research laboratory and seldom meet informally[59]. For a researcher working inside the island universe of a well-established specialty, a journey of discovery to another galaxy of knowledge can seem very daunting. Scientific knowledge thus tends to break up into self-contained problem areas, each evolving conceptually as if it were an independent discipline[60].

Fortunately, research specialties are not permanent epistemic structures. They originate as clusters of questions designed to be superseded by convincing answers. They are held together by theoretical paradigms [8.4], research technologies and other usages[61] that eventually fall out of use. However seemingly well entrenched, they are always vulnerable to a discovery [8.9] that outflanks or undermines them[62]. They grow too large and subdivide, or fail to reach consensus [9.2] and split into competing schools[63], or simply fade away for lack of interest.

The tendency towards fragmentation is limited by institutional factors[64] and by inter-personal competition. For example, the 'centre' of an established specialty becomes crowded with scientists doing 'mainstream' research. Many begin to realize that there are diminishing returns from following a fashion[65], and that there are more promising problems in 'marginal' areas[66], where they soon find that they have to take account of problems and techniques from 'neighbouring' fields. Thus, academic science as a whole is stitched together by scientists whose research careers cross the boundaries between the conventional specialties[67].

Moreover, this quasi-geographical metaphor understates the extent of these 'margins'. This is a subtle point that requires some clarification. To start with, one must remember that science is a body of *organized knowledge*[68]. In spite of their social and personal aspects, scientific specialties are defined *epistemically*. They are differentiated and catalogued in the archives [9.3], for example, in terms of their subject matter, not in terms of institutions or persons. Indeed, if the immense mass of existing scientific documents were not very thoroughly and precisely ordered in relation to their cognitive contents, it would be impossible to decide whether any new contribution was original.

The standard bibliographic practice is to catalogue the scientific archives along a single dimension. The literature of each specialty is labelled with a decimal number indicating where it might be found along an enormous notional bookshelf stretching from, say, *Modal Logic* (0.001) to *International Law* (999.999). This number not only gives an idea of the 'cognitive distance' between two specialties. More importantly, it encodes an overall *taxonomy*, analogous to the type of scheme used to classify biological organisms [6.3]. Each significant figure is associated with a level in a hierarchical classification[69]. Thus, for example, in the standard classification scheme for physics, the specialty in which I used to work was labelled '*72.15: Electrical conduction in metals*'. This was a sub-field of the field '*72.10–72.90: Transport properties of condensed matter*'. This, in turn, lay in the subdiscipline '*71–79: Condensed matter physics*', which was a major part of the whole discipline '*01–99: Physics*' – and so on.

This type of 'linearized tree' works well enough for *reference* purposes, but its defects become obvious for an archive that is to be *searched*. This is because it is classified on the basis of a very small number of criteria that have had to be chosen very arbitrarily. As a consequence two fields of research that might be classed as very distant by these criteria can actually be very close in other ways. On a typical university campus, for example, the Department of Anatomy is housed far away from the School of Engineering. And yet much research on the anatomy of the brain has such close conceptual links with micro-electronic engineering that they ought to be located in adjacent laboratories. Again, where does 'climate research' belong, as between such disparate disciplines as Economic History, Plant Physiology, Geochronology and Atmospheric Physics?

The relationships between various domains of knowledge cannot really be represented as a map in the conventional sense [6.4]. Deep tunnels of shared technique burrow beneath the disciplinary divides[70], and oceans of ignorance are traversed by satellite links of conceptual metaphor[71]. Such connections can only be properly represented by adding further classification criteria – that is, by going into more dimensions. Indeed, one of the objections to the notion of producing a unified account of all scientific knowledge [10.8] is that it would have to be charted according to an infinity of criteria. For this reason, even a simple 'sketch map' of the network of relationships between research specialties can quickly become too complex to be visualized in ordinary spatial terms.

Thus, the quasi-geographical metaphor is very misleading. It makes research specialties seem too 'compact'. In reality (so to speak!), they are what mathematicians call *fractal*. Their 'boundary regions' and 'trading zones'[72] are much larger, and far more convoluted, than their 'interiors'. A research specialty does not have a genuine 'core' where 'mainstream' research can flow on undisturbed, nor a 'centre' that is far from a frontier with any other specialty. Indeed, for many problem areas the existence of such a frontier is so disputable that an untutored observer would say that they merge or overlap without any discontinuity.

8.4 Disciplines and their paradigms

It is right to describe a scientific specialty as a 'problem area', since it is not so much a body of knowledge as a *research programme*. In principle, it merely encapsulates the current objectives of a cluster of loosely linked projects [8.2] with certain features in common. But if such a programme persists for a few decades then it becomes a cultural institution, handed down from generation to generation as an epistemic, technical and communal 'tradition'[73]. Much of the psychology and sociology of academic science is about how the actions and interactions of individuals and groups are shaped by research specialties: much of the history and philosophy of science can be written in terms of their emergence and conceptual evolution[74].

The scientific knowledge that comes out of a research specialty is typically very confused and uncertain. Nevertheless it is not produced entirely opportunistically or contextually, as a mere by-product of local laboratory life[75]. As we have seen, it is usually part of an evolving stream of similar knowledge. What is more, it is related to larger epistemic structures, from which it gets its meaning and to which it is designed to contribute. In other words, the 'problems' that span a specific 'problem area' are normally formulated and framed in the more general terms of a larger 'field' of science. This, in turn, is usually part of an even larger domain of knowledge – and so on, up to such broad academic categories as 'disciplines' and 'faculties'.

Thus, for example, the specialized research that I did on the theory of electrical conduction in liquid metals arose in the course of a career devoted mainly to gaining an understanding of the conduction of heat and electricity through 'condensed matter' – that is, solid and liquid materials. This field I saw as part of a much larger, more loosely organized activity, involving thousands of research physicists around the world, for the study of all aspects of the physical behaviour of matter in this state. Indeed, although I could not possibly be familiar with all this research in any detail, I needed to have a general idea of what was going on in order to make use of selected results in my own work. At the same time, I was a member of the academic staff of a university department of physics, where I gave lectures on other subdisciplines, such as advanced quantum theory and statistical mechanics, which were also part of the professional expertise that I applied in my research. And on occasion I exchanged research information with the chemists and mathematicians with whom we organized joint courses for students in the Science Faculty.

Like the higher-level taxa of biological systematics [6.3], these larger categories are defined very arbitrarily. Indeed, there is no established nomenclature for the various levels of academic classification[76]. How finely can an official academic discipline be subdivided before the lower taxa are better described as 'fields', 'specialties' or 'problem areas? In much of what follows I shall use the grand term 'discipline' to cover all such subdivisions, right down to the smallest scale.

Such categorial schemes may seem notional, but to the mind of the experienced research scientist, these are real structures in the scientific domain. They locate her, and her specialty, in successively wider communities, and relate her, and her work, to successively more general bodies of knowledge. As they get broader, they do not have to be mastered in detail by any single person. But awareness of them is part of the tacit knowledge required to do 'good science'.

Some of these subdivisions of knowledge are associated with well-entrenched academic institutions [2.9]. Several of the major disciplines have survived for centuries as distinct academic 'tribes'[77], often claiming cultural superiority over other less 'pure' forms of knowledge[78]. But they are not as compact[79], uniquely defined[80] or well-ordered as they often claim. The traditional linearized tree of the library catalogue shows only one dimension of the relationships between them, which are actually far more confused and complex than even a two- or three-dimensional 'map' would indicate [6.4, 8.3]. Nevertheless, they are the acknowledged 'owners' of substantial tracts of knowledge.

The concept of a discipline [3.9] thus has a strong hold on the mentality of every scientist, as well as a tight grip on their career. People do define themselves scientifically by the Russian-doll set of ever more specialized categories in which they believe they are enclosed[81], and come to see the scientific domain almost entirely along such mutually exclusive 'tunnels of knowledge' [82]. Indeed, this *déformation professionelle* can become a total mind set. 'Scientism' [10.8] sometimes narrows further, to 'physicalism' or 'economism', where the whole of existence is looked at through the blinkers of a single academic discipline.

Like their biological counterparts [6.3], the higher taxa of the academic ecosystem are largely social constructs. But they exist cognitively in the minds of individual scientists and are differentiated and classified epistemically. Do they have any significance simply as bodies of knowledge? 'The world' is a seamless web[83] [6.2, 10.8]. But in trying to represent various aspects of it scientifically to each other, we often get the feeling that it is easier to 'carve' at certain natural 'joints'. Are the distinct categories that we seem to find in our *knowledge* about the world as genuine as the distinct species of objects and organisms that we find in the world itself?

In fact, many of the subdivisions of academia are historical relics, no longer justified by present-day understanding. I once observed, for example, that exactly the same experiments – on visual perception in cats – were being done by biomedical researchers in university departments labelled variously as *anatomy*, *physiology*, *pathology* and *zoology*. But some disciplines are defined quite rationally in terms of their characteristic research objects. *Astronomy*, for example, includes all that we know about the universe beyond the earth, *palaeontology* covers knowledge gained from the fossils of prehistoric creatures, *criminology* is the study of the social phenomenon of crime – and so on. Other subdivisions are defined primarily by their method of research – e.g. *electron microscopy*, *statistics*, *palaeography*.

But there are also a number of fields of knowledge which we recognize as distinctive, even though they cannot be defined in terms of their research objects or methods. These range in scale from research specialties to major disciplines. How does one define the scope of *health econom-*

ics, for example, or *social history*, or *evolutionary psychology*, or *physical chemistry*? These are not about generally recognized natural entities or social activities as such. In each case, certain *aspects* of the world are being singled out for study on the basis of their *theoretical* significance.

The power of theory over research is encapsulated in Thomas Kuhn's metascientific concept of a *paradigm*[84]. Scientists come to see the world in terms of their models of it [6.10]. Their explorations and interventions[85] – that is, their research projects – are conceived as actions in the scientific domain [6.11]. This is the stage on which they play out their imaginative lives. This is the arena where they must actively display their originality and creativity [9.1]. This is the set of mental images and epistemic practices that unites them in a discipline.

Like many other important cultural concepts, the notion of a paradigm escapes precise articulation[86]. It is more than, say, an established scientific theory, for it is often a complex cognitive structure with significant social, psychological and technological dimensions[87]. Take 'the structure of DNA as a genetic code'. This is a cryptic statement of a well-founded theoretical model for a feature of the natural world. Indeed, it stands for a whole family of theories [6.4], in that to accept it implies acceptance of the numerous other biological concepts to which it is linked[88]. It is a cognitive 'frame' with 'slots' for standard 'exemplars' – that is, explanatory 'themas'[89] for typical phenomena – and standard methods for investigating them further[90]. Once taken on board, it suggests to the individual scientists a whole array of hypothetical questions that might now be considered do-able research projects – and so on.

The 'DNA paradigm' thus provides the consensual language in which the researchers that cluster around many different problem areas and disciplines [6.5, 8.8] can communicate and coordinate their work. As a result the new scientific discipline of molecular biology has emerged, where students are trained to think in this language and to show that they can apply it professionally in their research[91]. And as the paradigm extends its influence, cognitively and socially, out of academic science into applied medical research, it is becoming the basis for a whole range of useful inventions [4.8]. Indeed, the whole pharmaceutical industry, and other large societal enterprises [4.10], such as the Human Genome Project, have been set in motion by it.

The paradigm that holds sway over a scientific community is not necessarily associated with such a compelling model or theory. A discipline may be held together by a package of research practices[92], a conceptual language, or a universally shared view on what counts as scientific evidence in that discipline[93]. As Ludwik Fleck pointed out[94], a communal 'thought style' is almost inevitably adopted by members of a 'thought collective' who are committed to exchanging ideas or interacting intellectually. From a naturalistic point of view, a paradigm is more than a socially transmitted stock of knowledge. It is essential for the intersubjective communication [5.10, 5.12] of what Fleck called 'collective representations' – that is, scientific results and concepts.

But even an implicit paradigm can have a powerful cognitive influence. In particular, it may harden into a 'constellation of absolute presuppositions'[95], or a rigid 'categorial framework'[96] that is so closed to alternatives that it treats every contradiction as unthinkable, invisible or entirely exceptional[97]. The growth of watertight 'thought worlds' around social structures [10.4, 10.5] – art, religion, economic and political institutions, etc. – is a familiar human phenomenon[98].

Just how this happens is still widely debated. Even a well-established scientific paradigm, supposedly ruled by cognitive considerations, has a cultural meaning that demands a 'thick description'[99]. A philosopher might emphasize its role as a background of shared assumptions against which research claims can be judged with some degree of assurance[100]. But a social scientist would note the emotional attachment of scientists to quite local 'domain assumptions' which were built deeply into their minds during their research apprenticeship and which bond them to their scientific colleagues[101]. In other words, paradigms and disciplines evolve together as 'epistemic institutions' where the social and cognitive elements are closely intertwined.

This feature of academic science so contradicts the Legend that for long it was treated by metascientists as unthinkable, invisible, or entirely exceptional. It is peculiarly subversive in that it suggests that even the purest forms of scientific knowledge can never be purged of non-cognitive influences. It is difficult enough to protect science from 'external' societal interests [7.5 – 7.10]. Now we see that the most disinterested exercise of individual originality is always bounded by collective paradigms. In other words, 'internal' social factors enter into scientific knowledge at the very point of production.

Can it really be that most scientists are seriously blinkered in their choice of research problems? To put it crudely, are they so subservient to communal opinion that they seldom even think of undertaking projects that would challenge the established practices or the received wisdom of their community? Most scientists would vehemently deny this, both from personal experience and professional tradition. But many reputable metascientists assert, in effect, that it is the true position.

It cannot be denied, for example, that blind conformism – encapsulated in the story of the astronomers who would not look through Galileo's telescope – is not uncommon in science. Thus, a whole generation of geologists held back (or were prevented by their peers) from studying the implications of the notion of continental drift because it was so contrary to the universal belief that the earth's crust was rigid[102]. Similarly, many economists now complain[103] that it is impossible to do research outside general equilibrium economics, in spite of its manifest failures. Again, as I have hinted above, the obvious fact that research is a social activity was for long 'invisible' to metascientists because it seemed quite out of keeping with the established view that science was essentially a cognitive process [7.5].

Paradigms clearly have great power over research, right down to the smallest scale. This is not just a previously unacknowledged finding of historical and sociological research. It is part of the personal experience of every academic scientist who has tried to buck the trend of their specialty. Moreover, the same issues arise in the largest possible context – that is, where a whole human culture is the paradigm within which science has to evolve and seek credibility [10.4].

The question so crudely posed above will come up again in the analysis of conceptual change [9.6]. It is sometimes argued that a revolutionary new paradigm is necessarily 'incommensurable' with the one that it superseded[104]. But if our scientific thoughts are indeed ruled so firmly by paradigms, how can they ever change so radically? Is this really a piecemeal process[105], or are there, indeed, non-cognitive 'social' processes that disrupt the apparent continuity of the knowledge base in the scientific domain[106]?

As is well known, full-blown social determinism in the production of knowledge leads almost inevitably to metascientific *relativism* [8.13, 10.7]. For the moment, however, let us reserve our position on the validity of this epistemological doctrine, against which there are some very strong arguments. Logically speaking, for example, it defeats itself by self-reference[107]. Why should anyone be convinced by such a counter-intuitive proposition unless they already belong to the social group who cannot help but believe it anyway – and then, so what?

Moreover, looking at scientific paradigms from a naturalistic perspective, we often see them to be highly influential, but seldom absolutely dominant [9.5]. There is ample empirical evidence that scientists do, at times, honour the norm of originality by thinking the unthinkable, seeing the invisible, and taking serious notice of the exceptional. Such 'heretics' often have difficulty in presenting their research claims formally to the scientific community; but the traditional institutions of academic science are usually too weak to silence them completely without the support of more powerful external institutions such as the Church or the State.

8.5 Getting down to fundamentals

Scientific paradigms often claim to be *fundamental*. But as we have seen, they are never epistemically complete or coherent. Nor are they necessarily co-extensive with the disciplines they supposedly rule. Indeed, many fields of research are not dominated by a single conceptual scheme. Some disciplines, for example, are defined primarily by the life-world entities that they study, or by their research technologies. Until an all-embracing theoretical model is discovered – if ever – they persist in a 'pre-paradigmatic' state where any relevant facts or theories are given house-room [6.6]. It is often remarked, wistfully, that the human sciences seem unable to escape from this condition – or that when they do, as with classical equilibrium economics[108] or psychological behaviourism, they become oppressive orthodoxies.

Indeed, it is quite usual for a field of research to be spanned by competing theories, each claiming authority as a conceptual paradigm[109]. But this cognitive dissensus does not imply communal anarchy or sectarianism. Researchers doing similar work in the same laboratory can often be found to be subscribing to quite different theories[110]. In practice, the norm of scepticism [9.1] requires the co-existence of many schools of thought. But an experienced academic scientist becomes adept at shifting back and forth between competing candidate paradigms to gain a multiple perspective on her work[111]. The norm of originality would be vacuous if she could not, in the limit, develop a unique personal 'paradigm'[112].

Even where there is an acknowledged paradigm which everyone accepts in principle, it usually consists of a diversity of theoretical models bearing only a vague family resemblance[113]. For example, the theory of plate tectonics, which is fundamental to modern geology, cannot be formulated as a closed set of propositions. Notoriously, certain aspects of the Central Dogma that was supposed to govern research in molecular genetics[114] are now coming under scrutiny.

As earnest philosophers of science were quick to point out[115], a paradigm is a foggy concept that rapidly evaporates in the sunlight of formal analysis. Nevertheless, within its social context it has all the reality of a cultural institution[116]. That is, it functions as a fixed cognitive and methodological *frame*[117] for scientific work within a particular scientific discipline, subdiscipline, research field or specialty.

But academic disciplines and their subdivisions are not precisely defined or permanent. What is more, a 'fundamental' model that operates as an unquestioned paradigm in a particular subdiscipline may be known to be only one of a family of such models [6.10] in the parent discipline. For example, researchers specializing in the study of public attitudes regularly assume that these can be deduced from the answers to questionnaires and interviews, even though any such inference is disallowed by some of the more general theories current in social psychology.

The very notion of a hierarchy of increasingly more 'fundamental' paradigms is misleading [2.6]. It is often convenient [8.3] to produce a 'family tree' classification of knowledge and of research problems[118]. But the paradigms that shape research, like the disciplines, fields and specialties that they frame, cannot be mapped so neatly. They overlap, interpenetrate and interact with one another at every level, and from level to level, of supposed generality[119] [10.8]. It is easy enough to say that a progressive research programme typically has a 'hard core' paradigm surrounded by a 'protective belt' of more flexible ideas[120]. But when push comes to shove the distinction between the core and the belt is seldom obvious[121], and may only be apparent to the metascientist reconstructing the story 'rationally' in retrospect[122].

Here also, physics is not typical. No other discipline claims to be near to the ultimate goal of having a 'grand paradigm' capping the whole hierarchy. As soon as the master equation for the 'theory of everything' has been announced, then – so we are assured – all of physics will be reducible to it, and deducible from it [4.8, 6.5, 10.8]. Thus all the 'sub-paradigms' that rule such subdisciplines as atomic physics, nuclear physics, elementary particle physics, etc. will be seen as convenient mathematical approximations which we are free to accept or revise at will.

Even though the Grail of a Final Theory is no more than a dream, it has been the goal of a long history of successful research. Over the centuries, physicists have discovered and correlated a succession of ever more general Laws of Nature, from Newtonian mechanics to quantum chromodynamics. In spite of their promise, each of these successive theoretical systems has turned out to be not quite closed[123]. But except for the enormously super-macroscopic or minutely sub-sub-microscopic phenomena of cosmology and elementary particle physics, this goal has effectively been reached. For all practical purposes, the relativity theory of Einstein, and the quantum theory of 'the Copenhagen school' already operate in tandem as a fundamental 'grand paradigm' for the remaining physical sciences.

But the other natural sciences resist the doctrine of *physicalism*, which would subordinate them all to this one paradigm. This doctrine relies upon the unproven conjecture that all representations of the world can in principle be reduced to a unique, absolutely 'fundamental' theoretical model [10.8]. In any case 'being fundamental' is not an absolute epistemic quality [2.6]. It is simply a supposed relationship between one body of knowledge and another[124]. In the human sciences, such relationships are often matters of serious scientific dispute – for example, the place of sociobiology in the vicious triangle of biology, psychology and sociology. But the search for a closed theoretical system that might, say, encompass all these disciplines is no longer on the agenda of academic science

From a naturalistic point of view, paradigms are not permanent structural features of scientific knowledge. They emerge in the scientific domain [6.11] as clusters of theory and practice that strongly affect the direction of research. Hence they influence the further development of that domain – including their own evolution [9.7]. But there is no reason to suppose that they will converge on a unique final structure – a latter-day version of a 'Great Chain of Being'. They are useful primarily as epistemic scaffolding, not necessarily to be preserved as significant features of the world models that they help us to construct.

8.6. Normal science

In practice then, even a grand paradigm that frames a major discipline is no more than a structure into which can be slotted a variety of more or less distinct sub-paradigms[125], each with its own 'local' theories, conceptual languages, bodies of empirical data, research methodologies, etc. Research in these subdisciplines might well be gravely upset if the master paradigm were dethroned, but is otherwise largely unconcerned about how 'fundamental' it might be.

The fact is that much research can be described as *normal science*. Researchers try to solve problems that seem 'do-able' [8.1] because they are framed by well-established paradigms[126]. By committing themselves to certain theoretical authorities, to the reliability of certain instruments, to the imitation of certain examplars of 'good science'[127], they try to reduce these problems to *puzzles* requiring no more than the ordinary technical skills in which they are professionally trained[128]. They turn their research field into an exercise ground for a specialized cultural practice[129].

In effect, normal science limits the demands of research on the individual intellect and psyche. It is a medium through which it is easy enough to show originality[130] without losing the social approval of one's community[131]. It is a cognitive context in which to acquire tacit knowledge of a research specialty[132] and to learn and practise the general art of scientific investigation[133]. It is the rationale of laboratory life[134], ensuring that order emerges from region of disorder[135] and that communications from laboratories with different 'signatures' can be combined coherently[136].

The notion of normal science tends to present it as a *routine* activity, which could be performed quite adequately by well-drilled technicians [8.2]. After all, even the most 'exciting' research can involve long stretches of mechanical attention to detail. New planets are discovered only by interminable watching of the skies, new understanding of animal behaviour requires years of painstaking observation, new theorems emerge out of long periods of futile algebraic manipulation – and so on.

But this misses the the point. Remember that the converse also holds – that an apparently dull and ordinary natural phenomenon, such as the light given out by a 'red-hot' object, the brittleness of a crystal, or the growth of a mould on an agar plate, can be the starting point for a brilliant path of discovery [8.9]. The 'normality' of a piece of research is not measured in terms of the tedium of its methods or the homeliness of its objects of study. It is measured by the quantity of 'me-too' research in a specialty, or by the proportion of papers whose citations are drawn from the same closed list. It is a *mind-set* that can take hold of researchers in almost any field of academic science.

This raises an obvious question: is there a natural progression in the history of scientific change [9.6]? Does every research field typically pass from pre-paradigmatic anarchy to orthodox normality, with a limited phase of exciting discovery and paradigm formation in between[137]? Is it then taken over by increasing quantities of routine research whose results are of less and less relative interest or importance[138]? Does the paradigm eventually become so deeply entrenched that a veritable revolution is required to break its grip[139]?

The difficulty with any such periodization is that it assumes that research fields and paradigms persist as distinct entities throughout the process of change. As we have seen, this is seldom the case. They are never precisely defined, often co-exist as rivals and overlap, layer upon layer[140]. They speciate, merge or transform themselves internally. A 'phased development' model does not make sense in the human sciences, where paradigms never seem to triumph or die[141]. Even in the history of a discipline with a grand paradigm [8.5], such as elementary particle physics, major episodes of theoretical change may not be synchronized with those of experimental innovation[142].

What is more, there is little empirical evidence that research fields always decline into stultifying 'normality' after the enthronement of an over-arching paradigm. On the contrary, they often burst into new life. This is because they become more *finalized*[143]. Like most general terms used to describe the research process, this is an ill-defined concept, with ambivalent policy resonances [4.9]. But it indicates that research projects can be formulated within a more stable framework of concepts and techniques, and directed towards more specific ends. 'The art of the soluble', as practised intuitively by individual researchers, is made more open by being given a more explicit rationale.

Metascientists tend to see the advent of a master paradigm, such as classical mechanics, quantum theory, evolutionary biology, plate tectonics, molecular genetics or transformational grammar, as the End of History in its field. Most scientists see it as the beginning of a Golden Age of research. They are at last in a position to undertake serious research on the movements of the planets, the structure of atoms, the diversity of organisms, the causes of earthquakes, and the rest. To call such research 'puzzle-solving'[144] does not merely suggest, quite unjustifiably, that it is necessarily trivial, or frivolous, or contrived. It also denigrates the skills and insights required to formulate and tackle successfully the new range of scientific problems that now become 'do-able', either within or outside the original field[145].

Let us remember, after all, that the concept of a scientific 'problem' must include the intellectual background against which it is perceived. To propose a research 'project' [8.2] in a particular field, however fundamental, one must state its objective. This can only be done in terms of the current paradigm or paradigms in that field. In other words, almost all research is to some extent 'finalized', in that it is directed towards a rational end. This is just as true, for example, for the modern theoretical physicist seeking to establish the relationship between the four fundamental forces of Nature as it was for her nineteenth-century forebears, who were 'puzzled' to explain the peculiar connections between sensible heat and mechanical work. In each case, the research has a final purpose conceived in terms of what was by then well known.

Peer reviewers naturally tend to favour research projects whose objectives can be presented as attainable in principle – even if only by a heroic exercise of theoretical reduction. What really counts, however, is whether this objective is likely to be reached in practice, and whether the putative 'solution' has other epistemic virtues such as explanatory efficacy [10.1] or wider applicability [7.2]. In other words, finalization does not make research easier to do, or its actual results more certain. It just shifts the goal posts and marks out new fields of play in an endlessly challenging game.

Epistemically speaking, finalization indicates the availability of knowledge that can be relied on (up to a point!) in the production of more knowledge. *Reliability* is a fundamental scientific virtue [7.3]. It is the quality that makes science *useful*. Naturally enough, metascientists have given finalization a utilitarian slant, and discussed it mainly in connection with the possibilities of technological application[146] [4.10]. The study of nuclear fusion, for example, is presented as the prime example of finalized research, in that it is driven almost entirely by the more or less rational calculation that it will, in the end, show the way to an unfailing source of energy. Indeed, in cases such as this, a finalized science is scarcely distinguishable from a high technology.

But the usefulness of science is not limited to the world of everyday practice [10.6]. Improved reliability is a powerful operational quality within the scientific domain itself. Finalization is often a feature of the *scientific* application of a paradigm and of its wider extension. Thus, for example, the grand paradigm of molecular biology – 'the structure of DNA as a genetic code' – was discovered in the 1950s, but had little direct medical use until the 1980s. Nevertheless, as soon as it was established scientifically, this paradigm became the means by which innumerable research projects could be formulated and undertaken over an immense range of academic science. These applications, in turn, guided and stimulated further research on the basic DNA thema. In other words, molecular biology became a finalized scientific discipline long before it spawned an effective technology.

8.7 Who sets the problems?

Scientists greatly value personal autonomy in research. The academic norm of originality calls for individual freedom of problem choice. This freedom is hemmed in by material circumstances, historical opportunity, epistemic conviction, and above all, communal doctrine [8.1]. Nevertheless, academic science has always assumed that its research problems arise out of its own working. They are supposed to be generated naturally by the research process itself – through discovery, prediction, criticism, controversy and so on[147].

In principle, the finalization of a research field need not upset this tradition. Indeed, research activity is usually stimulated when a 'breakthrough' enables the members of a research community to undertake projects with a good prospect of achieving specific scientific goals. But it also enables people outside that community to propose research with a good prospect of achieving specific *practical* goals. Finalization thus opens the way to 'Mode 2' research [4.11], where problems typically arise 'in a context of application'[148]. Indeed, it is an essential factor in the transition to post-academic science [4.4], since it enables research to be institutionalized as a rational instrument of *policy* [4.9], on a larger societal scale[149].

Much policy debate is concerned with how to optimize the societal benefits of the scientific enterprise. For example, who should be involved in setting the research agenda[150] [7.6], what do the various actors bring to the process, and how should they allow for the intrinsic uncertainty of research [9.1]? Even very basic research does not take place in a power vacuum. It has to be supported financially and administratively by bodies whose interests go beyond the mere production of knowledge [7.9]. They naturally exercise these interests at the point of maximum leverage – that is, when research problems are being set. All policy talk about foresight, priorities, accountability, etc.[151] is really focussed on 'problem choice'.

In political discourse, moreover, the term 'problem' is often applied to a complex societal situation involving irreconcilable conflicts of interests, values and preferences between influential groups of people. Such situations can usually be characterized perfectly well in lay terms, and enquiry into them is a normal function of responsible citizenship. Post-academic social scientists are in duty bound to take on such problems [7.10], despite the work that has to be done to transform them into research projects, and the knowledge that they cannot hope to have consensual outcomes[152] [9.2]. In effect, although scientific enquiries of this kind are now an indispensable instrument of equitable government, in terms of problem choice they are often nearer to 'the art of the possible' – i.e., politics – than to 'the art of the soluble'.

But we still have to look at how problems are actually formulated and turned into projects[153]. In particular, we have to think about the *epistemic* effect of trying to get a handle on the finalizability of research in order to direct it towards non-scientific goals[154]. What is actually new about the type of knowledge produced in this new mode?

It is obvious, for example, that post-academic science has only limited respect for *strategic autonomy* in research[155] – that is, freedom for individual scientists or small groups to define their own long-term research programmes. Indeed, it often requires highly qualified researchers to work together on problems which they have not posed personally, and which they may not even have chosen collectively as a team. In general, the science that then gets done is technically 'better' and more 'relevant' than if it were left entirely to the idiosyncratic judgements of individual scientists.

But the elimination of wasteful effort on seemingly ill-conceived projects is to some extent offset by a concealed epistemic cost. It means that a few wild conjectures never get a chance to show their hidden capabilities, which are just occasionally revolutionary. Thus, the effect of 'collectivizing' problem choice must be to limit the range of *variation* of research projects.

This is significant, because academic science can be considered an 'evolutionary' process[156], whose long-run efficacy depends on having a highly diversified stock of variants, as well as a highly selective environment [9.7]. In post-academic science, problems are socially pre-selected, on hypothetical grounds, before they are even tackled or can be tested by their outcomes. What will be the long-term effect of the introduction of this 'Lamarckian' factor on the advancement of knowledge[157]?

It must be emphasized, however, that finalization is not equivalent to predictability [8.11]. For example, in a finalized discipline such as physics, very elaborate and expensive research projects can be undertaken with a reasonable expectation that they will work as planned[158]. It does not follow, however, that these projects always produce their intended results. Sometimes, as in the notorious case of the 'War on Cancer'[159], they far outrun the basic understanding needed to actually solve the problems at which they were directed. In other cases, such as the unfinished saga of the attempt to harness nuclear fusion, the goal recedes as research advances towards it.

Accountants and managers regard 'failed research' as totally uneconomic. But 'negative' results are an essential part of any evolutionary process [9.7]. In any case, precisely because every research project is unique, it almost always spins off 'positive' scientific information. Indeed, elaborate instrumental facilities, such as particle accelerators or satellite observatories, typically make discoveries [8.9] of a kind that were not at all anticipated when they were planned. The history of such projects demonstrates the impossibility of deliberately 'constructing' knowledge to suit even the most pressing of human needs [8.13].

Although post-academic research projects are not always very large, 'Big Science' [5.7] is usually carried out under Mode 2 conditions. Even the purest and most fundamental scientific discipline then becomes a 'problem context' for the application of established knowledge[160]. The skills and imagination of very large numbers of scientists are thus entrained in *technical systems*[161], whose organizational features appear to be much the same, right across the R&D spectrum. One of the major characteristics of post-academic science [4.11] is this structural convergence between the academic and industrial research traditions[162].

The 'post-industrial' mode of R&D that has thus evolved deviates markedly from the academic ethos [4.11]. The single-minded pursuit of practical ends generates knowledge whose nature, form and epistemic quality depend less on its actual mode of production than on the situation in its 'context of application'. Thus, for example, the Human Genome Project is devoted to the production of 'maps' whose validity will be judged in terms of their medical utility, whilst the equally expensive and systematic effort to explore the ocean floor is designed primarily to produce maps that are of value in anti-submarine warfare or oil-prospecting[163]. In each case, the social organization of the research involves a different mixture of academic, governmental and commercial institutions, some of which are still activated by the notion of 'doing good science' [164]. In each case, however, the epistemic status of the final product is entirely pragmatic [7.3, 10.7], and is not meant to be judged by traditional academic standards, which are often very different[165].

Nevertheless, it would be wrong to equate post-academic science with 'commissioned' research, where very detailed questions are set by authorities quite outside or far above the laboratory frame. Mode 2 problems are typically formulated in 'hybrid fora' where the researchers involved may have more or less weight, depending principally on how close the problems are to practical application[166]. Indeed, it seems that most scientific workers feel satisfied with a reasonable degree of *technical* autonomy[167] – that is, freedom to take an independent approach to a scientific problem that has only been posed in principle by others. In practice, this may not be much more of a constraint on individual originality than the informal curb of peer opinion in a traditional academic specialty.

Post-academic science is not, of course, indifferent to research achievement. It is always on the lookout for technical virtuosity, and awards managerial authority to outstanding performers. Indeed, it tends to elevate them into 'stars' for whom the mass of other scientific workers simply gather information[168]. At the same time, however, Mode 2 inhibits the individual exercise of the highest form of scientific 'creativity' – the construction of soluble research problems – by redefining it as a group phenomenon. This seems to be the case in high energy particle physics, for example, where 'collaborations' are not just the research tools of their famous prize-winning leaders[169]. Again, in space science, the original authors of the ideas that finally gave rise to brilliant experimental projects can only be found by tracing them back through elaborate hierarchies of committees and workshops, project proposals and feasibility reports.

Even personal success in formulating fundable research proposals [8.2] may simply reflect a shrewd eye for what is currently regarded as 'good science', or even purely social skills in constructing networks of allies [5.13] in and beyond the relevant research community[170]. Non-cognitive factors of this kind have always been influential in academic cycles of credibility and credit[171]. Their influence is far greater in the more highly structured social world of post-academic science.

The pragmatism of Mode 2 orients research towards specific problems. Instead of pursuing the traditional scientific goal of mapping the scientific domain as a whole, it takes a looser, more experimental approach, where the problem itself provides the focus of continued effort. This approach works well in the study of natural and artificial phenomena in problem areas whose contexts are partly universal and partly local. Thus, for example, very practical, down-to-earth questions about the mechanical properties of household plastic materials such as poly-thene led to the discovery of a whole class of semi-crystalline structures whose theoretical existence had scarcely been conjectured.

In effect, post-academic science makes progress by focussing on the regions around specific problems. Its dense communication networks encourage thorough exploration of limited areas, and of the pathways between them. Disciplinary frontiers are overrun in order to construct detailed local maps of potentially useful domains of knowledge and prac-tice. For example, research on the practical problems of growing perfect crystals for computer chips has become a coherent specialty combining concepts and techniques from a number of different subdisciplines of physics, chemistry and engineering – and contributing new knowledge and methods to each of them. In the end, this striving for 'local' under-standing, without preconceived notions of what might require explain-ing or be acceptable as an explanation [10.1], may even be more effective in closing the gaps in the knowledge map than a single-minded pursuit of general intellectual unity [10.8].

Very often, of course, a problem arising in the context of application cannot be solved without attention to more fundamental issues. It is quite clear now, for example, that the 'War on Cancer' – the attempt to produce a vaccine against viral cancers – could not have succeeded without a much better understanding of basic cellular mechanisms[172]. The whole scientific enterprise is built upon positive epistemic feedback between theory and practice[173]. Ideally, then, difficulties in applying basic principles should always trigger off basic, academic research. But instead of advancing head on against entrenched paradigms or concep-tual enigmas [9.5], post-academic science can often be satisfied with rea-sonably reliable pragmatic solutions to its problems, whether or not these solutions have a broader theoretical basis.

Most scientists believe, I suppose, that all these local maps will even-tually turn out to be pieces in a grand jig-saw, ready to fall neatly into their places in the scientific world picture [10.8]. But the construction of a reliable representation of a local reality usually involves the development of a correspondingly local language[174] [6.6]. The more elaborate this rep-resentation, the more difficult it is to uncouple such a language from its context, and to use it in representing the realities of other problem areas.

As a consequence, post-academic maps of knowledge may well be meticulously detailed and systematic in their coverage, but they will still be divided up into specialized domains [6.11], characterised by mutually incomprehensible technical discourse about artificial research objects[175]. Moreover, these domains, like all research specialties [8.3], will always be in a state of flux, and often overlap one another. Novel solutions will be found for long-standing problems. Novel problems will arise in new contexts of application. Paradigms, techniques and specialized skills will be continually segmented and recombined into new configurations.

Post-academic science accepts that its representations of various aspects of nature are purely conventional and does not insist thay they must converge and combine into a unique map. It does not spurn the genuine results of 'reducing' a theoretical system to one that is more fundamental, but is not driven by *reductionism* as an metaphysical ideal [10.8]. Indeed, it regards formalized schemes of thought [6.6] less as strengthening skeletons than as potential barriers to understanding. In the language of post-modernism, post-academic scientists do not reject such 'grand narratives' out of hand. They merely take a sceptical attitude towards their claims.

In other words, *pragmatic finalization* [8.6] is favoured above *explanatory unification* [10.1]. This may not, of course, be a genuine metascientific distinction. It may simply describe the changing social role of scientific disciplines as they become more mature theoretically and more effective technologically[176]. Indeed, there is much to be said ethically for spelling out the practical goals of research [7.9] rather than referring vaguely to the material benefits from knowledge produced 'for its own sake'.

The trouble is that it is difficult enough for an academic scientist to decide on purely cognitive grounds what research to do at any given moment [8.1] without including the variety of social goals and interests that might eventually be affected by the results. It is also doubtful whether the manifold cultural consequences of scientific knowledge can be catalogued or foreseen. The urge to make scientists more 'socially responsible' in their choice of problems may not have the desired effect of producing knowledge that is more likely to be used beneficially[177].

8.8 Interdisciplinarity

In 'normal' academic science, research problems are formulated and tackled within the bounds of an established discipline. But the 'contexts of application', where post-academic science finds its problems [7.9], are not so bounded. They are almost always *transdisciplinary*. Post-academic science repudiates the metaphor of the university as a notional brain, whose permanent modules are academic faculties and departments, each dealing tidily with its allotted specialty. It is not inhibited from taking on any well-posed problem, basic or applied, that arises naturally in the regions where traditional disciplines interact or overlap.

In rhetorical moments, academia welcomes a broad view. Academics often expatiate on the virtues of *interdisciplinarity*, conjuring up visions of a community of scholarly saints, marching forward together under the banner of a final theory. The actual trend is in the opposite direction. Even the most fundamental fields of academic research have become increasingly specialized, diversified and fragmented [3.9, 8.3]. There are thus many different ways, in principle, in which this trend might be bucked[178]. And yet, the label 'interdisciplinary' attached to a research centre has come to mean little more than that it is the opposite of 'basic' in its research interests[179].

Post-academic science is not merely transdisciplinary. It is defiantly post-modern in its *pluralism*[180]. It welcomes wide definitions of knowledge, and decentred diversity, without fear of possible inconsistencies. The 'context of application' inevitably introduces 'trans-epistemic' factors, such as human values and social interests[181] [7.5]. The knowledge that it produces is not organized around theoretical issues, and is not automatically subject to clear rules of coherence and credibility. It may combine cognitive and non-cognitive elements in novel and creative ways – witness cognitive science itself. The interpenetration of university-based and industrial research in fields such as biomedicine exemplifies the formation of hybrids with research cultures which do not share the traditions of academic science.

For scientists who believe in the Legend, this pluralism can be very distressing. How is it possible to do 'good science' under such conditions? How can we formulate new ideas without a unique set of general rules by which to judge them? Where are we to place new data without a system of categories derived from a unique paradigm? Even from the naturalistic perspective of this book, the epistemology of post-academic science is not at all transparent. How much can be considered well established in a body of knowledge that mixes theories and practices, ideas and data, cognitive and non-cognitive elements, from a number of distinct research communities[182]? Post-academic science concentrates on reliability in specific applications: is its general reliability seriously compromised by this post-modern pluralism and incoherence?

This pluralism and instability in the domain of knowledge is reflected back into the social dimension. Post-academic scientists, being students of 'problems' rather than of 'subjects', have to find their way through a more open landscape of career paths[183]. Interdisciplinary projects require *multidisciplinary* research teams, where cooperation moderates the tribal antagonisms that differentiate disciplines[184] [3.9, 8.4]. 'Venture research'[185] calls for skills and traits that are not recognized in 'normal' academic science. Socially as well as intellectually, the post-academic scientific world mirrors the confusion and insecurity of the 'post-industrial' socio-economic order[186].

Nevertheless, successful research still requires time and sustained effort. 'Big Science' projects, such as nuclear fusion experiments, take years to plan and carry out[187]. As the maxim affirms, creativity is usually based on an infinite capacity for 'taking pains' – that is, for undertaking vast amounts of routine work over long periods of time [8.6]. Post-academic science has deconstructed the academic ideologies that legitimated the disciplinary frameworks in which individual researchers could feel secure[188]. But new socio-epistemic structures are sure to be erected, to provide the stable environment that scientific originality needs to show its worth.

Indeed, the 'context of application' [7.9] is not an open space, without frontiers[189]. On the contrary, it is already closely occupied and subdivided amongst a variety of industries, government departments and technological professions. The problems that activate post-academic science are often deeply rooted in history, and are typically 'owned' by well-established institutions, such as pharmaceutical companies, arms procurement agencies, associations of engineering and medical practitioners, environmental protection commissions, economic advisory councils, and so on. This elaborate social structure is associated with an equally elaborate epistemic structure, where the 'problem areas' are differentiated much more arbitrarily, and are often even more narrow and specialized, than they are in academic science. In other words, post-academic science is not free from the processes by which new academic disciplines emerged in the past[190], and post-academic scientists are not necessarily all that free to cross the new intellectual and professional frontiers thus created.

In many respects, therefore, the advent of post-academic science is only another phase in the eternal dialectics of research. 'Disciplines' stand for stability and uniformity. 'Interdisciplinarity' is a code word for diversity and adaptability. The advancement of knowledge in all branches of science is driven at all levels by the tension between these opposing forces[191]. The idea that research will henceforth be undertaken by mutable configurations of Mode 2 researchers[192] is very appealing to the restless, enterprising, self-winding scientist – especially if he has a tenured post [3.10]. But disciplines, research communities and other specialized epistemic institutions are still needed to provide the peer groups and 'authorities' to review project proposals and their findings, and to establish that reliable knowledge has indeed been produced.

Indeed. from a cognitive point of view, 'interdisciplinarity' is one of the major sources of mental creativity[193]. Original ideas seldom come entirely 'out of the blue'. They are typically novel combinations of existing ideas [9.7]. To 'make the connection', one has to cross the boundaries between supposedly distinct paradigms – that is, between distinct disciplines. On the fine scale of everyday research practice, of course, where paradigms and disciplines are usually indistinct and impermanent, this is no big deal. But our intuitive measure of the originality of a great scientific idea is the height and thickness of the barrier between the paradigms and disciplines that it has transcended. The call for more interdisciplinarity in research is really a plea for more scientific originality! No wonder this call is so difficult to heed.

In effect, interdisciplinary research is *abnormal* [8.6]. It is *radical* in intention[194], if not always *revolutionary* in outcome[195]. It deliberately tackles problems that are not within the scope of a single, well-established paradigm. But this is not really so uncommon even in basic academic science, where it is not driven by thoughts of application. Indeed, 'good science' is typically a mixture of 'normal' and 'abnormal' epistemic practices, often on a day-to-day scale[196]. A lot of the time, I treat my current research problem as a 'puzzle' where well-tried standard methods will surely yield useful but unsurprising results. But I continually worry about the soundness of these methods, try to look at the problem from other points of view, and am tempted to think the unthinkable about what is really going on. And even in rare moments of leisure – for like all scientists I am an addictive workaholic – I find myself wondering whether it is possible to combine these apparently incommensurable attitudes of mind.

'Abnormal' research is thus as much a feature of academic science as its supposed antithesis. The question is: what triggers it off? Any research specialty can provide an endless stock of 'do-able', if not very original problems [8.1]. Why should an established scientist undertake research that will take her out of her area? It is not enough to say that good scientists are psychologically motivated by curiosity [2.7], or that they deviate adventurously from the communal consensus in search of the social rewards of discovery[197] [8.9]. What is there in the *epistemic* situation that can make such a decision perfectly rational, even for quite a cautious person?

The immediate answer is that scientific disciplines are very seldom dominated by over-arching paradigms that reduce all their problems to 'puzzles' . This is rarely true at any level, right down to the narrowest problem area [8.6, 9.6]. The fact that a field of research is covered in principle by an unchallengeable general theory does not mean that it is free from uncertainty or controversy. Conflict between co-existing sub-paradigms is much more usual than consensus [9.2]. Competing 'schools of thought' may even accumulate so many adherents that they coagulate socially into distinct subdisciplines. Nevertheless, socio-epistemic boundaries within a relatively small research community are highly convoluted, and many researchers remain uncertain where they stand [8.3]. They are then impelled by the norm of scepticism to undertake research designed to resolve the 'interdisciplinary problems' thus brought to light [9.1].

We shall return later to the dynamics of scientific change [9.6]. But even a 'revolutionary' paradigm that sweeps all before it usually leaves a number of 'disconfirming instances'[198] in its wake, and as it is extended into new fields it is sure to reveal further *anomalies*[199] – that is, phenomena that it cannot explain [8.9]. This is just as true in the microcosm of a narrow problem area as it is for a major scientific discipline. What is more, an active research community will often put such anomalies high on its research agenda, even if truly provocative enigmas[200] seldom reveal themselves to its collective vision.

8.9 Discovery

Science is always on the move. Researchers choose their problems and formulate their projects in an ever-changing epistemic environment[201]. New knowledge is continually moving the goalposts of their game plans. Indeed, this is a self-activating process [9.8]. We have seen that academic scientists do research in the hope of making *discoveries*. But every discovery is a catalyst for further research.

'Discovery', like 'research', is a characteristic feature of academic science[202]. But it has to be understood as something more than a culture theorist's term for a sequence of social events involving the private or public announcement of a research result, the formal registration of a research claim in the scientific archives, and its validation, interpretation and retrospective attribution by a research community[203]. These social practices do indeed accompany scientific discoveries. They are important elements in the communal acceptance of new items of knowledge [9.3]. But the conventional meaning of the word has a specific *epistemic* significance in scientific contexts.

In effect, several ordinary dictionary meanings[204] are intertwined. On the one hand, 'to discover' means 'to find after study or search', signifying that a scientific discovery is an outcome of the intentional action of research. On the other hand, it means 'to learn about or encounter *for the first time*', with the associated meaning of being *the first person* to do so. The phrases I have italicized convey the sense of *novelty* which is inseparable from the whole concept.

The essential point is that a scientific discovery is not just a validated research finding. It is an *unexpected, unforeseen, surprising* finding[205]. For example, the possibility that the mass of an atom was concentrated in a small nucleus was scarcely imagined until this was discovered experimentally by Ernest Rutherford. On the other hand, I do not think it is correct to say that Carlo Rubbia's team at the CERN particle accelerator 'discovered' the W and Z particles. The elaborate experiment that they carried out in 1983 was carefully designed to detect just such entities, whose existence had already been widely predicted on theoretical grounds [8.11]. It would have been much more surprising if their search had *not* had this happy outcome.

Indeed, because they surprise even their makers, discoveries are often simply treated as 'eureka' episodes[206], to be accounted for retrospectively in terms of mental phenomena such as *gestalt* switches, perceptions of anomaly, bisociation and so on.[207]. This is a significant feature of the scientific life, to which we shall return. But science is much more than DIS-COVERY writ large. The research process cannot be mapped on to its psychological dimension and condensed into this single word.

In ordinary scientific usage, moreover, discoveries are deemed to be *empirical* [5.2]. Only 'facts' can properly be said to have been 'discovered', as distinct from 'theories', or 'models', which are 'proposed', or 'constructed' [8.13]. Such distinctions are impossible to define logically, and are not always applied consistently. For example, Francis Crick and James Watson are usually said to have 'discovered' the structure of DNA, even though what they did was to construct a double helix model that was consistent with the experimental data. The recognition of a significant pattern in the natural world – for example, a 'Law of Nature', such as the Periodic Table of the Elements – counts as a discovery[208]. And yet, Charles Darwin is seldom said to have 'discovered' biological evolution, even though the theory that he proposed explained vast quantities of known facts.

Nevertheless, in most research contexts, 'factuality' is regarded as an essential element in the concept of discovery. A discovery claim might, for example, have been misconceived or faked. Even a 'golden event' showing the existence of a new elementary particle, captured for eternity in a nuclear emulsion or bubble-chamber photograph[209], must be tested by independent replication or triangulation [5.8] before it is accepted as authentic by a research community. This notional link between discovery and empiricism is thus deeply entrenched in the scientific culture, and cannot be left out of a naturalistic epistemology. For most scientists it implies the independent existence of the entities that the discoverer claims to have found[210]. In effect, a doctrine of *realism* [10.7] is packed into a single word.

For this reason, metascientists who reject realism are bound to dispute the conventional distinction between 'discovery' and 'invention'[211]. This distinction is to a large extent arbitrary [8.13]. Every discovery is shaped by human perceptions and intentions, and has to be assessed in relation to its social background. The concept of discovery thus touches on fundamental epistemic issues, to which we shall return.

For the moment, however, our concern is not with how discoveries are made, or what they contribute to knowledge. It is with their influence on the on-going research plans of scientists. This influence is powerful because it is so paradoxical. Obviously, being unexpected, a discovery is 'abnormal' in its effect[212] [8.6]. It unframes a puzzle, disconfirms a hypothesis, or discloses a phenomenon that challenges a paradigm. It opens windows on new fields of ignorance[213], suggests connections between supposedly distant disciplines[214], or reveals distinctions that were commonly disregarded[215]. It thus creates an 'anomaly' requiring further investigation, much of which will have to be interdisciplinary [8.8].

On the other hand, making discoveries is a regular feature of scientific research. A scientific discovery is usually the outcome of a research project addressed to a specific scientific problem [8.2]. It arises typically out of a carefully devised investigation to solve a particular 'puzzle', test a particular hypothesis [5.6], or explore a particular domain of nature[216]. It gets its meaning from the paradigm under which this problem was formulated and from the knowledge that the project was designed to produce[217]. However unexpected it may be at the time, a scientific discovery can be seen as 'normal' in its larger context[218]. Some enthusiasts have even tried – not very convincingly – to simulate it with a computational algorithm[219].

An exciting scientific discovery is thus much more than an astonishing piece of information about the world. It is information that is relevant to the prior beliefs of the discoverer, and thus to the existing system of beliefs in her research community[220]. However disconcerting it may be, it reports an empirical 'fact' that has to be taken seriously [5.2]. The body of knowledge already shared by that community is suddenly seen to be imperfect or incomplete without it[221].

Indeed, a scientific paradigm is typically an epistemic network [6.4] where any significant hole or flaw is likely to have noticeable effects at other nearby nodes or links. These are often the public clues that lead several researchers independently into projects that explore the network or test its integrity in this neighbourhood. These projects are never, of course, precisely identical in goal or method[222], but they often result, almost simultaneously, in essentially the same discovery[223] [3.6]. Although the existence of true *multiple discovery* is disputed by some meta-scientists[224], many scientists have personally experienced situations very like it[225] and interpret these as evidence for the 'reality' of scientific knowledge [10.7]. In effect, it illustrates the duality of scientific paradigms, as over-arching social institutions and as underlying representations of the uniformity of nature.

I am not suggesting, of course, that scientific discoveries are socially determined, or in any serious sense predictable. If they were, then they would not be discoveries! Just because research often has a sophisticated rationale it does not follow that discovery must have a compelling logic. Computational simulations of scientific inference are not (as yet, anyway) very convincing.

In any case, our naturalistic perspective cannot ignore the *serendipitous* element in most accounts of scientific discovery. When scientists talk about their research they always emphasize the fortuitous circumstances through which they achieved success – the happy chance of accidentally leaving a Petri dish uncovered, or using a contaminated chemical reagent, or leaving a photographic plate wrapped in black paper near the discharge tube, or whatever.

Such anecdotes inevitably relate to mental events, which are particularly vulnerable to the tricks of memory and unconscious motive[226]. They are all too easily re-interpreted metaphysically, as theologians have amply demonstrated by arguing that everything that happens to us in our earthly existence can be understood as the working of Divine Providence – or, as it may be, of blind Fate. Nevertheless, there seems little doubt that true serendipity[227] – the accidental discovery of something not sought for – does play a major role in the production of scientific knowledge[228] [5.2, 5.6, 6.1]. What is its epistemological significance[229]?

The key point is that serendipity does not, of itself, produce discoveries: it produces *opportunities* for making discoveries. Accidental events have no scientific meaning in themselves: they only acquire significance when they catch the attention and interest of someone capable of putting them into a scientific context. Even then, the perception of an anomaly is fruitless unless it can be made the subject of deliberate research[230]. In other words, we are really talking about discoveries made by the *exploitation of serendipitous opportunities* by persons already primed to appreciate their significance.

Strictly speaking, therefore, there is nothing special about the epistemological status of a serendipitous discovery. It will have to be communicated to the relevant research community, subjected to communal criticism, put through the standard tests, and accepted as valid – or not – in the usual way. But because serendipity is completely unpredictable in its incidence and outcome, it introduces a factor of *blind variation* in the production of knowledge. This factor is strongly favoured in *evolutionary* theories of scientific change [9.7]. A serendipitous observation involves a wild leap outside the limits of what was until that moment supposed[231], and thereby enables science to advance into domains of understanding that were not previously imagined.

When scientists list the conditions for scientific serendipity[232], they usually treat it as if it arose naturally out of *curiosity*. This is not so much a personal trait or an attitude of mind as a virtue associated with a social role [2.7]. Many traditional features of academic science favour and celebrate serendipity. Academic scientists are usually highly specialized in their interests [8.3], but often have in mind a much larger agenda of problems than they are actively working on[233]. What is more, having sensed an opportunity, they are free (in principle!) to take time off to look into it more closely, and to balance the risk of finding nothing against the possible reward for making a significant discovery.

For this reason, it is a matter of some concern that post-academic researchers[234] have much less freedom to undertake relatively modest investigations without formal approval [8.6]. In effect, they are all deemed to be doing 'normal' science, which has no place for serendipity. Research grants and contracts tie them into finalized projects from which they are not officially permitted to deviate to chase after presumed wild geese[235]. Nevertheless, research remains as uncertain as ever in its outcomes, and must surely continue to be ruled by 'opportunistic logic' in its detailed operations[236]. The 'tinkering' which is so characteristic of the research culture is essentially the practical exploitation of serendipity[237], writ small.

8.10 Hypotheses

Up to now, we have been considering the norm of originality primarily in relation to the planning of research. Many people would say, however, that the supreme expression of scientific genius is in making novel but fruitful *hypotheses*. The scientific domain is mapped and held together by networks of *theories* [6.4]. But how do theories come into existence, and acquire their scientific force? They are often said to be 'discovered', but this cannot be strictly true. A scientific theory is essentially a mental construct [8.11]. However closely it may relate to empirical 'facts', it cannot be 'discovered' without at least a pinch of the magic ingredient of 'invention'.

The notion of a 'theory' is too primitive for fruitful analysis [6.1]. From a naturalistic perspective, a scientific hypothesis is simply an embryonic theory. Indeed, as in Faraday's famous simile, it is very like a new-born baby[238], requiring to be nurtured, strengthened and tested before it is accepted as 'full-grown' by a research community. The *justification* and *validation* of theories is a social process governed primarily by the norm of scepticism, which will be discussed in the next chapter.

In scientific practice, however, the 'context of discovery' cannot be separated epistemically from the 'context of justification'[239]. In effect, a hypothesis that did not already indicate why it was proposed and how it might be tested in some way[240] would be like a baby born without limbs, clearly incapable of growing to self-reliant maturity. Then again, who knows exactly how it will turn out? The mere process of testing a hypothesis focusses scientific attention on a particular aspect of nature[241] and often produces new research findings to which it has to adapt as it develops[242].

But the 'baby' metaphor can be misleading. Scientific theories do not necessarily emerge as well-formed hypotheses or retain their individuality for long. They are often little more than structural elements in paradigms – that is, temporary nodes in shifting networks of concepts, models, 'facts' and techniques, linked to various features of the natural and social worlds [8.4]. What we ought perhaps to say[243] is that each linkage in this assembly starts life as a hypothesis, and remains to some degree 'hypothetical' even when it has been accepted as 'reliable' and 'well established' by the research community.

A scientific theory is often presented in the relatively simple form of a 'law of Nature' [6.4]. In effect, this states that there is a certain type of relationship between certain types of natural objects, organisms, properties, phenomena, etc. In many cases, this relationship is experienced as a 'regularity', or 'pattern', in some aspect of the world [6.1]. Indeed, much research is specifically directed towards the discovery, creation or recreation of circumstances where just such a pattern can be perceived[244]. When this research succeeds, it induces the hypothesis that these circumstances can be *generalized* – that they have revealed particular instances of a relationship that holds *universally*. For example: 'I have been watching ravens. All the hundreds of ravens that I have seen were black. Therefore *all* ravens are black.'

This mode of reasoning is often stated formally as the principle of *induction*. It is basic to scientific theorizing: and yet it is obviously logically incomplete. What is more, as philosophers have discovered since David Hume pointed this out more than 250 years ago[245], no amount of elaboration can make it sound. Like pattern recognition, it is a natural cognitive process which cannot be reduced to a formal algorithm.

Indeed, induction is effectively an extension of pattern recognition [5.12, 6.2] into the time dimension. It is natural to try to make sense of a spatial regularity, such as a grid of equally spaced lines, by intuitively extending it out of its local frame to infinity in all directions. Similarly, it is natural to give meaning to a regularly occurring event, such as the 24–hour periodicity of sunrise, by intuitively extending it back into an immemorial past and forward into an endless future.

Let me make it quite clear that this is not a philosophical cop-out. The process of *inductive inference* is not 'irrational'[246]. On the contrary, from a naturalistic perspective[247] it stands out as one of the principal pillars of scientific rationality [6.8]. It is a *universal* human capability, honed by millions of years of biological evolution[248]. By linking *theory* with *experience*, it makes a direct epistemic connection between entities in the *scientific domain* and representative images of the *life-world* [10.2] Thus, it is a characteristic *cognitive* process, involving the cerebral manipulation of *mental models* [8.14]. Yet it is *intersubjective*, in that it can be communicated, understood, perhaps disputed, perhaps agreed, between individuals [5.10], and eventually made *communal* property as an archival text [9.3].

The forms of reasoning used by research scientists to interpret their results are not different in principle from those used generally in non-scientific contexts[249]. They vary in detail from field to field[250], and from context to context, for the setting for serious scientific thought, talk and action is not limited to the research laboratory, private study, lecture room or conference hall[251]. In practice, the tacit element of inductive inference in such reasoning is simply taken for granted.

But *formal* scientific discourse is different [3.3]. The norm of communalism obliges every researcher to present her hypotheses within their supporting scaffolding of inferences, so that they can be fully assessed and eventually shared by the research community. This obligation diffuses back into the research process, shaping the formulation of problems and projects [8.2]. Scientists are not always sound logicians[252], and scientific hypotheses cannot be completely constructed or justified by a process of logical deduction [6.6]. But the intersubjectivity of elementary logic [6.6] is a vital communal resource in producing 'public' knowledge[253]. Logical argument is thus a major ingredient in *abduction* or *retroduction* – the rationale of scientific inference[254] – which plays such a crucial role in the epistemic practices of science.

This feature of science is well known. Hypotheses that are well-structured logically and/or mathematically can often be inferred from existing theories, and are relatively easy to test for internal consistency. But our concern here is with their 'external' consistency – that is, how they relate to the empirical 'facts' they claim to represent. Here it seems that strict logic is unhelpful[255].

Take the familiar case of a universal 'law of Nature' generalized from an observed regularity. As we have seen, this type of inference is not logically compelling. Moreover, it is not logically symmetrical. On the one hand, the observation of yet another black raven, yet another white swan, is uninformative. Inductive generalizations are unaffected by confirmatory instances. On the other hand, a single deviation from the hypothetical pattern obviously renders the whole inference invalid. One white raven, one black swan, completely disconfirms the 'invariant colour' hypothesis and expunges it from the communal archive[256].

A scientific hypothesis is not a full-blown theory: it is a *suggested* theory. In due course, no doubt, it will have to stand up to the most challenging tests that sceptical analysis can devise. To begin with, however, a hypothesis is epistemically fragile. Logic is unhelpful because it cannot strengthen it in affirmation, and is too sweeping in negation. This is because formal logic makes no allowance for the ever-present element of uncertainty. When we say that a situation is 'hypothetical' we mean that it is conditional on circumstances which may or may not occur. In scientific discourse, as in ordinary life, a hypothesis is an idea which may or may not turn out to be valid. It is a glimpse through the veil of ignorance that always obscures our knowledge of the future.

Why then bother about the epistemic status of an unproven hypothesis? Why not say that it has no scientific standing as 'knowledge' unless and until it is securely established? The trouble is that no scientific hypothesis ever becomes *completely* secure and above suspicion. This policy is equivalent to the ancient but essentially vacuous philosophy of total scepticism[257]. It would leave all theoretical science in limbo – along with most of the 'facts' with which it is associated.

What is at stake here is our notion of *belief* [10.5]. This cannot be dissociated from *action*. A *scientific* hypothesis (like a new-born baby!) calls loudly for *scientific* action[258] – critical comment, searching tests, confirmatory observations, 'more research', etc. This action is shaped by our epistemic evaluation of the hypothesis – is it scarcely credible, rather doubtful, more or less proven, or what?

What is more, hypotheses seldom come on stage alone. An experienced research scientist is usually aware of several more or less serious theories competing for acceptance in the same field [8.5]. In formulating problems and planning research projects, she not only has to suspend disbelief sufficiently to keep them all in mind[259]: she also has to arrange them in some sort of order of 'credibility'. In accordance with the principle of Occam's razor, for example, she might decide to test the simplest, most likely hypothesis before looking into one requiring the introduction of more complex, speculative ideas. Or, in an adventurous mood, she may knowingly risk complete failure in following up a speculative idea that might lead to a big breakthrough.

Scientific research is rational action under conditions of uncertainty. Although this uncertainty sometimes comes near to the indeterminate Japanese 'mu' state[260] – 'neither yes nor no' – it is seldom utterly blank. The scientific domain of every researcher is structured notionally into what might be called 'levels of belief'[261]. When scientists talk about 'credibility' in relation to the work of a colleague [5.7, 7.3, 7.10] they mean more than her personal sincerity or honesty (which is scarcely even questioned). They are referring to the degree of belief that they attach to her theories and to the 'facts' she presents in their support. In other words, scientists are accustomed to labelling hypothetical entities in terms of their 'degree of credibility'. These labels are seldom made explicit, and are far too vague and variable to determine the course of research on their own, but they undoubtedly play a part in all scientific action, private and public.

This layered belief structure is often revealed in formal reviews of research fields. Indeed, apprentice researchers quickly learn a code of polite euphemisms for publicly indicating extreme disbelief in a wild hypothesis[262]. In their private communications, scientists are notoriously less inhibited. They also emphasize the strong element of gambling in their work [263]. In the language of horse-racing, they are expert punters, with a good eye for form. This shows up in anecdotes of private wagers on the outcome of a highly contested investigation[264]. In the conduct of their careers, they usually back favourites, but are occasionally willing to stake their whole reputation, at long odds, on their personal assessment of a rank outsider.

The practical logic of scientific inference is thus essentially *probabilistic*[265]. I am not referring here to the need for statistical analysis – especially in biological research[266] – to allow for sampling and measuring variability in empirical data. Nor am I suggesting a direct connection with techniques, such as the Delphi method, for forecasting future scientific and technological developments by aggregating the prophecies of expert scientists and technologists[267]. I simply mean that the basic rationale of many familiar epistemic practices in academic science can be formally analysed in terms of *subjective probability*[268]. It can make metascientific sense to think of the 'credibility' of a hypothesis as a number between 0 and 1, and to calculate how this number changes in the light of further information, such as the outcome of a new experiment or observation.

Before going further, it must be said that the very idea of subjective probability is a challenge to rigorous epistemological analysis. Nevertheless, it is a standard component of everyday reasoning, where its use is consistent with the mathematical axioms of the *probability calculus*[269]. In particular, it provides a natural field for the application of *Bayes' Theorem*, which is an algebraic formula that covers the process of inductive inference.

The basic formula is so simple in form that I risk offending anti-mathematicians by writing it down and explicating it:

$$\mathbf{P}(p|q,H) = \mathbf{P}(p|H) \times \mathbf{P}(q|p,H) \div \mathbf{P}(q|H)$$

It concerns the probability for the occurrence of an event p – for example, that my hypothesis about invariant avian colouration will eventually turn out to be valid. At a certain moment, I estimate this to be $\mathbf{P}(p|H)$, indicating that all I have to go on is a body of information H – e.g., that so many thousands of white swans have been seen in the past. Just then an event q occurs – e.g., a black swan swims by. This knowledge leads me to change my belief in my hypothesis. I change my estimate to $\mathbf{P}(p|q,H)$, standing for 'the probability of p, given q and H'.

The theorem states that the change should be proportional to the ratio of two other probabilities. The first of these, $\mathbf{P}(q|p,H)$, is the probability that q might have been expected to occur, given *both* p *and* H – i.e., that my hypothesis had actually been shown (by other means) to be quite valid. This is divided by $\mathbf{P}(q|H)$, which stands for the probability of q occurring, under the same conditions H but *irrespective* of the occurrence of p – i.e., whether or not it is true that all swans are the same colour. The rest is just arithmetic.

Under appropriate conditions[270] – which I shall not try to spell out – this formula is easy to prove, and exact. But these conditions are often wildly unrealistic. Moreover, all the terms in the formula are purely notional. How could one possibly assign a number to the 'probability' of a certain event occurring under vague general circumstances which cannot be specified and quantified? What should be included in 'H'? No wonder that strait-laced metascientists will have nothing to do with it[271]!

Bayes' Theorem can clearly be applied to many instances of scientific inference. But it does not solve the basic problem of induction. Indeed, its direct applications are often very misleading. For example, according to the *Law of Succession*, the probability of yet another 'success' after n successful trials is $(n + 1)/(n + 2)$, and thus tends to near certainty as n increases. But this apparent confirmation of the basic principle of inductive generalization only holds under formal mathematical conditions of 'blank ignorance' which bear no relation to reality[272].

Nevertheless, *Bayesian reasoning* is an effective *heuristic* device in scientific research. This may be because it gives the correct answer whenever an inference can in fact be drawn by strict logic. If, for example, we were testing the hypothesis that *all* swans are white, then the formula shows that the sighting of a single black swan would reduce its probability to zero. On the other hand, if the event q had been the observation of yet another *white* swan, then the formula indicates that this would be a poor reason to make a significant change in our prior belief.

What is more, Bayesian reasoning supports a number of more complex principles of scientific inference that are often considered to be purely 'intuitive'[273]. It helps to explain, for example, how a 'surprising' discovery can often give considerable credence to a disputed hypothesis, or why '*ad hoc*' hypotheses should be treated with some suspicion. It justifies the high value that scientists attach to *consilience* – that is, the 'jumping together' of diverse hypotheses and 'facts' to produce a coherent theoretical system[274]. It can even explicate Ockham's razor, i.e. the apparently aesthetic preference for 'simplicity' and 'economy' in scientific theories and arguments.

Scientists do not regularly deploy *formal* probabilistic arguments in support of their hypotheses. But like everybody else, they employ informal Bayesian reasoning in the same spirit as they use logical inference. Indeed, like everybody else, they are as imperfect as 'Bayesians' as they are as 'logicians'[275] [6.6]. In any case, the art of drawing inferences and assessing hypotheses cannot be expressed entirely in Bayesian terms[276]. In practice, as any post-modern critic would immediately point out, evidence is often 'made' to conform to a preconceived hypothesis, thus inverting the Baconian notion that scientific hypotheses are typically inferred from the evidence[277].

Nevertheless, informal probabilistic argument is a vital feature of the scientific culture. Phrases such as 'it is likely that . . .', 'the weight of the evidence is against . . .', 'these results suggest that . . .' are used freely in research reports, review articles and other forms of public scientific discourse. Admittedly, there is no way of defining 'objectively' the *a priori* probability of a singular event such as the eventual validation of a hypothesis. But then, by frankly accepting this probability as a *subjective* concept, we make allowance for the variable *individual* element in all scientific practice, whether in skilled observation or imaginative cogitation[278]. What is more, informal Bayesian reasoning is so 'universal' that it can be communicated and accepted *intersubjectively*, and hence shared communally [5.10]. It is thus a natural element in any *collective* representation of scientific knowledge, right up to the level of an established fact of life [10.6].

8.11 Prediction

To qualify for consideration as scientific knowledge (however tentative), a hypothesis has to be admitted into the scientific domain of a research community. As we have seen, a hypothesis is not just a 'mini-theory': it is a theory carrying a credibility tag. The numbers written on this tag are very indistinct and are deciphered differently by different members of the community. Nevertheless, peer reviewers and other reputable authorities often agree broadly on the 'probability' of a well-known hypothesis.

Indeed, one of the general characteristics of an academic disciplines is its 'credibility threshold' for hypotheses. Cosmologists, for example, regard it as part of their epistemic culture to entertain – and publish in their communal archives – conjectures whose *a priori* probability is almost zero. As Lev Landau remarked, 'they are often in error but never in doubt!'[279]. In the adjacent discipline of astrophysics, however, not everything goes, as can be seen from the reception of Halton Arp's heretical views on the interpretation of redshifts[280]. And at the other extreme, anatomists insist on overwhelming evidence for any hypothesis that gets into *their* official literature.

The norm of scepticism ensures that much of the action in science is systematic controversy over the credibility of 'facts' and theories [9.1]. Bayesian reasoning [8.10] can never be rigorous, but it is a highly effective *rhetorical* weapon in such conflicts. In particular, it usually gives a clear indication of how a *new* finding affects the credibility of a hypothesis. For example, that anomalous black swan will surely delight the opponents of 'invariant avian colour law', and draw a clutch of highly disputable justifications from its proponents. Friends and foes alike will agree, however, that this observation greatly reduces the *a priori* probability of that hypothesis, and does indeed call for a lot of explaining away.

We are thus entering the traditional heartland of the 'scientific method'. The principle that every hypothesis should be thoroughly 'tested' [5.8] simply means that conscious efforts should be made to change its credibility. Guided by Bayesian reasoning, a scientist can estimate the contribution of a *potential* discovery [8.9] to the evidence for or against a particular hypothesis, and thus be motivated to do research in that direction. In other words, Bayesian reasoning is often the underlying rationale in the choice of problems, the design of projects, the formulation of models, and many other epistemic practices in science.

The most convincing test of a scientific hypothesis is surely the successful *prediction* of a previously unobserved empirical 'fact'. This not only provides evidence of its *practical* reliability, it can also have a dramatic effect on its *theoretical* status. This effect is discounted in formal logic, where only *disconfirming* instances that *refute* hypotheses are deemed to be significant[281]. But it is an elementary exercise in Bayesian reasoning[282] to show that the surprising corroboration of an apparently unlikely conjecture can greatly increase its credibility. In other words, there is nothing wrong with the conventional notion that predictive *verification* [5.8] can be just as important as *falsification* in the production of scientific knowledge[283].

The psychological effect of a successful prediction[284] is obviously strengthened by the fact that one can rule out any possibility that the hypothesis has been contrived precisely to get the observed result. But the Bayesian argument does not really require that the 'prediction' should actually have been made in advance of, or in genuine ignorance of, the observation that confirms or disconfirms it[285]. It applies also to *retrodiction*, or, as we typically encounter it, *explanation* [10.1], where the essential element of surprise resides in the unlikeliness of the proposed connection.

For example, physicists are often pleased to find that the magnitude of an observed phenomenon can be explained theoretically by an elaborate mathematical calculation [6.7]. Such calculations are often 'opaque', and cannot easily be engineered in reverse to infer their initial assumptions from their unobvious conclusions. Thus, it is customary to say that Neils Bohr used his quantized orbit hypothesis to make a very precise 'prediction' of the atomic spectrum of hydrogen, even though this was a well-known physical phenomenon that had been measured accurately many years previously. In effect, the surprising information that valid empirical data can be deduced from a doubtful hypothesis forces us to take that hypothesis much more seriously.

In the right conditions, this *hypothetico-deductive method* can be so compelling that it is sometimes supposed to activate the whole scientific culture. Indeed, it is argued[286] that a scientist should not even entertain a hypothesis from which it is not possible to deduce an observation that might fail to be confirmed. In other words, *potential falsifiability* is held to be the much sought-after *Good Housekeeping* certificate of genuine science.

Interpreted as a methodological maxim[287], this is unobjectionable. Research communities have no place for theories that cannot be *tested* in some way, and a test that cannot possibly be failed is no test at all[288]. But the ordeal of successful direct prediction is too severe and demanding. It would exclude from science all theoretical inferences about the past, such as hypotheses about the origins of life, the earth and the universe. It would not only eliminate all historical disciplines: it would also shut out most of the human sciences, where verifiable prediction is usually so uncertain [6.11, 7.9], and doubtful hypotheses slowly acquire credibility from the success of the practices they inspire[289]. In spite of their apparent inevitability in retrospect, most technological innovations are unforeseen[290]. Even in the biological sciences, the predictability of many phenomena is very limited[291] – indeed, in the case of evolutionary processes [9.8], impossible in principle[292]. In such circumstances, theories can only be considered refuted by reference to elaborate statistical strategies such as the 'non-significance' of the evidence against an assumed null hypothesis[293].

What is more, failure to confirm a carefully calculated theoretical prediction does not entirely refute the hypothesis from which it was deduced. A scientific hypothesis, however tentative, cannot be treated as an isolated proposition. It arises from, is framed by, and is linked to, a wider theoretical environment. No experiment, whether it 'succeeds' or 'fails', is absolutely 'crucial'[294]. What is put to the test is always a whole network of accepted concepts and supposedly established 'facts' [6.4, 6.11], any one of which might be the culprit[295]. What seems unchallengeable may prove to have been a fantasy supported by ramifying chains of facile 'corroboration'[296]. Although experimenters often make the mistake of ending an investigation prematurely when the anticipated results seem to have been confirmed[297], overhasty rejection of a 'falsified' initial hypothesis may be equally counterproductive[298]. And so on, beyond formal analysis.

The point I am making is not that scientific knowledge is built on sand. It is that issues that are usually supposed to arise only later, in 'the context of justification'[299], play an important role, from the very beginning, in 'the context of discovery'. It is not just that these two 'contexts' are intermingled in the day-to-day life-world of research[300]: they are also interwoven epistemically. Scientists try to differentiate in principle between theoretical 'features' and observable 'instances'[301], and even separate 'theorizing' and 'experimenting' into distinct academic disciplines[302]. Nevertheless, just as scientific experiments can only be carried out rationally in the light of the theories they might justify [5.6], scientific theories can only be postulated rationally in terms of the experiments (or other systematic empirical investigations) that might justify them.

The stormy marriage between theory and evidence is the dialectical rationale of research[303]. It must be emphasized, however, that tenable scientific theories cannot be produced just by thinking about how they might be justified. 'Testability' is only one of the epistemic standards by which hypotheses are judged as they come into mind or are presented by others. Theoretical ideas are considered plausible on the basis of a diversity of tacit, even contrary, criteria. Scientists sometimes try to express their theoretical preferences in terms of 'generality', 'specificity', 'simplicity', 'parsimony', 'complexity', 'rigour', 'flexibility', 'symmetry', 'incongruity', 'communicability', 'subtlety', 'accuracy', 'scope', 'fruitfulness', or just 'elegance' [304]. But these are essentially *heuristics* – helpful 'guides to discovery'[305] – not recipes that necessarily guarantee a truly satisfactory product.

From an evolutionary point of view [9.7], the art of research is knowing where to look, and recognizing the value of what you have found, in the absolutely immense 'search space' of possible observations and ideas. Thus, maxims[306], rules of thumb, research strategies, methodological principles, phenomenological theories and other informal 'short cuts' in this space[307] are indispensable features of all scientific paradigms[308]. According to the Legend, they should never be mistaken for strict laws or regulative principles[309]: in practice, such distinct frontiers are impossible to police in the changing landscape of epistemic values where science is actually conducted[310].

8.12 Hypothetical entities

Where do scientific hypotheses come from? This is a question that has been explored endlessly by historians of science. But retrospective analysis of specific cases has not given a general answer. There is no guide to the fruitful notion that may come unexpectedly to mind. On this point, history supports Paul Feyerabend's famous anti-establishment slogan: 'anything goes'[311].

Nevertheless, an experienced scientist usually has a considerable repertoire of promising ideas and 'facts' that can be put together in various combinations to make many different testable theories [8.10]. The task of disposing of those that don't work is usually easy enough: the hard part is to think of a combination – preferably the only one – that really does the job[312]. Doing research often seems rather like solving a jig-saw puzzle, where all the pieces are lying face up on the table, waiting to be picked up and fitted together by the diligent and perspicacious scientist as she makes imaginative hypotheses about the picture that seems to be emerging[313]. Indeed, it is sometimes suggested that 'artificial intelligence' might be used to automate and speed up this tedious process[314].

The jig-saw puzzle metaphor is apt enough at certain stages in the theoretical development of certain sciences. It applies, for example, when fragments of information point to the existence of a hidden mechanism or structure – a universal force, a double helix, a crust of floating plates, a symmetry principle – whose form will surely explain so much. But this metaphor not only commits us prematurely to a strongly realistic epistemology [10.7]. It also implies an unusually well-defined epistemic context. Even in 'normal' research [8.6], the question may be – what puzzles are we trying to solve, anyway, and which pieces belong to which puzzle[315]?

What is more, as we go from the physical sciences through biology to the human sciences, the 'pieces' become more plastic, so that they can often be forced to fit together almost any which way[316]. Human scientists are thus frequently frustrated in their efforts to prove decisively the reality of the 'deep' structures that they have postulated to explain the highest truths[317]. I am not suggesting that such structures might not exist. But we have found no reason to think, as is sometimes asserted[318], that theories involving 'unperceived' entities are the quintessence of science. The task of the theorist in the human sciences is not usually to arrange the known or potentially discoverable 'facts' so as to reveal the hidden monads – the genes, the atoms, the quarks – that run the whole show. It is to construct a reliable and meaningful map of some aspect of reality out of a limited set of stable and significant features selected from the booming, buzzing confusion of the life-world [6.5].

Nevertheless, scientific domains contain many significant entities that are not only quite unlike everyday objects but are also not directly observable by everyday means [6.11]. Scientific knowledge often involves *hypothetical* entities – atoms, genes, quarks, black holes, and so on – which are supposed to govern phenomena in the life-world but which lie outside ordinary perception. This raises widely debated issues. Quantum theorists insist that 'hidden variables' can be excluded on technical grounds[319]. Some extreme *positivists* and *operationalists*[320] have gone so far as to insist that it is 'unscientific' to postulate invisible entities[321], as if these were hangovers from the days when people believed in ghosts, or as if they were mental crutches[322] that would be joyfully cast aside when we had reached a healthier state of understanding [10.8].

One of the characteristics of the social sciences is that they have not succeeded (though not for want of trying) in revealing an invisible world of entities and concepts behind the societal life-world. The realm of 'theory' explored by some post-modern scholars in the human sciences is so disconnected from empirical 'facts', and so far from communally agreed, that it is scarcely credible as a scientific domain. However fancifully named, the categories used by most social scientists are not different in essence from the familiar ones of class, conflict, money, etc.[323]. At most, they create categories, such as 'inflation', or 'exclusion', which conveniently define phenomena that are readily observable in ordinary life [10.2, 10.7]. Nevertheless, for a strict positivist, this is not the point. The most down-to-earth social theorist cannot do without the hypothetical entities – intangible institutions, past societies, long-dead people – required to make historical sense of the present day[324].

The Legend has it that even when such entities are considered to be 'real' – which is interminably disputed amongst philosophers – they are essentially 'scientific'[325]. This means that they enjoy a different mode of existence from everyday objects such as tables and chairs or cats and dogs. To make sense of this distinction, however, one must first look at the various types of entity that actually figure in life-world knowledge [10.6]. These are not limited to relatively solid and enduring objects of 'meso-cosmic' dimensions[326]. They include, for example, long-dead historical persons such as the Emperor Napoleon[327] and intangible cultural entities such as the plays of William Shakespeare. It is not obvious by what criteria these should be distinguished from comparable 'scientific' entities, such as Tyrannosaurus Rex[328] or the Second Law of Thermodynamics[329]. Again, the proceedings of an immaterial social institution such as the House of Commons are customarily treated with the same practical respect as the stone-work of the Palace of Westminster in which they occur [10.7]. Similarly, the arcane events that are interpreted as an experiment in particle physics have to be taken as seriously as the software and hardware of the vast instrument at CERN where they are generated.

What is more, we are long past the days when the dictum 'seeing is believing' implied visibility to the naked eye. Yet the metaphor of 'throwing light on a subject' is as cogent as ever[330], in that the shared human capacity for visual pattern recognition [5.12, 6.2] continues to play a central role in the production of 'public knowledge'[331]. But the lifeworld nowadays includes what we 'see' through eyeglasses, microscopes and telescopes, not to mention remotely controlled video-cameras and baggage-inspection X-ray machines. In spite of the complexity of the instrument and the indirectness of the path of the image from the object to the eye, a virus whose structure can be 'looked at' through an electron microscope can scarcely be regarded as 'hypothetical'[332].

Consider this distinction from the scientific side. The history of science is full of cases where an otherwise unobservable entity has been the key postulate in explaining an observed pattern of phenomena[333]. 'Atoms' could explain the rules of chemical combination, 'genes' could explain the rules of biological inheritance, 'quarks' could explain the rules of high energy particle transformation.

In the course of time, as an explanation gains credibility, so do its constituents. Within a few years, the inferred entities are treated by scientists as normal items of scientific thought[334]. They are manipulated mentally [8.14] and experimentally to interpret or produce other phenomena, until, in the end, they are firmly knitted into the network of accepted theory. It is interesting to listen to experimental physicists, for example, blithely describing an explosion of particle tracks entirely in terms of the interactions between quarks – even though nobody has yet succeeded in observing a single quark on its own[335].

Sometimes, of course, such entities have later been made 'visible' by a brilliant exercise of technical virtuosity. In recent years, for example, the scanning tunnelling electron microscope has shown us single atoms sitting quietly on solid surfaces, just as they have been imagined. But such exercises are of no great epistemological significance[336]. Research is not the sort of detective story that needs a dramatic denouement where the villain is revealed in his true colours.

Good scientific practice requires that hypothetical entities should not be given more complex properties than are needed for them to operate in the particular theoretical context where they were conceived[337]. As time goes on, however, they are often used in other contexts and acquire a variety of other properties. In this way[338], they become less and less 'hypothetical' – that is, more and more credible – until, so to speak, we realize that they also are active characters in the narrative.

What is more, the career of a scientific entity does not necessarily stop when it is no longer in dispute. It may then emerge from the scientific domain and become part of the life-world [10.7]. This is the history of many of the objects and concepts that are taken for granted in everyday modern life. Microbes, extinct species, vitamins, radiation, energy, anticyclones, schizophrenia, etc. left the orderly care and intellectual jurisdiction of the scientific community for the rough and tumble of common use. This *reification*[339] – transformation into a 'thing' – does not seem to have been studied empirically, but the diversity of current meanings for words like 'energy'[340] suggests that it is seldom a sharply defined process. In other words, a naturalistic account of science would confirm the post-modern insight that boundaries such as these are much more permeable than the Legend has taught us to believe.

8.13 Constructivism

In exploring the norm of originality, we have mainly assumed that academic science is activated by individual scientists seeking personal recognition for their contributions to public knowledge [3.8]. But an account that ignores collective effects is clearly incomplete. Science is a *social* practice, undertaken through a whole variety of formal and informal institutions – research groups, networks, learned journals, specialist communities, and so on[341]. Indeed, the norm of scepticism operates almost entirely through such institutions [9.3].

In this sense, science can properly be portrayed as the 'social production of important public stories'[342]. But should one go further, and describe these stories as social *constructs*? This question rouses strong passions. *Constructivism* is appealing as a direct counter to naive scientism, but is often presented just as dogmatically [1.2].

These two views of science have become competing 'strange attractors' for every metascientific issue in the neighbourhood. Working in this neighbourhood, I am continually aware of the danger of leaning too far towards one side or the other, and falling into the corresponding academic black hole. That is why I have deliberately avoided any attempt to survey the literature that has developed on each side – especially where this consists mainly of Byzantine tracts against rival sects within the same closed intellectual universe.

And yet here (to change the metaphor) is precisely where a just and honourable peace process is needed, to put an end to the 'science wars'. I could, perhaps, have postponed discussion of this question until nearly the end of the book, and then issued a declaration of non-partisanship. But it is important to introduce it here, because it opens up a number of interesting areas of analysis, and yet often leads into what many would see as a semantic (or ideological?) dead end.

The point is that when academic science is treated holistically as a 'social institution' – which it undoubtedly is – the controversy tends to move up into the sociological stratosphere. The argument turns on whether, quite generally, everything that is 'produced' by any such institution ought not to be described more properly as 'constructed' by that institution. In other words, the question is answered, one way or the other, by reference to an over-arching sociological principle which is far above empirical demonstration or even, I suspect, complete explication. I am not suggesting that Olympian debate over the notion of 'pan-sociologism' is academically fruitless, but only that it can never be concluded in a way that is likely to satisfy either side in more specific conflicts.

Our naturalistic approach requires us, rather, to take a worm's eye view. We must look inside academic science and study the way in which knowledge originates and is 'processed' as it becomes established. It is quite clear that this processing takes place, from the word go, within a social context[343]. This is not to diminish the vital role of the individual human mind [5.9, 8.14] in the initial conception and formulation of problems, projects, inferences and hypotheses. But we cannot perceive the simplest object without 'putting a construction on it'[344]. Even the highly personal trait of originality has to operate within a nested hierarchy of institutional frames. Does this mean that what comes out of such operations is not only 'shaped' by this context [7.5], but is actually *constructed* socially?

One point is clear. In standard English, 'to construct' is nearly synonymous with 'to assemble'[345]. At every stage in their development, scientific ideas are 'put together' from parts drawn from the whole intellectual and practical environment of their producers. Famously, Darwin incorporated his knowledge of animal breeding into his evolutionary hypotheses, and Einstein brought the everyday experience of weightlessness in a falling lift into his theory of General Relativity. Even the strictly 'scientific' elements in such theories have 'social' origins, since they come from bodies of knowledge shared and attested by various scientific communities. And the process by which these parts are assembled is not completed until the whole 'construct' has been presented for communal acceptance.

Interpreted thus, science is surely 'socially constructed'[346]. Why then all the bother? To my mind, some of this arises from the multiplicity of meanings for the word 'construct'[347]. The thesaurus lists other near synonyms, such as 'to design', 'to manufacture' or 'to fabricate'. These remind us that 'construct' is an active verb, normally used with a suggestion of *intentionality* – that is, a mental act in which we direct ourselves to something in some way [348]. It implies that the social machinery that 'produces' knowledge does not run automatically, but is guided, albeit by an invisible hand, towards rational ends.

Academic scientists strenuously deny the existence of any other guidance than the objective nature of the world as it is. They thus find this imputation of a hidden purpose very perplexing. The list of synonyms extends into a domain of ambiguity where there is plenty of space for misunderstanding[349]. For example, does 'construction' include 'fabrication'? Professional scientists take great pride in being 'honest seekers after truth'. The bare assertion that a scientific discovery has been 'constructed'[350] is thus very offensive. In everyday language it immediately suggests that the work in question is not authentic and hints at bad faith [9.4].

Constructivists continually insist that that this is not what they are getting at. But their repeated claim that they are simply using a technical sociological term, without cynical moral implications, comes close to sounding disingenuous. This is not to deny that cases of deliberate fraud or partisan misrepresentation sometimes occur in academic science. But there is no evidence that these deviations from the ethos are so widespread as to require 'unmasking' by a generalized insinuation of systematic professional hypocrisy.

A less provocative interpretation is that the guiding hand is invisible because it is the unconscious instrument of general social forces, such as gender, ethnic and economic interests. In other words, 'constructivism' focusses on the means by which the norm of disinterestedness is unwittingly subverted. It implies that scientists are influenced subliminally to construct theories that favour the interests that sustain and bound their social environment [7.5, 7.6].

Nevertheless, having peeled away the marginal meanings (including the entirely different reference to 'children's personal constructs' in educational theory[351]) we are still left with a significant core. To put it simply, constructivism challenges the Legend by raising the basic question whether scientific knowledge is 'found' or 'made'. From a naturalistic perspective, this may be viewed as an empirical question about scientific practice. But for the traditional philosophy of science it is often the keystone in an argument for or against scientific realism [10.7], and is therefore given great epistemological weight.

Looking back over previous chapters, we cannot avoid the conclusion that the naturalistic answer is that scientific knowledge is both found *and* made.

On the one hand, academic science gives great weight to the empirical 'findings' of research [5.2–5.8]. These are not always entirely novel or unsuspected, but many of them certainly are. Very elaborate technical and social procedures are used to check, counter-check, replicate or otherwise authenticate these before they are accepted communally as 'facts'. Scientific knowledge thus contains much that has been 'found'.

On the other hand, in its striving for universality, academic science requires the formulation of theories [6.4–6.11]. These, again, are not always novel or unobvious, but many of them certainly are. Very elaborate technical and social practices are employed in testing and criticizing these before they are accepted communally as reliable 'laws' or 'explanations'. Scientific knowledge thus contains much that has been 'made'[352].

In the face of these well-known features of scientific practice, constructivists argue that what scientists call 'facts' – even 'objects'[353] – are all essentially 'constructs'. They can never be fully authenticated without the exercise of human agency. They are often hypothetical entities [8.12], whose existence is inferred by tenuous chains of reasoning from the outcome of carefully contrived experiments, incorporating a great amount of theory[354]. This is perfectly true, and often overlooked by scientific positivists.

But this total defence of constructivism cannot be accepted beyond a certain point. Taken to its logical conclusion [10.7], it leads to a metaphysical idealism where everything in the world, 'scientific' or otherwise, is a 'construct'[355]. Along the way, it glides over the natural phenomenon of contingency – the unpredictability, the uncertainty of outcome – that is central to scientific discovery [8.9].

Conversely, anti-constructivists would have it that what scientists call 'theories' are the causal relationships or intrinsic properties that are 'found' along with the material entities – often previously hidden – that underlie our direct experience of the world. For example, they are often tracked down and confirmed in astonishing detail by elaborate experiments that deploy great quantities of other hardware [5.5, 5.6]. In the course of time, initially diverse and arbitrary hypotheses are whittled away or seen to converge until an unequivocal map of this aspect of nature is revealed. This also is true, even though it is often ignored by constructivists.

But this attempt to crush constructivism is also not very helpful beyond a certain point. It merely expresses a metaphysical utopianism where nothing in the world is other than the way that science will ultimately discover it to be [10.6–10.8]. We cannot wait until that joyful day. For all practical purposes, all the knowledge coming out of science must continue to rely heavily on concepts and techniques made by people as fallible and venal as ourselves.

These arguments and counter-arguments do not just cancel each other out. They reinforce the view that science is a genuine amalgam of 'construction' and 'discovery'. Like the rest of life, it marries intention with contingency. These are often logically distinguishable[356], but their union is too close and intimate, at the level of mundane practice, to be dissolved by philosophical edict. Embarrassing as this may be for the doctrine of scientific objectivity and value neutrality[357] [7.1], it is in the nature of scientific knowledge to be a mosaic of 'made' and 'found' pieces, right down to its deepest foundations.

This is not to decry metascientific enquiry into the sources of particular scientific ideas. It is obvious, for example, that the balance between 'discovery' and 'construction' can vary markedly from discipline to discipline. Generally speaking, the natural sciences have evolved in an epistemic environment that favours 'discovery', and have adapted to its rigours by rejecting all problems tainted with 'subjectivity'. The human sciences, by contrast, survive through the methods they have developed for making sense of the baulky products of personal intention and/or social action[358]. But as we see in intermediate disciplines such as psychology and geography, this contrast is one of academic practice, and is not based upon a sharp dichotomy of epistemic principle.

At this point, the reader may ask how I can take such a timid, tepid, aloof attitude towards the conflicting claims in this fierce, hot battle. Why don't I come down off the fence, or at least set them out in detail? Let me be quite straightforward. The whole purpose of this book is to get a clearer view of the nature of science, over which there is such bitter conflict [1.2]. What I am saying here is that neither side can make anything like a convincing case for exclusive occupation of the disputed territory. The strength of science is tolerance of difference of opinion [9.2]. The arguments either way are quite insufficient to justify epistemic cleansing of a culture where 'discovery' and 'construction' have cohabited tacitly for centuries. On the contrary, they point towards a positive analysis of how this partnership operates, and can be made even more fruitful.

Until the 1960s, metascientific thinking was dominated by the Legend, which depicts all the action in science as 'discovery'. The broad intellectual movement loosely covered by the term 'constructivism' (or its near associate, 'relativism' [10.7]) came as a welcome challenge to this one-eyed vision. At last it was realized[359] that in many fundamental respects science should be seen as a social institution, organized to yield a social product.

This insight was resisted because it was thought to threaten mortally the practical realism of the scientific culture[360]. As we shall see, this threat is greatly exaggerated[361]. Nevertheless, constructivism, even in the 'weak' form[362] that I am presenting here, has opened up a whole new dimension of epistemological enquiry. It is now clear that the nature of scientific knowledge could not have been established simply by 'philosophical' (or even 'psychological') analysis of the actions and beliefs of isolated individuals. *Sociological* analysis is also required to take account of their *collective* labours and representations.

The 'sociological turn' in science studies has influenced almost every page of this book. It prompts the naturalistic perspective on scientific practice as a cultural form [1.5] and the epistemic exploration of the institutional norms of the scientific community [3.2] – the least discussed yet most potent sociological feature of academic science[363]. But it should not be interpreted as a turn away from material or rational considerations[364]. In particular, centuries-old questions about free will and determinism[365], or about the rational grounds for belief in science[366], are not answered at a stroke by proclaiming that everything is 'social'[367]. On the contrary, it is the apparently spontaneous convergence of the testimony of many independent witnesses [5.7] that gives solidity to our individual experiences[368]. As I have been emphasizing in this section, the 'world' is ever with us: we live together mainly by our *shared* perceptions of it, and not entirely by self-serving fantasies[369].

Nevertheless, the constructivist critique strikes to the heart of traditional notions of scientific originality[370]. Even the most novel hypotheses normally emerge from, and are sustained by, strong networks of 'established knowledge' [8.1, 8.10]. New science almost never appears out of an epistemic 'empty quarter'. It usually arises in an intellectual territory that is already colonized – often multiply – by academic disciplines, research specialties, theoretical paradigms and other social structures [8,4]. It must also make good its claims to a socially acceptable place in these structures – which may be impossible if it involves displacing a paradigm that is very deeply entrenched[371]. In other words, even as an item of abstract theory, it has to be, or become, 'socially structured' if it is to fit into the overall corpus of 'public knowledge'[372]. Even the most conventional histories of scientific discovery can be read instructively in this light.

Some constructivists – especially those who emphasize the influence of established social interests such as male domination [7.5] – interpret this as conformism. Science, they would say, is a dogmatic system which only accepts new findings that accord with its already-received doctrines. But remember that these doctrines were not laid down by a higher power. They are the communally accepted and ordered outcomes of previous research. Their social authority derives from the universality with which they are respected in practice by active scientists.

A new research finding, once accepted, becomes part of the very structures that gave it meaning in the first place[373]. In doing so, it changes, positively or negatively, the credibility of the items it was intended to test [8.10]. Indeed, even by formulating a 'problem' a social scientist may effectively generate 'interests', such as communal values and preferences [7.10], which were previously unrecognized by those who were found to hold them[374].

At the same time, however, a new finding strengthens the knowledge structures that are 'taken for granted' [10.3] in designing and interpreting this test[375]. For example, the interpretation of a satellite image may be

strongly disputed by the parties to a geophysical or ecological controversy, each trying to build it into their own theoretical construction of
what actually happened in a certain region. Yet by their agreement that
such images do constitute valid evidence in this dispute, they add further
social authority to a research technology which was itself 'constructed' in
the course of similar disputes a few years back.

As Otto Neurath famously remarked[376], scientists behave like 'mari-
ners who have to rebuild their ship on the high seas, without ever being
able to strip it down in dock and construct it afresh from the best available
components'. Faulty planks are removed, one at a time, mended or
replaced, and put back in approximately the same place. Each step in this
process of continuous self-assemblage[377] is strongly determined by the
structure around it, and yet, in time, a completely new structure is produced.

In more sophisticated sociological terms, scientific knowledge
changes and grows [9.6] by *structuration*[378]. This is the means by which, it
seems, social structures typically reproduce themselves[379]. Even the
most knowledgeable and self-conscious individuals are strongly constrained by custom and precept to perform the roles and carry out the cultural practices required of them as members of a particular society [10.4].
But their doing so re-enacts these customs and re-validates these precepts, thus sustaining the very structure which constrains them. This is
usually thought of as a means by which cultural traditions can be preserved almost intact for many generations.

In the structuration of science, however, every researcher must not
only integrate the work of their predecessors and rivals into their own[380].
Each such cycle of 'deconstruction' and 'reconstruction' also requires the
addition of genuine elements of originality and scepticism. This is all
that is needed to turn it into an *evolutionary* process, giving rise, in due
course, to novel epistemic structures which were never 'constructed' as
such [9.7, 9.8].

8.14 What do scientists have in mind?

Scientific originality reveals itself socially, but its wellsprings are individual. *People* are at the junction points of the interpersonal networks that we
call scientific communities, research teams, communication systems,
academic institutions, etc. [5.10]. These nodal elements are not just

repeater stations and information packet switching centres. They receive messages from many sources, combine them, transform them, and transmit them in new forms. A naturalistic account of science must surely acknowledge that a great deal of the action is *mental*: it goes on in the separate *minds* of individual scientists [5.9].

Metascientific attention has mainly concentrated on the very small proportion of cases[381] where the new messages were sufficiently novel – and also proved later to have been sufficiently interesting[382] – to indicate high 'creativity', or even 'genius'. Strictly speaking, even the most creative scientist cannot be considered an 'idea generator', in isolation from the intellectual and social networks of which she is an active element[383]. In spite of much folk theorizing and anecdotal evidence to the contrary, the talents and commitment required to do 'good science' do not seem to be associated with any particular psychological make-up[384] [2.7]. Nevertheless exceptional scientific creativity is obviously fascinating and has attracted much study.

Unfortunately, the results of these investigations are inconclusive. From a computational point of view, for example, creativity appears to involve the systematic exploration and transformation of enormous 'conceptual spaces', guided by heuristics such as 'generalize', or 'specialize', or 'consider the negative'[385]. But there is little evidence that scientists (like chess masters) actually use anything like such brute force algorithms in their creative thinking[386].

Again, the role of 'sceptical misfits'[387] and other socially marginal researchers who 'break through barriers of conformity' and 'make new connections' is often stressed[388]. But it turns out that specialists who have built up large repertoires of relevant 'themas'[389] [6.10] are often just as innovative as 'migrants' who bring in new habits of mind and new ideas from other fields[390].

What we really learn from such studies is that exceptional scientific originality is merely an extreme form of a normal human capability. Millions of scientists, usually quite ordinary people, operate quite adequately as individual nodal elements in scientific networks. From day to day, they acquire information from their research instruments, from their colleagues, from the scientific literature, from the life-world, and from recalled experience. They regularly select and recombine this heterogeneous material, and assemble it in essentially new forms which they transmit to other scientists. What they do 'in their heads' as a matter of routine in everyday scientific practice may not always seem very clever, but no form of 'artificial intelligence' can yet match it [6.3, 6.6]. Although some scientists seem to talk like robots, the Turing test would soon suss out a real computer masquerading as a human researcher.

Needless to say, various practical tools of thought, such as writing, measuring, calculating and computing, have great enlarged this capability in particular directions. But the nature of science is still largely shaped by the natural human capacity for *cognition* [5.9]. This does not mean that our present inquiry must come up with a model showing how it is that people can 'understand', 'remember', 'imagine', 'assess', 'infer', 'attend to', 'communicate', etc., etc., the multiplicity of things that they do. Science employs these familiar cognitive operations in the living forms they are given to us, not as the skeletal simulations of cognitive science or the fossilized abstractions of high philosophy. Even a systematic formal analysis of what human minds actually do to the information on which they operate would take this whole inquiry back to square one, since it would inflate a whole universe of epistemology just to contain its infinitely elaborated explications.

In our naturalistic account, then, we do better for the present to rely on the folk psychology embodied in common usage than on the findings of neurobiology[391]. In particular, we need to look at the modes in which information is normally 'cognized' in the course of scientific practice. For example, we know that these are subject to biological constraints. The normal human brain is limited in the rate at which it can take in a coded message, in the number of nested clauses it can grasp as a whole in a grammatical sentence, and – except perhaps for a few mathematical weirdos – in the number of space-like dimensions in which it can imagine a hypothetical structure [6.4]. These limits can be stretched by training and experience, but can only be transcended by sophisticated devices that transform the desired information mechanically into intelligible forms.

On the other hand, the human capacity for *symbolic* communication is unrivalled. This takes place typically through the medium of *natural language* [5.12, 6.2, 6.6, 6.10]. In effect, the brain receives information in a coded form, where the words in no way resemble the entities they represent. These messages are evidently decoded mentally, combined, manipulated and transformed, before being recoded for onward transmission. As we have seen, under appropriate conditions this establishes the intersubjectivity that makes scientific knowledge a communal resource.

This 'information technology' model of cognition would seem to apply to a whole range of specialized symbolic and technical 'languages', including algebra [8.7]. One of the fundamental features of the scientific culture is that research communities develop these languages and learn to use them amongst themselves [5.1, 5.12, 6.6, 6.11]. Physicists talk 'physicsese' about the interactions of electrons, quarks, etc. Chemists talk 'chemistese' about valence bonds, oxidation states, etc. Sociologists talk 'sociologese' about ascribed roles, social constructs, etc. And so on. In each case, individuals understand the special words that represent the particular facts and concepts commonly used in their research community, and are able to use this understanding to make new statements in the same language.

So far, so good – but what actually goes on when a scientist uses her 'understanding' of previously received scientific statements to produce a new scientific statement[392]? Can we say anything useful about the mental operations that intervene between hearing 'there is a blip on the output indicator' and exclaiming 'that must have been a top quark'?

In some instances, perhaps, these internal mental operations are conducted entirely in a regular symbolic language such as algebra. There are, presumably, logicians and mathematicians who are able to manipulate their formulae mentally according to the rules they would have to obey in a formal proof. It is well known, moreover, that every competent user of a natural language can generate any number of novel statements that are grammatically correct transformations of one another[393] [6.6]. But a computer that performs verbal operations without any sense of what they are about cannot replace a genuine scientific mind[394]. (Or is that how we get the portentous strings of gobbledygook that are apparently accepted as scientific discourse in some corners of academia?)

In other words, 'mentalese' – the private medium of thought – surely cannot be just 'other words'[395]. It must, of course, be translatable into a public language[396]. But it seems to me[397] that the elements of mentalese must also be structurally similar to the entities they represent, so that most mental operations automatically have counterparts in the 'external' domain to which they refer[398]. To think creatively, we have to be able to manipulate our thoughts about specific objects, relationships, concepts, people, etc. according to the same rules that those objects, etc. would have to obey 'outside our heads'[399]. Given 'here is a bird' and 'here is a cage', I cannot afford to pause to consult a dictionary before manipulating these terms mentally so that they combine naturally into the new message, 'bird in cage'. This obvious life-world possibility must already be inherent (somehow!) in the way that I represent these entities in my thoughts.

To put that more concisely, 'mentalese' must have a major *iconic* element[400]. This is borne out by the role of visual communication in science [5.12, 6.2]. Diagrams, pictures, abstract maps and material models play a much greater part in scientific practice than is allowed for in the Legend[401]. A great deal of scientific information is transmitted and received in modes that can be transformed directly into mental images. The patterns, connections and other structural relationships that show up in such images – their implicit 'logic' – can reasonably be attributed to corresponding features of the real world[402].

What is more, this is not just a static process of 'recognition'[403]. The powerful role of 'visualization' in scientific thought[404] has long been appreciated amongst naturalistic metascientists[405]. This power is sometimes considered a sign of genius, but it is only an enhanced version of a natural human capability[406], linked to our neural machinery for sensori-motor coordination[407]. Without conscious effort – sometimes independently of language[408] or even of normal vision[409] – people are able to 'imagine' transformed versions of an original scenario – the hand rotated to fit into the glove, the landscape viewed from another angle, the grotesque mask as the face of a living demon, the sequence of events run backwards in time[410].

Indeed, recent developments in cognitive science[411] suggest that much of our ordinary thinking is performed through the subconscious manipulation of *mental models*[412] [5.12, 6.4, 6.10, 10.1], rather than through the processing of symbolic codes. For example, when we encounter the word 'in', we automatically project an schematic image of an entity into a schematic box, and interpret the situation in terms of that visualized relationship. Thus, our ability to understand abstract logical consequences or contradictions is linked to the mental mechanisms by which we grasp the spatial relationships that we encounter in the real world[413]. In effect, the Venn diagrams used by logicians to illustrate propositions such as 'all *s*'s are *p*'s', etc., etc. have hidden counterparts in everyday 'mentalese'.

This is the ability we use to build up an internal 'map' of our immediate environment[414] [6.5, 6.11]. We can thus, for example, envisage and plan our movements in the rooms of our house, the fields of our farm, the streets in our town. Psychological research has shown that many other animals can do the same. It is obvious that we are here dealing with a product of biological evolution. Organisms acquire brains that enable them to adapt to the complexities and changes of their environment by acquiring detailed knowledge of it[415]. Indeed this basic cognitive mechanism is so distinct that it can be specifically affected by damage to a particular region of the brain[416].

Our sense of the reality of the life-world is connected with the realization that our own internal 'map' is so similar to the corresponding 'maps' used by other people that these must all be just variants of a general 'map' that we seem to hold in common [10.7]. Indeed, quite unsophisticated people have no difficulty in making and understanding a material map that represents this tacit social construct. What is more, once one has learnt the conventions of a system of cartography – the symbols that stand for roads, forests, rivers, etc. – one can 'read' a public map of an unfamiliar region, and locate oneself in it. The whole art and science of making and using maps depends upon this general human capability, even though people differ in the ease with which they can actually carry it out in practice.

Now take this perfectly ordinary cognitive phenomenon one stage further. Recall one's first visit to a foreign city. To start with, one has to consult the street map at every turning. But gradually, with further experience of its reliability in practice, this public map is 'internalized'. That is to say, we learn to use our memory of it to guide our movements, just as if it were our own private 'mental map' of our environment[417]. We may even report back to the publishers that a particular road has been given a new name, and that the map should be amended accordingly. In sociological terms, a two-way structuration process [8.13] develops between an individual 'personal construct' and a collective 'social construct'.

It is thus very clear why so much scientific knowledge is communicated and stored in the form of pictures, diagrams, maps and models [5.12, 6.10]. These are not just powerful media for the transfer of information into, out of, and between human minds[418], and thus for the 'networking' required for a collective product. They seem to be the media of cognition itself. A scientific theory presented in the form of a model is already well formed for reception and manipulation inside the 'black box' of the living brain. Even before it is fully incorporated into the public 'map' of a scientific domain[419], it becomes a tentative part of the private 'map' which every scientist carries around 'inside their head'.

Scientists must have such 'maps' to guide them in their research[420]. In some scientific disciplines – particularly physics[421] – these have become so extensive and elaborate that they give the impression of constituting a comprehensive, coherent, collective 'world picture'. But this is an illusion [10.8]. Such maps are not only grossly simplified[422] representations of different aspects of the life-world, as variously experienced by diverse research communities. They are also very seldom as consensual as they claim [9.2]. Despite their public credentials, their private versions vary markedly from individual to individual[423]. This is particularly true in the human sciences. Each of us builds up from infancy a mental model of our social environment [10.3]. Although this model must, of course, be a variant of a locally current version of social being, it will inevitably be highly personalized. Yet from it we have to give meaning to the abstract collective representations of a science such as sociology.

The map metaphor for individual cognition[424]does not solve the riddle of scientific originality. A scientist concentrating on a serious research problem cannot read off the answer from any existing 'map'. She has to have in mind a diversity of other 'facts' and concepts, of varying degrees of credibility. They may be represented mentally by a heterogeneous collection of free-floating 'schemas', 'scripts', 'frames', 'directed graphs', 'relational databases' and other 'list structures'[425], clustered into loosely linked, open-sided 'microworlds'[426]. These do not yet – may never – fit neatly together. Perceiving in them a map, and inferring from them a model, is the name of the game[427].

But cognitive science is beginning to demystify 'intuition'. Constructive imagination is not irrational, even when it cannot be followed step by step in a logical sequence[428]. A long-standing acquaintance with a particular field of knowledge makes one familiar with its underlying structures[429]. Expert scientists not only know what is going on in the domain of their research; they also map this domain internally with complex features which are usually quite different from those shown on the official maps[430]. These features often have previously hidden connections. Not suprisingly, the recognition of such a connection – sometimes below the level of conscious attention – is seen by the outsider as an 'inspired guess', or a mental 'short cut'. Scientific intuition of this kind obviously cannot be taught as such, but it is a natural product of basic education, wise training and reflective experience in the practice of research[431].

Scepticism and the growth of knowledge

9.1 The agonistic element

The final norm of academic science [3.7] is *organized scepticism*[1]. This sounds like a philosophical doctrine, but is not a call for total doubt[2]. The metaphysical notion that we cannot *really* know *anything* is not incompatible with being a scientist, but is so general and abstract that it has no more impact on scientific practice than it does on other aspects of life [8.10, 10.5] . Again, scepticism has psychological overtones, favouring a 'questioning' attitude, akin to 'curiosity' [2.7]. This attitude is as necessary to scientific progress as personal 'creativity', although it must not be confounded with a conservative stance that automatically rejects every new idea[3].

But its real force is sociological. 'Scepticism' is a code word for those features of the scientific culture that curb 'originality'. Personal trust is an essential feature of scientific life [5.7]. But scientific communities do not accept research claims on the mere say-so of their authors. The active, systematic exercise of this norm by individual researchers is what, above all, makes science a communal enterprise. 'Peer review' is the key institution of the scientific culture[4].

We have already noted a number of the ways in which this norm indirectly shapes scientific knowledge. To be considered 'scientific', a 'fact' or theory has to satisfy a number of general epistemic criteria, such as reproducibility, logical consistency, independence of the observer, etc. These are essential conditions for communal acceptability. Researchers are well aware that their research claims will be subjected to highly sceptical appraisal. They therefore design their research projects, from the beginning, to produce results that satisfy these conditions[5].

Again, many scientific practices are motivated by the desire to strengthen (or weaken!) the credibility of particular observations or hypotheses [5.6, 8.10]. Indeed, research is performed in an all-pervasive atmosphere of scepticism. Scientists are driven to quantifying data, replicating experiments, seeking empirical facts to refute conjectures, testing theories against their predictions, and so on [5.8], just to survive[6].

In principle, *organized* scepticism is the responsibility of the 'gatekeepers' to the formal literature[7][9.3] . The editors and referees of reputable scientific journals take this responsibility seriously, and regularly reject contributions that do not satisfy their standards. Although these standards are often very vague, and vary widely from journal to journal and from discipline to discipline, they set benchmarks of credibility in the research communities they serve. An empirical fact or theoretical inference cited from a peer-reviewed scientific publication may be presumed to have stood up to at least this degree of expert scrutiny, even though it is not guaranteed to be 'true'.

Indeed, as all scientists are fully aware, much of the knowledge codified and stored in the official archives [9.3] is far from convincing. The critical process, being distributed over a whole community, is unsystematic and of variable quality, so that very weak results often slip through the net. Many mistakes and misconceptions [9.4] are not challenged until they affect the interpretation of other research[8] – and even then may never be publicly corrected[9]. Thus a great deal of what is asserted, quite sincerely, in the official literature will later turn out to be quite wrong[10].

The decision to accept a research claim for publication does not make it immune from further scepticism. Indeed, the credibility of an item of scientific knowledge could never be settled by any such decision, however officially pronounced. That is one of the reasons why journal referees are customarily anonymous. They have to take decisions affecting the interests of their peers under conditions of such extreme uncertainty and potential bias[11] that they are bound to be open to question on some point or another. Every experienced research scientist has a stock of anecdotes about the disputes that can arise between would-be authors and zealous referees.

Indeed, such disputes are all part of the action. Thus, authors often revise their papers in the light of comments by sceptical referees[12]. The most important manifestation of the norm of scepticism is not selection but *criticism*[13]. When scientists communicate with one another, publicly or privately, formally or informally [3.3], they do not just transfer information from mind A to mind B; they engage in *dialogue*[14]. The scientific culture is an institutionalized context for *argumentation*[15]. A scientific community is an agonistic field, where researchers cross verbal swords over the significance of each other's claims[16].

This argumentation takes place in many different fora, ranging from research group coffee clubs to international congresses. It may involve face-to-face conversation, personal correspondence[17], electronic networking[18] [5.13], public debate, or printed exchanges of opinion in a scholarly publication[19]. Although such exchanges are often ostensibly triggered by the publication of a particular book or paper, this may be no more than a distinctive episode in a long-running saga of informal and formal dispute[20]. In Big Sciences such as high energy particle physics, these critical interactions mainly occur inside research teams, before their findings are disclosed[21]. In the human sciences, where researchers work more on their own and consensus is not the only show in town[22], they usually take place in print, typically as scholarly reviews of recent books or articles.

Scientific knowledge, then, is as much the product of argument as it is of observation or cerebration. But what does that signify epistemically? Certainly, research claims cannot be accepted as communal property without a series of critical negotiations. Surely, the revisions produced by such negotiations make research findings more 'valid'? But since the ultimate standard of 'validity' is communal acceptability, this is a truism.

The norm of scepticism does not impose an independent epistemic requirement. It merely strengthens and makes more explicit the criteria by which research claims are assessed in any particular research community[23] [6.8, 8.11]. Indeed, when scientists argue they often make direct reference to general metascientific principles – objectivity, empiricism, logical consistency, 'falsifiability', etc. – which would otherwise be taken for granted[24].

Bitter experience reminds us, however, that the value of criticism depends enormously on the spirit in which it is offered and taken. Constructive argument requires shared commitments to goals and basic principles, respect for the other's standpoint, the capacity to rehearse opposing views, and above all, desire for mutual understanding rather than victory[25]. In principle, these virtues are fostered by the various social practices associated with the norms of communalism, universalism and disinterestedness. Since their origins in the seventeenth century[26], these practices have been cherished by scientists as the operational framework of a 'republic' or 'commonwealth' [3.1] where their intellectual disputes can be staged without external interference or internal social strife. It may well be, indeed, that scientific communities remain somewhat more open – for arguments of their particular kinds – than most other arenas of public debate [10.5]. But is this all that we can reasonably expect, and all that we require, of science as a social institution?

One must agree with many observers of the scientific culture[27] that its standards are far below those of an 'ideal speech community'[28]. For example, numerous cases can be quoted of scientists who have behaved 'unprofessionally' – even as formal referees[29] – in order to get their ideas accepted, or at least widely discussed[30]. But even an ill-posed and inconclusive debate about an undecidable conjecture[31] can make a positive contribution to the scientific enterprise. This is often how new methods and values are diffused[32], boundaries marked out[33], orthodoxies challenged[34], methodological assumptions uncovered[35], agreements and disagreements about basic principles clarified[36], unforeseen problems revealed[37], or research efforts redirected[38]. Noisy public dissension is more often a sign of vitality than of social or intellectual chaos[39].

Nevertheless, the real test of an epistemic institution is how it handles an intellectual disagreement that has become socially polarized into a *controversy*. It may be that most such controversies are just social magnifications of grubby personal rivalries[40]. But there is a less cynical interpretation. According to the Legend, disputes should be about findings, not persons[41]. In reality, however, scientists identify themselves personally, and are identified by their peers, with the ideas they have originated. Just as the reputation of a scientist declines if her ideas lose credibility, so their credibility declines if they are not wholeheartedly defended by their progenitor[42] [7.3]. The promotion or rejection of a new scientific idea inevitably becomes a cause in which other scientists are entrained.

In Big Science, where research projects are long and research teams large, various procedures have evolved to paper over, or play down, interpersonal and intergroup conflicts, whose eventual outcome may be presented by a rapporteur speaking for the whole 'collaboration'[43]. A multi-authored scientific article is usually the polished product of informal, rough, reciprocally critical exchanges between its principal authors[44]. But passionate, even dogmatic advocacy is as essential to science as cool, rational analysis[45]. Once a disagreement has gone public, it goes against nature – human nature – to suppose that people should not line themselves up on one side or another, form opposing 'schools of thought', support 'friends', try to thwart 'opponents', and so on. To describe such phenomena as 'pathological'[46] denigrates human solidarity, and sets an unattainable standard of communal or cognitive 'health'.

Scientific controversies are often referred to as 'games'[47], but the rules are sometimes so ill-defined that military metaphors seem more appropriate[48]. And as in many real wars, their historical origins are sometimes so trivial and accidental that the apparent bone of contention – for example, the sex life of the whiptail lizard[49] – means little to outsiders[50]. To give an idea of their diversity, here is a sample of the topics that came up when I searched electronically for 'controv...' in my notes for this book: the origins of hydrocarbon fuels[51]; the value of the gyromagnetic ratio of the electron, and the nature of cosmic rays[52]; the authenticity of reports of cannibalism[53]; the dating of the human settlement of the Americas, and the effects of diet on human behaviour[54]; Africa as the original source of AIDS[55]; the connection between HIV infection and AIDS[56]; the theory of biological evolution by natural selection[57]; the place of the Devonian in the geological record[58]; the oblateness of the sun, and the role of hidden variables in quantum theory[59]; respiratory chain systems in mitochondria[60]; the chemical transfer of memory, ether drift, cold fusion, Pasteur and the origins of life, gravity waves, and the missing solar neutrinos[61]. These are just some of the issues on which genuine differences of scientific opinion have erupted into open controversy.

The trouble is that these examples of controversy have so little in common. If you study them more closely, you find that some are short and sharp whilst others rumble on for centuries; some are about general principles whilst others are about specific observations; some are about the credibility of marginal phenomena whilst others are about the scope of central theories; in some situations a single protagonist faces the world, whilst in others a research community is evenly split into contending 'schools'. In some cases – especially in the human sciences – controversies that seem purely 'academic' actively engage powerful material interests or value systems in society at large; in other cases, extra-scientific affiliations are concealed under a cloak of missionary zeal for 'good science'[62].

The course of every scientific controversy thus depends on its personal and institutional environment. And as more knowledge accumulates around the point at issue, it usually tends to become more and more localized epistemically. In the end, everything seems to hinge on the credibility of particular 'facts' and theories in one or more very small regions of the scientific domain[63]. Thus, only someone thoroughly familiar with those regions can make out what is going on – with the attendant risk of themselves being drawn into the battle[64].

Controversies are so prevalent in academic science that they cannot be said to breach its social system[65]. What makes them instructive is that they reveal and define 'normal' scientific practices by stretching them beyond their customary limits. Unfortunately every controversy is 'abnormal' in a different way. It presses on different norms and activates different psychological and social forces amongst its participants.

In my view, scientific controversies, like other features of the scientific culture, cannot be understood if they are approached entirely empirically. Like other social phenomena, they need to be located within the sort of analytical framework that I am trying to present in this book. But the fact that the study of scientific controversies has played an important part in developing this framework does not mean that these should be the sole focus of metascientific analysis.

Indeed, just as obsession with controversy often distorts the lives of individual intellectuals[66], so it distorts our image of the culture in which they live. The effect is to replace the myth of a super-rational community that 'knows no force but argument' by an equally specious myth of a sordid power struggle between protagonists who 'know no argument but force'[67]. Emphasis on the sadly familiar, quasi-military manoeuvres and quasi-political ploys that are characteristic of all social conflicts diverts attention from many practices that are peculiar to science. As we have seen, some of these practices foster competition and social fragmentation, whilst others favour cooperation and social cohesion[68].

There is an alternative perspective in which even the most violent scientific controversies can be seen as taking place within a pluralistic environment where there is tacit agreement on fundamentals[69], where dissenters and dissenting activities are systematically tolerated[70], and where even the most radical and persistent heretics are not ostracised[71]. These contrary perspectives are not mutually exclusive: they are two sides of the same coin.

From a naturalistic point of view, the most striking general feature of scientific controversies is the language in which they are conducted. Although an academic community can be quite as disputatious as any ancient Greek *polis*, a modern scientific debate no longer resembles a political or legal *agon* in the *agora*[72]. Many of the traditional modes of polemic – invective against persons, accusations of unworthy motives, appeals to authority, dramatic figures of speech, and so on – are very rarely used in public. Indeed, except perhaps in book reviews, they would be considered quite pathological and almost certainly counterproductive[73]. A knowledgeable audience can, of course, hear the anger and scorn behind the euphemistic exchanges. Everybody knows that when a scholar writes 'it is not clear that...' she really means 'what utter nonsense!', and so on[74]. Lists of such 'interpretations' are standard items of professional humour amongst scientists.

Although not enforced by a Speaker or Judge, this public courtesy obviously facilitates open criticism[75]. An affectation of disinterested humility [3.5, 5.1] is an essential part of the scientific role[76]. But even in a heated controversy, scientists use the austere impersonal language of formal scientific communication, and systematically understate their claims and counterclaims. Can they really be so inhuman that their words are untinged with mental manipulation and propaganda[77]?

Paradoxically, this unassertive, low-key tone is itself a mode of *rhetoric*[78] – that is, a metalinguistic form[79] skilfully designed to *persuade*. Look at it this way. For a research scientist, the object of the exercise is to produce findings that are acceptable to a research community. The final stages of acceptance involve publication in the scientific literature [9.3]. Any debate about the credibility of a research result must eventually revolve around how it might be presented in the archive[80]. By formulating her claims – or counterclaims – as if they had *already* been accepted, she has gone a long way towards persuading us that they are indeed acceptable. We begin to think of them as 'taken for granted' items of knowledge [10.3] even before they have been through the mill of communal criticism[81].

In any case, how else can scientists communicate clearly with each other about their ideas except in the language of their culture? 'Scientific' concepts are defined in terms of their operations in the corresponding 'scientific' domain [6.11]. This is a world from which all 'subjective' features have been excluded. 'Scientific' entities can only be discussed 'scientifically' after they have been 'objectified' – that is, taken out of the life-world contexts where they were found, and to which they supposedly refer[82] [5.3, 6.1]. This process of separation is, of course, open to debate[83]. But from then on people can argue about them, in private just as in public, only in the language where they have meaning, that is the technical language of a research community[84].

Thus, the language of public controversy is often a valuable source of information about the epistemic principles of a branch of science. It is noteworthy, for example, that such controversies are seldom about the absolute 'truth' or 'validity' of a particular research finding – whatever that might mean! They almost always involve competing claims to the same territory on the knowledge map[85]. In other words, they are usually about the *relative credibility* of apparently incompatible hypotheses [8.10].

Thus, a scientific controversy is typically a display of 'Bayesian rhetoric'. Each side strives by every means to produce arguments that raise the credibility of their own theory and lower the credibility of their opponents'. Many of these arguments are specious, although not necessarily 'dishonest'[86]. Scientists, like lawyers and politicians, learn to 'wriggle' argumentatively (as Darwin put it) [87], retreat into technicalities[88], exaggerate the expertise of their peers[89], treat as well established their own previous interpretations of doubtful propositions[90], and so on. As in other arenas of public debate, much of the action is devoted to anticipating, detecting and exposing such devices as they occur in specific cases.

Nevertheless, scientific reasoning differs from other forms of argument because it is fundamentally Bayesian [8.11]. In most cases, for example, it is more powerful in attack than in defence. That is why disputing scientists often search diligently for small items of evidence against hypotheses they disagree with, rather than making the larger effort to obtain evidence confirming their own ideas[91]. Although Bayesian logic does not make any distinction in principle between 'refutation' and 'corroboration'[92], this asymmetry is very evident in the practical rhetoric of scientific controversy.

9.2 Consensus – or just closure

According to the Legend, science is particularly good at putting an end to controversies. Overwhelming empirical evidence, or irrefutable logic, will eventually settle the matter, one way or the other, beyond all reasonable doubt. On this view, the final objective of research is a paradigm governed by a closed theory[93]. Indeed, a disputed knowledge claim should not be considered 'scientific' unless it is capable – in principle at least – of being tested to 'closure'.

But science is not really like that. As we have seen, it is a Bayesian system where there is no way of terminating scepticism and criticism. This is particularly difficult in the human sciences[94]; but even the most comprehensive 'falsification' is never final [8.11]. Experiments [5.6] advance the argument but are never definitive[95]: logic [6.6] has moral force[96] but never covers all the eventualities: even the goodness of the fit between theory and experiment is in the mortal eye of the beholder[97].

The openness of academic science to criticism is one of its major social strengths. Public controversies are health warnings against dogmatism [10.5], and often draw attention to unsuspected risks. Of course, 'good science' is usually pushed aside when genuine cognitive uncertainties are enlarged and enlisted by contending societal forces [7.9]. But this is the price that post-academic science must be willing to pay for its place in the post-modern world.

Nevertheless, the fundamental inability of science to arrive at complete certainty is a grave social embarrassment. It seems to license people like Creationists to dismiss its most secure findings on the grounds that they are 'controversial'. Even more seriously, it tempts scientists themselves to withdraw from active responsibility for the immediate consequences of their knowledge claims. When pushed for an opinion, they can say that there are some differences of opinion which can only be resolved by more research. The testimony of many an expert witness has been discredited by a reluctant admission that the moon could just conceivably be made of green cheese, for all that we *really* know about it scientifically.

In fact, the demand for closure comes mainly from outside the academic world. Many people use scientific knowledge in taking practical decisions [2.4, 7.2]. They would like to feel that this knowledge is absolutely reliable – unlike the distressingly uncertain information on which they mostly have to guillotine debate and take action[98]. Nobody expects a group of lawyers, politicians, theologians or doctors to have identical expert views. But any outward sign of disagreement amongst scientists is taken as a grave weakness.

This is a major source of confusion in public debates on social issues involving science[99]. It is often suggested, for example, that such 'transepistemic' issues[100] [7.10, 8.2] could be resolved more easily if there were a 'Science Court' to determine the 'facts', as agreed by the scientific experts on both sides. But matters of fact are seldom what is really at stake. Such an institution would, of course, exclude the very people with the most relevant expertise, such as economists and other social analysts – scholars whose 'scientificity' is continually being called into question because their professional opinions on life-world issues are often so contradictory[101] [7.10].

We arrive here at another paradox. A science can be distinguished from other modes of knowledge production by its striving for the maximum area of *consensus*[102]. But this consensus must be a *voluntary* one, achieved under conditions of free and open *criticism*. These conditions are enabled and regulated, more or less effectively, by the CUDOS norms [ch.3]. The norms of communalism and universalism envisage eventual agreement. But the norm of 'organized scepticism', which energizes critical debates, rules out any official procedure for closing them. Consensus and dissensus are thus promoted simultaneously[103].

This paradox is resolved by the cunning of *habitus*[104]: social practices are so arranged that the contradiction never has to be faced. For example, it is no longer thought appropriate to set up an official tribunal – even under the aegis of an august National Academy – to settle a scientific controversy[105]. Good science is not made by majority verdicts[106]. Discredited claims are never *killed*: they simply *fade away*[107]. Although their supporters usually learn to live with the dominant view, they are not expected to recant in public[108]. Indeed, some of them soldier on for decades, refusing to admit defeat[109]. If they are as eminent as Linus Pauling, Fred Hoyle or Peter Duesberg they may even continue to be given space for their 'heresies' in the formal literature, despite the fact that nobody seriously believes them[110]. The custom then is for other authors to ignore them completely, or to cite them perfunctorily, without bothering to express negative opinions that would only re-ignite a fruitless controversy[111].

The epistemic factors that operate in these situations are extremely varied. Ideally, they ought to make any alternative seem 'incredible'. But this can never be absolutely assured. The 'crucial experiment' that seems to clinch the matter for all time may be a rational reconstruction visible only in the historical rear mirror[112]. The bottom line for the 'truth' of a scientific proposition is its unquestioned acceptance by a research community[113]. But the general criteria by which this is eventually achieved cannot be defined precisely. In practice, they are interpreted flexibly, and vary considerably from discipline to discipline[114]. Every working scientist soon discovers that the archival literature is actually full of inconsistencies left over from disputes that were simply abandoned because they could not be argued to a convincing conclusion[115].

Research communities do have socio-cognitive processes by which 'reasonable agreement' is reached on the acceptability of disputed research claims[116]. But these processes can be so prolonged and can require such delicate social negotiation[117] that they are often almost invisible. A change of communal thinking may thus become established tacitly[118], even though the points on which agreement is supposed to have been reached are so vague that they cannot be listed definitively[119].

Thus, a coded comment in an authoritative review or comprehensive treatise may be the only public signal that a particular research finding is now beyond dispute. By this time, however, research designed to refute it will have ended[120], and almost everyone will be treating it as 'well established', with only an occasional reference to its chequered past[121]. In the 1960s, for example, most geologists began to use Alfred Wegener's ideas on continental drift as if quite oblivious to the storm of controversy that had surrounded them for half a century[122]. Modern biologists do not refer to the vast literature on the successive stages by which Charles Darwin's evolutionary hypothesis eventually became part of the unquestionable canon of biological theory[123]. These doubts have now become matters of academic history, for which natural scientists care much less than their humanist colleagues[124].

Scientific communities do reach effective consensus on many of the facts and theories within their competence. Henceforth, these facts and theories may well be considered as cogent and reliable as any knowledge so far granted to our tiny human minds [10.7]. But this is a practical, working consensus, not a solemnly affirmed doctrine. It is expressed through the way that researchers use this knowledge to breed new experiments, new techniques and new theories[125]. It is incorporated into the scientific domain of their thought and action, so that a quark seems almost as substantial as a golf ball and a DNA sequence more meaningful than a telephone number.

But even in the natural sciences, this operational consensus may conceal a wide range of views on what it is that is actually agreed. A scientific concept cannot be formulated unequivocally [6.6]. The pronouncements of the 'established authorities' and the officially sanctioned definitions in the textbooks are conventional fictions that may be more superficial than they appear[126]. Their formal language simply cannot convey all the uncertainties, ambiguities, *caveats* and 'anomalies' within which even the most secure scientific ideas are embedded.

Amongst physicists, for example, there has long been complete,

unchallenged agreement on the validity of classical mechanics, thermodynamics, relativity theory and quantum theory. And yet the official scientific literature continues to accept numerous perfectly serious papers
and books offering alternative, often highly contentious accounts of how
these mathematically well-posed theories should be understood.
Formalists argue that these are merely different interpretations of a consensual core, but their every attempt to get agreement on how to define
such a core rigorously merely leads to further dispute. The situation in
the biological sciences is, of course, even less definite. Almost every writer
on evolution, for example, produces a different list of what 'every biologist believes' – especially when they want to reconcile this with the words
of Charles Darwin or more recent worthies[127].

This is not just a problem with grand theories [10.8]. It applies right
down to the smallest details of fact or theory. The members of a specialist
research community will usually insist that there is a consensus amongst
them on a great number of points. Yet their individual lists of these are by
no means identical! Indeed, it is not uncommon to find several apparently inconsistent theoretical models peacefully co-existing within the
same subdiscipline[128], not as rivals but as partial representations of distinct aspects of the same whole.

As we approach the human sciences, we find more emphasis on refining
the terms of debate than on consensus as such[129]. The academic map of a
discipline is quite often subdivided into 'traditions' and 'schools' which
offer directly competing interpretations of the same basic material[130].
There is a tendency, moreover, for each of these to turn into a 'verification
circle' for its own views, rather than working for convergence with other
views through sincere 'falsification dialogues' with its rivals[131].

Given the nature of the subject matter, this pluralism and polycen-
trism are not unexpected[132]. It is not just a matter of keeping criticism
and controversy within manageable bounds[133]. A scientific account of
human behaviour is necessarily hermeneutic [5.11]. That is to say, its
meaning can only be established intersubjectively, and shared by
members of a research community, through empathy – their common
experience and understanding of the human condition. Such meanings
are so vulnerable to alternative interpretations that it is very difficult to
bring any argument to a thoroughly convincing conclusion[134]. The
attempt to formulate widely acceptable generalizations merely generates
abstract theoretical systems – in classical economics, for example – that
have lost touch with the life-world[135].

What is surprising, rather, is that the perceived consensus in some human science disciplines is not all that small[136]. The reason is, perhaps, that what the members of a research community feel they share is not a doctrine but a *paradigm* [8.4] – a whole bundle of epistemic practices that they believe to be efficacious in the production of reliable knowledge. In some disciplines, the preferred approach is experimental: in others, theory is dominant. In some fields, only quantitative instruments and mathematical formulae are considered trustworthy; in others, hand-written texts and eye-witness testimonies are accepted.

The shared practices characteristic of a scientific specialty are not just methodological. Commonalities of problem choice, of educational tradition and of relations with practice span the culture of a specific discipline and shape the outcome of research. Thus, the notion of a 'consensus' is not entirely objective. In summoning up thoughts of what is shared in a research community, it evokes a feeling of solidarity, a spirit of mutual trust [5.7], which surely plays a vital part in the credibility of science[137].

9.3 Codified knowledge

Over all, academic science is a social institution for the *collectivization* of knowledge [4.6]. Individuals are encouraged to acquire information and understanding, and to transmit them to the 'archives', where they become a *communal* resource. As I have tried to show throughout this book, there is no unique 'philosophical' formula for what should count as 'scientific knowledge'. But within the scientific culture – and to some extent in society at large – this term is conventionally restricted to the accumulated contents of this archive[138].

Of course, working scientists possess individually a great deal more knowledge of a scientific kind than gets into the official literature [5.8]. Nevertheless, public archival knowledge is not just a collection of carefully selected and smoothed lumps of personal knowledge, codified for easy reference. It is a distinct epistemic *institution*, with a number of crucial functions[139]. And like any such institution, it has attracted more than its fair share of criticism and cynicism.

Ideally – and this is as far as the Legend goes on the subject – the scientific literature is a repository of valid scientific knowledge. This knowledge is presented in *didactic* form, ready-made for use in research, practical application, education or edification[140]. Thus, scientific progress [9.8] can be directly measured by the growth of the archive, both in quantity and quality, and science policy [4.9] can be based upon *sciento-metric* analyses of the numbers of authors, citations or research sites that clusters of papers have in common[141].

In reality, the actual material stored in the archives is extremely heterogeneous in style and substance. In particular, the research findings claimed in the primary literature are of varying credibility [8.11]. This cannot be determined from the way they are presented. Thus, competition increasingly drives researchers to publish 'preliminary' findings – typically in the form of 'letters' or short contributions to 'conference proceedings'[142] – which are not always later justified[143]. But the other current trend, towards longer and longer papers, conscientiously reviewing all the previous literature and ostensibly integrating a few new results into an accepted theoretical model[144], does not necessarily produce more reliable knowledge. In either case, specialist expertise is required to assess its credibility in the light of what is currently believed[145].

Expertise is also required, of course, to interpret the turgid, technical jargon. The notorious opacity of the official scientific literature is neither surprising nor remediable. This literature is designed for the professional use of research communities, not for public edification[146]. This is why academic scientists usually find it so hard to make their ideas intelligible to the public at large. They are accustomed to writing as if addressing the most knowledgeable and critical experts in their field. This is entirely different from the job of the science journalist, which is to inform people who know almost nothing of the subject[147].

A century ago, when this mismatch was less marked, most working scientists could explain their work reasonably clearly to non-specialists. Not any more. Even in the human sciences, where some scholarly books find still wider audiences[148], most of the official literature is now just as recondite as it is in the natural sciences. And the well-known but relatively few scientists nowadays who are good at 'popularizing' science [3.3] are not given much academic recognition for their peculiar skills. In effect, they are seen as actors in the layer of specialized institutions, professions, social roles, publication media and styles of communication that has developed to fill the widening knowledge gap between 'science' and 'society'[149] [4.11].

'Expository science'[150] has a range of societal functions, from solemnly advising governments to lightly entertaining the general public. In recent years, it has escaped the embrace of the scientific community to become a social institution in its own right. Instead of garnering scientific knowledge reverentially from the formal research literature, journalists, political activists and even educationalists try to get it at first hand, from the mouths of the researchers who have produced it. Academic scientists naturally deplore the weight thus given to research claims that have not necessarily been cleansed by peer review. But much about the nature of science, and especially the innate rhetoric of scientific argument, is conveyed more effectively in an unscripted TV interview or policy debate than by a carefully prepared written text[151].

Indeed, research specialists often find they can only communicate with one another across disciplinary frontiers in an informal 'expository' mode[152]. But this is only one of the media through which knowledge is diffused within and between research communities. In addition to peer-reviewed reports of original research findings, the official archives of science themselves contain a great variety of 'secondary' literature where these findings are conscientiously catalogued, abstracted, surveyed, criticized – or, most significantly, ignored.

What is more, these documents are linked together by *citations*[153]. In principle, all research results are original, and can stand up for themselves. But in spite of their repudiation of the voice of authority[154], scientists must always rely heavily on previously published facts, concepts and methods. All such prior knowledge has to be authenticated by reference back to an archival source. In effect, this requirement closes the epistemic circle around the norm of communalism [ch.5]. Research scientists must not only contribute all their findings to the communal store but also draw from it all the scientific knowledge they use in their research.

The etiquette of citation is not precisely formalized, and varies from discipline to discipline[155]. But it is an obligatory practice that performs a number of different functions throughout academic science. For example, citations have a rhetorical role in the fomulation of project proposals [6.2] and the presentation of research results. They not only indicate possession of the expert knowledge required of a member of a research community. A show of humble deference to past achievements – and especially to eminent achievers – also enlists moral support for new claims[156].

Citations operate sociologically to 'recognize' researchers individually for their contributions to knowledge [3.8]. Indeed, one of the features of post-academic science is the notion that the standing of a researcher amongst her peers can be assessed by counting 'hits' in the *Science Citation*

Index[157] – an annual catalogue of all the *new* papers that have cited each *previous* paper stored in the archives. This attempt to quantify the 'impact' of research findings [5.4] has many very obvious defects[158]. Nevertheless, it shows that the measure of the credibility of an item of scientific knowledge is its actual use in scientific practice [9.2].

In fact, citation indexes were originally invented to facilitate *information retrieval*[159]. The archives of modern science are of Byzantine complexity. Even the most elaborate classification schemes [6.3, 8.3] cannot provide pointers to every scrap of knowledge that might be applicable in a particular context. But items that are highly relevant to a present-day research report can sometimes be found by going back to the earlier papers that it cites, and then looking at the subsequent papers that have cited *them*. Thus, for example, a pharmacologist working on a particular drug may go back to the paper where the chemical formula of the compound was first determined, and then use a citation index to search for other references to this compound, whose uses may be much more diverse than he had supposed.

The issue here, however, is not whether these assessment or retrieval techniques are effective in practice. Nor is it whether citations are reliably comprehensive, appropriate, objective, worth counting, etc., etc. It is that by linking the scientific literature backwards and forwards in time, citations make indirect connections between all the antecedents of a discovery, and also between all its consequences. In effect, clusters of 'co-citations' are material manifestations, within the archives, of the conceptual networks of theories, data, techniques, etc. that span scientific domains [6.4, 6.11]. For example, when researchers in different subfields of physics all cite Heisenberg's paper on quantum mechanics, they show that all their different theories belong to a network of ideas in which Heisenberg's theory is a major node. Again, we need not stop to ask whether the study of such clusters is really the best way of getting to know about the emergence and evolution of specialties, disciplines and paradigms. The essential point is that the scientific archives have this intrinsic *textual* structure that parallels the *epistemic* structure of the knowledge they contain.

But of course it is their epistemic structure that makes the scientific archives more meaningful than a mere stockpile of 'facts' and concepts. As we have seen, a new research problem, project or finding is always presented within a context of explicit or tacit knowledge [8.1]. A scientific paper customarily begins with a 'review of the literature', depicting this context and locating the new work in it. Concepts, hypotheses, empirical data, uncertainties, theoretical critiques, credibility debates, etc. are drawn from the archives and rationally articulated in a form designed for acceptance as the current consensus[160]. This 'structuration' process [8.13] both preserves and reconstructs what it claims to be describing. At the same time, it offers to the reader a 'mental map' [8.14] of a specific locality of the scientific domain.

Unfortunately, such maps are often highly idiosyncratic. In the *primary* literature of a research specialty, almost every author seems to be using a slightly different version. The role of the *secondary* literature [3.3] is to set out the common features of these maps and to suggest how discrepancies might be reconciled – or, very often, conveniently ignored. A bottom-up hierarchy of 'progress reports', review articles, treatises and textbooks then extends this process over wider and wider domains of knowledge[161].

This secondary literature merges seamlessly into 'expository science' – for example in encyclopaedia articles. Nevertheless, it is still part of the formal communication system of academic science. *Synoptic* texts by responsible scholars are extensively cited by researchers, and are as vital to the scientific culture as primary research reports. And yet their production is not directly fostered by the academic ethos. In the natural sciences, wide-ranging review articles and monographs are valued as communal resources, but are not counted as 'original' contributions to knowledge. In the human sciences, where research findings often appear first in book form, the writing of textbooks may be financially rewarding, but earns little professional credit.

This is largely a symptom of specialization [3.9, 8.3]. It is rare for a research scientist to get to know enough, outside a very narrow field, to understand and write about a wider area from first-hand knowledge. And suppose that she has gone outside her acknowledged sphere of expertise, what 'peers' are there to review her work, except nit-picking specialists in those other fields into which she has foolishly ventured? There is very little career incentive to spend years on a project which is likely to expose one to damaging criticism and loss of personal academic credibility.

But there is a broader epistemological issue. A 'map' covering a wide area cannot be precise in every detail [6.5]. Even a general overarching paradigm, such as Darwinian evolution or quantum mechanics, has to be interpreted in terms of particular facts. The intellectual cartographer must select the features to be included, and depict them in broad brush strokes that gloss over their complexities, uncertainties, and heterogeneous sources[162]. In other words, synoptic writing involves a great many personal choices, construals and simplifications which are not indicated clearly to the reader, and whose subjectivity is irreducible.

For this reason, the different research communities that might be covered by such an overview are reluctant to accept it as communal property. They simply do not have the critical resources to eliminate the personality of the author and render it completely 'objective'. Indeed, in the human sciences, research claims based on subjectively tainted 'secondary sources' are often considered highly questionable. The customary method of presenting the knowledge in the archive on a wide canvas is a 'symposium' where individual specialists are invited to survey their own bits of the action. The resulting publication usually reads like a motoring atlas, where there seems no connection between the roads at the top of page 53 and those at the bottom of the adjoining sheet, on page 77!

Nevertheless, in spite of their deficiencies, the scientific archives are a remarkable social institution. Getting knowledge into print was obviously a practical prerequisite for a scientific culture[163]. Printed 'inscriptions'[164] are indispensable for sharing accurately and storing cumulatively a mass of 'memories' far exceeding anything that we could have or hold as mortal individuals[165]. Indeed, the technology of the written word is the vital factor. Symbolic codification in a standard form is so essential for scientific communalism that it is difficult to imagine anything like academic science in a non-literate society.

There is a deeper factor. The emergence of *language* transformed human cognition by enabling us to represent our thoughts explicitly to ourselves in strings of words. More recently, our cognitive capabilities have been further transformed and enlarged, without biogenetic change, by what seems no more than a convenient cultural device. The invention of *writing* permits thoughts to be frozen in time, scrutinized at leisure, and stored in new forms. It provides each one of us with an 'external memory field' that we use as an additional workspace for our personal cognitive operations[166]. In learning to read and write, the literate individual extends and adapts her personal 'mental domain' [8.14] so that it is able to accommodate the abstract theoretical structures typical of modern science[167]. Conversely, these structures are shaped by the cognitive capabilities released by this powerful social technology[168].

Let me emphasize that *all* human action requires a 'mental map' of the physical and social environment [10.3]. Even in the most traditional non-literate society, this 'map' is usually so much the same from individual to individual that it can be treated as a cultural invariant. But as anthropologists have amply demonstrated[169], it does vary markedly from society to society [10.4]. If there is any meaning to the concept of 'the world as it really is', then this should be termed *World 1*, to distinguish it in principle from any of the various ways in which it is represented in *World 2* – the world in which people mentally enact their experiences[170].

The scientific archives contain a collectivized, symbolic representation of World 2. The information codified in books, libraries, maps, computer databases, etc. belongs to a distinct epistemic 'world', since it exists independently of the memory or cognitive powers of any particular individual[171]. But this *World 3*, as Karl Popper called it[172], obviously includes more than science [10.5]. Academia is concerned with other organized bodies of knowledge, such as religious creeds, political doctrines, legal codes and literary texts[173]. It is very hard to demarcate these from less systematic forms of 'public information'. As experience with the Internet all too clearly demonstrates, World 3 extends seamlessly into all corners of modern life. It ranges from computer algorithms to cookbooks to poetry, from international treaties to newspapers to party manifestos, from engineering blueprints to advertising posters to stained-glass windows, from recordings of folk music to symphony concerts. Indeed, as anthropologists and archaeologists skilfully demonstrate, public meanings can be 'read' in the artefacts, buildings, institutions and social practices of every culture, past and present. In general terms, World 3 is a universal natural product of language-using humanity[174].

Symbolic social interaction is obviously of enormous epistemological and cognitive significance. Its evolutionary emergence, current modes and enduring mysteries are the subject-matter of an immense literature, spread through all the human sciences. But is there any more to it, from our point of view, than an enhanced technology for the capability that makes science possible – intersubjective communication [5.10]? This capability is too fundamental to be explained in more primitive terms. At the deepest level of principle, it is the same whether scientists communicate face to face, by word of mouth or blackboard formula during a provocative research seminar, or through the wall of death, in the pages of a century-old book stumbled on whilst browsing the dusty shelves of an obscure library.

From a naturalistic point of view, of course, the information technologies of writing, printing, depicting, photography, electronic data-collection and archival retrieval are essential to academic science [5.12]. These not only enable vast quantities of 'facts' to be acquired and processed for communal use. By presenting abstract theories and general classification systems as World 3 entities, we separate them from the multiplicity of individual minds and particular circumstances where they were conceived[175]. As 'knowledge without a knower'[176], such entities seem to be free of subjective influences, and can even evolve independently of any external forces[177]. Although complete 'objectivity' is unattainable, it remains one of the principal goals of the scientific culture. Many World 3 operations, such as the elimination of errors and inconsistencies in the scientific literature and its consolidation around general paradigms, clearly work towards this goal[178].

The scientific archives are orderly areas of cultivation in the forests of World 3. Metascientists still have much to learn from studying the sources of their epistemic fertility[179]. But they should not be worshipped metaphysically[180]. To treat their contents as autonomous obscures their human origins and connections. In particular, it ignores the dynamic interaction between World 3 and World 2.

The scientific domain within the mind of each researcher [8.14] is very largely an internalization of the supposed communal consensus, just as this consensus is supposed to be a representation of what is in the minds of all researchers. But this correlation is always imperfect, and is always being disturbed by the practices prescribed by the norms of originality and scepticism. Active research scientists are continually contributing novel findings to the archives or incorporating its changing theoretical structures into their own thoughts. The *cognitive* power of science stems from this synergy between individual and collective modes of 'knowing'[181].

More pragmatically, the formal archives are no more than repositories for the research findings accepted as publishable by research communities. Quite apart from their many uncertainties, errors and omissions, they do not include the unpublished contents of science – that is, the tacit, informal, technical and life-world knowledge required to write or read a scientific communication [5.8].

To put it hypothetically: even the wisest little green creatures from Alpha Centauri could not make sense of the scientific literature on its own. To interpret it they would have to tap into the World 2 scientific domains in the minds of specialist researchers – that is, they would have to enter into dialogue with experts, like other lay persons. But they would also need direct experience of how we humans ordinarily live and do things here, on 'the third rock from the sun'. This is self-evident for the human sciences, where empathy is the key to all understanding [5.11]. It applies right through the natural sciences, up to the boundaries of pure mathematics if not beyond.

There is a more profound objection to giving the World 3 version of science a privileged epistemological status. The notion of 'knowledge without a knower' is not just a paradox that can be resolved by reference to potential readers, temporarily frozen communications, and so on. It is an antinomy that implicitly defines the concept of 'knowledge'. The terms that cluster around that concept – terms such as 'belief', 'truth', 'reality', etc. – cannot be interpreted without a context of *action* [10.5]. This context may be hypothetical, but it is what gives them meaning[182]. By regarding the actor as superfluous to the act of 'knowing', we would exclude further discussion along these lines. In effect, we would be licensing a metaphysical position equivalent to the naive positivism from which we are trying to escape[183].

9.4 Getting things wrong

Scientific practice engenders a mind-set towards 'getting things right'. People assume that the measure of 'good science' is infallible correctness. Scientists feel that the worst thing that could happen to them professionally would be to make a mistake and to have it found out[184]. The sociological norms and epistemic principles of academic science combine in requiring them to deplore errors, and to detect and eliminate them as vigorously and rapidly as possible[185].

And yet the formal research record is full of erroneous 'findings', even about easily reproducible 'facts'[186]. In principle, these should generate obvious theoretical and/or empirical inconsistencies[187]: in practice, such inconsistencies often go unnoticed for long periods, and sometimes even open up false paths to genuine knowledge[188]. The persistence of published mistakes – usually minor, but occasionally astonishingly large – is one of the curses of scientific life.

But non-scientists should not be too scornful about the failure of scientists to live up to their ideals of perfect accuracy. It is all too easy for the non-specialist to disparage the difficulties of doing 'good science'. The norm of originality requires research to be conducted at the very margins of technical feasibility [8.1]. Those elaborate protocols and incredibly sensitive instruments [5.5] have not had the benefit of thousands of hours of road testing and routine use. The quest for CUDOS [3.8] requires researchers to undertake 'hopeless' experiments, to generalize from ill-prepared or non-representative samples, to infer causation from correlation, and generally to take other methodological short cuts[189] simply to get anywhere at all. If a research report had to be copybook perfect before it could be published, then what would it be copying? Only a philosopher[190] can indulge in the 'rational reconstruction' of what was, at the time, a thoroughly opportunistic, serendipitous, unpredictable enterprise.

One of the obvious sources of false information in the scientific archives is, of course, deliberate *deception*. The scientific culture depends fundamentally on personal honesty and mutual trust [5.7]. For this reason, deliberate *plagiarism* – the expropriation of genuine research results – is almost as reprehensible as their *fabrication*. And yet, in spite of peer review and other safeguards, it is relatively easy to get fraudulent research claims into the literature, and to profit from them careerwise for a while. Such cases, when found out, stimulate much institutional turmoil and public comment. The contrast between their condemnation as instances of grave social deviance and the relatively lenient sanctions applied to those who perpetrate them tells us a lot about the internal sociology of research communities.

But what is the *epistemic* influence of a scientific fraud? Does the fact that it is deliberately designed to deceive make it more damaging than an unintended error? This is a matter of dispute. On the one hand, it is held that this is a major issue, especially in modern biomedicine, where the empirical results of research are often represented by very complex – and expensive – datasets[191]. Certainly, nobody should doubt that people have been directly harmed by the falsification of test data, usually to bolster the claims of a practical procedure or influential theory[192]. Nor should one underestimate the corrupting effects of intellectual crimes such as Cyril Burt's faked data on the intelligence of identical twins, or the 'Piltdown Man' hoax[193].

On the other hand, it is argued that dishonest research claims, like other 'wrong' results, are eventually weeded out by the normal workings of the scientific culture[194]. A skilfully fabricated 'discovery' – i.e. convincing 'findings' that conflict with communal expectations – may earn temporary esteem[195], but will eventually be unmasked[196]. In any case, scientists generally assert that the incidence of downright fraud is very low, and that it is inconsequential by comparison with the unconscious bias, partisan manipulation of data and controversial misrepresentation that already distort 'what is known to science'[197].

It may be that such unprofessional practices as the 'massaging' of data, and the 'cooking' of results, are considered less acceptable nowadays than in the days of the zealous amateur, even though the 'authorities' have not the will or competence to deter them[198]. But a random 'audit' to determine the prevalence of minor modes of scientific 'misconduct' would not only be extremely expensive[199], it would also fall into a legalistic morass. Research communities do not drawn precise boundaries around 'acceptable ' epistemic practices [5.8, 8.11]. The criteria for 'falsified science', like those for 'good science', are field specific[200]. Highly publicized cases of outright fraud provide little guidance in the many situations where even the question of intentional deceit can only be answered in relation to the precise personal, institutional and cognitive context[201]. Indeed, the metascientific study of 'deviance' seems to teach us less about human motivation than about the unstated epistemic norms of research[202] and the on-going social structuration of knowledge[203] [8.13].

But the norm of scepticism is not just a reminder that scientific knowledge is a human product, inevitably permeated with error. Although it affirms the importance of seeking out and correcting such errors by the best available means, it is not just an injunction to 'get things right', as if addressed to the printers and proof-readers of a charmingly erratic newspaper. Nor is it an over-riding metaphysical doctrine of vacuous generality [10.7], asserting that nobody could ever 'really know' anything, however hard they tried[204]. At its deepest, it is essentially a *moral* principle, directed against the sin of pride. Having negated the possibility of obtaining completely sure and certain knowledge through science, we invert our metascientific values by affirming *fallibilism* as a positive virtue[205].

But this is not a total inversion. It does not mean that all supposed 'truths' are equally 'relative', and so 'anything goes'. On the contrary, it is a naturalistic stance that underlies our whole concept of rationality[206]. It urges us to seek the very best knowledge we can, whilst insisting that what we think we have found should not be treated as quite *incorrigible*. The normal strategy of research requires us to trust ninety-nine per cent of the items in an established network of theory and 'fact'. We do well to

'believe' these items – that is, act unhesitatingly as if they were unques-
tionably the case[207]. Nevertheless, we also do well not to put them per-
manently on an altogether higher plane of credibility than the few
doubtful items that we choose to challenge by our research. The whole
argument of this book is that there is no way of ensuring that we have 'got
things right' scientifically – which means that sometimes, who knows,
we shall find that we have 'got wrong' one of our best-loved beliefs.

9.5 Mysteries, marvels and magic

Just as folk psychology holds that scientists are full of 'curiosity', so folk
epistemology holds that scientific knowledge is full of 'mysteries' and
'marvels'. The scientific archives do indeed contain many reports of inexplicable observations and many theories built around counter-intuitive
concepts. What is more, academic science is surrounded by a penumbra
of *parascience*[208], where such mysteries and marvels are elaborately interwoven. What is the metascientific significance of such epistemic structures as astrology, telepathy and homeopathic medicine? According to
the Legend, these are *pseudosciences*, whose typical features exclude them
from the official canon. Does our naturalistic model of science uphold
this distinction?

Let us accept at once that this is a region of the scientific culture where
rationality and good faith cannot be taken for granted. All science, as we
have seen, is vulnerable to human frailties, cognitive and moral.
Academic science limits the effects of individual frailties by systematic
communal appraisal. But it has no regular mechanisms for distinguishing between fact and fancy, or between obsessive self-delusion[209] and
systematic deceit[210]. In situations where these might be suspected,
worldly investigations and mundane considerations of mental stability,
venal interests, potentialities for fraud, etc. are entirely appropriate, even
if they are not scientifically 'symmetric' with the research claims whose
credibility is at stake[211].

But having assured ourselves of the sincerity and sanity of their proponents, do we have other general grounds for refusing to consider such
claims worthy of scientific attention? This question usually arises when
they are submitted for publication in an archival journal. Reputable scientific journals do not publish all the communications they receive, even
from established scientists. Papers are rejected for a number of reasons,
such as technical defects, improper presentation, or irrelevance to a

particular research specialty. But are there general criteria by which an editor can immediately judge the substance of a communication to be 'pseudoscientific', and politely return it to the author without even submitting it to peer review? And if it does go out to referees, on what basis can they argue that it is not just 'bad science', but 'not science at all'?

Here is one of the major frontier posts on the supposed demarcation line around science[212]. But many of the items that the frontier guards would like to exclude are not illicit within science. For example, the acceptability of an unorthodox 'contribution' should not depend on its form but on what seems to be its 'substance' – although in practice it is not always easy to tease this out and present in a 'scientific' style. Again, it does not have to be submitted by a 'professional scientist' with a PhD. Indeed, in pursuit of intellectual equality and universality, some journals remove the names of the authors and their institutions from the typescripts that are sent to referees.

Some of the more zealous gatekeepers would like to exclude 'irreproducible' observations. True, it is not easy to assess the credibility of a report of one-off event, or of results obtained with a unique instrumental technique[213]. But science has to be concerned with a whole variety of singular events, including some that are not even observable, such as the origin of life[214]. What is more, the criterion of 'reproducibility' becomes very questionable once one tries to define it carefully – especially where human beings and other biological organisms are involved[215] [5.8].

What often makes a singular event mysterious is that it seems 'inexplicable' [10.1]. But this also applies to many well-attested and replicated scientific phenomena and even to the efficacy of much of our technology[216]. An unexplained event can be tolerated and even valued as an 'anomaly' [8.9]. That is to say, it can spark off an investigation that may end in changing the established framework of explanation – or, just as likely, in exposing a straightforward technical error. Such cases are the staple of the history of science, from the final accreditation of reports of 'stones falling from the sky' to the discrediting of the notion of 'anomalous water'.

The strategic issue for the researcher is how far to go in questioning the conventional wisdom[217] – or, to put it another way, to decide what paradigm has been 'falsified' by the anomaly [8.11]. Doubtful hypotheses are the meat and drink of the scientific culture [8.10]. But unfettered theoretical speculation provides as little nourishment as plodding orthodoxy. Scientists range temperamentally from 'speculophobics' to 'hypothophiliacs'[218]. But research communities tend to ignore an idiosyncratic conjecture[219] – until it is later honoured with a Nobel Prize. Indeed, such conjectures usually come from 'low status' scientists who reckon they have no reputation to lose, or else from those whose name is already so big that they can risk putting it on the line for a weird idea[220].

Science has no special difficulty with speculative theories that hypothesize hidden entities – genes, viruses, atoms, quarks, and so on [8.12]. But the explanatory power of such entities is sometimes attributed to 'marvellous' properties that contravene all our previous notions of the fundamental nature of things [8.14]. Their supposed behaviour is *counter-intuitive* – in effect, not possible for the sorts of entity with which we have previously had to deal. The outstanding case of such a theory is, of course, quantum theory, which assigns to electrons and other 'elementary particles' a number of properties that seem quite inconsistent with our most basic ideas about causation and localization.

This is not the place to enter into the epistemology of quantum theory. For the best part of a century, its paradoxes have engaged the most acute scientific and philosophical minds. My own impression is that we have only made clearer that what happens on the 'microscale' is not amenable to 'normal understanding'[221] – that is to say, to reasoning based solely on direct life-world experience of 'mesoscale' objects such as tables and chairs[222] [10.1, 10.3, 10.6]. On the other hand, the work of physical scientists throughout this whole period has shown, just as clearly, that they can learn to 'think quantally' [8.14] to produce reliable knowledge concerning observable phenomena[223]. Despite its mysterious logic, quantum theory is a coherent, internally consistent theoretical paradigm, with a firm empirical base[224], which is shared uncontroversially and used fruitfully by an immense research community. By naturalistic criteria, this is more than sufficient for a secure place within 'science'.

For centuries now, science and technology have been producing elaborate material and epistemic systems whose workings are so efficacious, and yet so difficult to comprehend, that they have seemed like magic. The strangeness of quantum physics and its cosmic-scale partner, relativistic physics, lends further support to the popular image of science as necessarily esoteric and 'unnatural'[225]. This image obviously give social power to the shamans of high science and helps to protect their practices – and especially their funding – from lay criticism[226]. Nowadays, however, it is beginning to make their research seem less of a societal enterprise deserving public support[227]. What is more, it encourages wildly speculative theorizing[228], as if illogicality and obscurity were positive virtues. There is no substitute for the work that has to be done to make a novel scientific idea acceptable to a research community.

Much scientific knowledge is described as 'trained common sense'[229], or 'common-sense knowledge writ large'[230] [10.6]. Distance from 'common sense' is not a sign of being truly 'scientific'[231]. But the results of research have always gone beyond the generally accepted ideas of the day [9.6], and often forced their revision[232]. As Max Planck's seminal contribution to quantum theory shows[233], potentially revolutionary ideas do not always come from groups in conflict with orthodoxy[234]. Nevertheless, an open mind towards the 'impossible' is an essential feature of the 'scientific attitude'[235].

But this does not lay upon every scientist a duty to explore and explain all the mysteries of life. The formulation of researchable problems requires as much skill as their solution [8.1]. It is difficult to get a handle on a singular, irreproducible 'anomaly', or an outrageously speculative hypothesis, that is totally incoherent with what we already know or believe[236]. If it resists all attempts to locate it in the established network of concepts or 'facts' about the world [10.5], it can begin to seem not only 'impossible', but also so 'meaningless' that it drops out of scientific consciousness[237].

The fact is that academic science is not well suited to the study of 'mysteries'. But that does not mean that such matters should not be studied at all. In a pluralistic society this is the role of 'alternative' science, in various forms. This is not easily differentiated from 'orthodox' science[238], except by being produced by 'alternative' institutions[239]. In other words the distinctive features of parascience are social and psychological rather than epistemic.

Thus, rather than striving to acquire communal CUDOS [3.8], para-scientific researchers are typically motivated by a desire to justify a system of preconceived beliefs. These beliefs are sometimes recognizably religious[240] [10.5] but are often highly personal and idiosyncratic. This zeal shows up in a number of ways, such as reports of barely detectable effects, *ad hoc* responses to criticism, and constant reiteration of outlandish theories[241]. What is claimed is generally so uncertain, fragmentary and heterogeneous that little progress can be made by submitting it to peer review or other modes of collective assessment[242].

In other words, a particular item of knowledge cannot be dismissed as 'pseudoscience' simply because of its subject matter, or on the basis of internal criteria of coherence, plausibility, etc. Even what appears to be complete nonsense may contain a germ of truth. It is for each of us, as free citizens, to decide whether it deserves closer attention, by 'scientific' and other means.

But its initial credibility must depend on what we know about the social processes by which it was produced, the degree to which it is attested by a trustworthy community, and so on. The 'strong programme in the sociology of scientific knowledge (SSK)' requires that all knowledge claims should be treated 'symmetrically', without metascientific bias[243]. Ironically, it is precisely in their sociological dimensions that the asymmetry between academic science and its parascientific cousins is most pronounced and significant[244].

9.6 Epistemic change

Whatever else we may say about scientific knowledge, it *changes*[245]. What is more, it tends to *grow*. This does not just apply to the scientific community, which has always been a growth industry [4.7], and the archival literature [9.3], which is amassed automatically. It applies also to the conceptual substance of what is thought to be well established.

In practice, 'knowledge' [1.6] is not the sort of entity where 'growth' can be differentiated from 'change'. Much 'new' science may well be little more than a minor modification of what was previously known[246]. But the relentless barrage of 'original' research claims is only partially offset by 'sceptical' rejections, revisions, refutations and deletions. Quite new experiments breed out of the results of old ones[247]. Theoretical concepts are found to make sense far outside their original contexts[248]. Ideas flow in – or are carried inside the heads of researchers – from the outermost peripheries of a subject to its inner core[249]. Established specialty frontiers are redrawn or traversed in great leaps across the academic map[250]. New disciplines, new fields of research, new methods and new research traditions continually emerge, in novel epistemic, technical and cultural combinations[251].

Much of this turmoil is shaped by social interests, economic and political events, practical developments and other factors outside science [7.5, 8.7]. But most metascientists now accept that the course of conceptual change, like the course of technological innovation to which it is linked[252], cannot be explained solely in terms of such forces[253]. Academic science continually explores, interacts with, and adapts to its socio-economic environment [9.7]: but it is also driven by forces arising deep within itself[254]. This inner dynamism, operating, so to speak, in every cell of its body, is a natural consequence of its social structure and is intrinsic to its being. Academic science can be milked, and to some extent harnessed [4.8], but it becomes a very different animal – if it survives at all – when controlled in detail. That, surely, must be clear from all that I have said so far in this book.

Look at it another way. Until recently, most historians treated science as a self-contained activity, drawing only occasional ideas from society at large. This sort of account is now seen to be seriously faulty and incomplete. Nevertheless, suppose it were valid. What course would one expect science to take if it were driven entirely by 'internal' forces functioning according to the Mertonian norms? What is the strength of these forces, and what would be the outcome if they acted alone?

In fact, we have already seen them at work 'locally', within the scope of a research specialty, during the scientific career of an individual researcher. The essence of the Mertonian scheme is that it postulates no 'long-range' forces. Its norms operate almost entirely on a microsociological scale. But on this scale they largely dominate the action. As we have seen, they play a major role in generating and shaping a whole range of typically 'scientific' practices and phenomena, such as observation, experimentation, theorizing, publishing, claiming discoveries, peer review, public controversy, priority conflict, archival storage, source citation, field specialization, paradigm formation, etc.

We need not repeat our account of how these familiar microsociological features of academic science arise, nor emphasize their dynamic role in conceptual change. The question here is: how might a system driven by such 'short-range' forces behave on larger temporal and epistemic scales? In particular, what are we to make of the notion[255] that the history of every major science is punctuated by epistemic discontinuities in which the meaning of every item of knowledge is radically changed? Are there *scientific revolutions,* which truly 'turn the world upside down'? Does it sometimes happen, as was said of plate tectonics, that scientists must really 'forget everything that has been learned in the last 70 years, and start all over again'[256]?

Scientific change on every scale necessarily involves the supersession of what had previously seemed stable paradigms [8.4–8.10]. Usually, such revolutions do not occur simultaneously over a very wide region of knowledge, so that major paradigms continue to hold together by *intercalation,* like a wall whose bricks are replaced one by one[257]. Often the change occurs almost imperceptibly, by a long succession of minor adjustments[258]. In the human sciences, where argumentation is usually so inconclusive[259], a revolution may look more like a swing of fashion than a true break with the past.

But sometimes, especially in the natural sciences, epistemic change is occasioned by a remarkable discovery, or a long-sought 'breakthrough', with startling cognitive consequences. There is no highest level of generality – or, to invert the metaphor, no deepest 'foundation' – that can be guaranteed to be immune from such changes[260]. When a widespread network of 'established' concepts is thus put into question, a mind-boggling 'gestalt switch' may be required[261], apparently shaking the bedrock of scientific rationality[262]. Even that card-carrying iconoclast Albert Einstein could never accept the probabilistic aspects of the new quantum theory that he had helped to bring to birth.

The psychology of abrupt, massive, cognitive change is clearly of great interest. So are the historical and sociological features of such scientific earthquakes – the retrospective recognition of precursor tremors, the actions and reactions of individuals, the new landscapes created, the threatening aftershocks, the slow return to 'normality', and so on. A radical scientific revolution may in fact be accepted quite smoothly and quickly[263] or it may be spread untidily over a number of turbulent years as its opponents are 'converted' or silenced by death[264]. Innumerable empirical 'facts' and phenomenological theories will still prove reliable, regardless of how they might now be re-interpreted[265]. In the long run, however, such details are of little significance by comparison with the apparently unbridgeable epistemic gap between the old regime and its successor.

The real question is whether such a gap is ever 'unbridgeable'. As we have seen, a paradigm shift at any level is usually reflected in a major change in the corresponding language of thought and communication. In general, if a gap opens up, it can be bridged by discourse in the higher-level language of a more inclusive paradigm [8.5]. But suppose that there is no 'higher'-level language, no more 'fundamental' account, out of which to construct such a bridge. Suppose, as in the case of the Copernican and quantum revolutions, a crack opens up in the solid ground of everyday reality [10.7]. Then, surely, the two scientific languages, and all their conceptual correlates, are *incommensurable*. There is just no way of matching meanings across the historical divide of the revolution[266].

For many metascientists this is a telling argument. They even go on to apply it to other epistemic gaps, such as between modern science and 'alternative' knowledge systems. But in spite of its use to lambast the Legend, this argument is confused[267], since it rests upon the very assumption that it attacks: it assumes that an absolutely over-arching paradigm, a truly fundamental 'world system', is a meaningful concept [10.8].

One solid philosophical achievement of 'post-modernism' is its critique of the very notion of a comprehensive 'metanarrative'. Naturalistically speaking, moreover – and in spite of assertions to the contrary – the natural and human sciences are no exception to the anthropological observation that no human society has ever succeeded in establishing such a 'total' system amongst its members [10.4]. Indeed, if such a cultural form could exist, we would be unaware of it, for there would be no window of partially shared conceptual schemes[268] through which we could communicate with its practitioners.

'Incommensurability' is a dramatic but essentially misleading notion that takes too seriously the rhetoric of the participants in processes of conceptual change[269]. The defenders of a 'classical' system of ideas exaggerate its completeness, coherence and universal credence[270] [9.2]: the attackers call for a comprehensive 'revolution' to break the hold of a stagnant but well-entrenched consensus[271]. But each side is too absolutist in its epistemic claims. In reality, radically new views often reveal why the old ideas were so sound[272], so-called revolutionaries are often just obstinate dissidents from current orthodoxies[273], and ancient and modern intellectual regimes often co-exist peacefully as mappings of different aspects of the world[274]. It is only to the Whig historian, conscious of all that has been achieved, that political and cognitive revolutions look so total.

9.7 The evolutionary analogy

The antithesis of revolution is *evolution*. As we have already noted, scientific paradigms often change very slowly, almost imperceptibly, over periods of years or decades, until they are quite different from what they once were. Should one say that all conceptual change, however abrupt it may seem on longer time scales, is really always continuous? Historians are adept at revealing the threads of remembered experience and tacit knowledge that connect new ideas with the past[275]. But that cannot be strictly true, since it would make any change impossible. Like Zeno's hare that could never catch the tortoise, no original scientific idea could ever be born, let alone win the assent of a research community.

The naturalistic solution to this logical paradox is to treat conceptual change in science as an 'evolutionary' process analogous to Darwin's account of the evolution of biological organisms [6.3]. This analogy comes easily to mind, and was suggested independently by such influential nineteenth-century metascientists as T.H. Huxley, Karl Marx, William James, Ernst Mach and – even before Darwin – William Whewell [276]. Nevertheless, it has only begun to be discussed systematically in recent years[277].

In effect, science evolves by an on-going, never-ending cyclic process of 'Blind Variation and Selective Retention' – BVSR[278]. A great variety of hypotheses and empirical research claims are put forward by researchers [8.10], but only a few of these are accepted communally and passed on as 'established' scientific knowledge. The essence of the analogy is that research claims, like the variant progeny of living organisms, are produced 'blindly', without foreknowledge of their likely 'fitness', which is only determined subsequently by 'selection' – that is, by differential 'survival' in a critical environment and 'reproduction' in the archives [9.3].

This *selectionist* account of epistemic change[279] in science is obviously consistent with the general argument of this book. The norms of 'originality', 'scepticism' and 'communalism' are put into operation as processes of 'Variation', 'Selection' and 'Retention' respectively. Like the notion of 'falsifiability'[280] [8.11], but with a wider scope, it stresses the role of criticism in optimizing the reliability of scientific knowledge. It is also a 'sociological' model, in that this process of 'selection' is not limited to self-criticism. The prescription that an individual scientist should check her findings exhaustively before making a research claim is secondary to a pervasive social function exercised publicly by a research community.

The evolutionary analogy provides numerous insights into the dynamics of epistemic change. A great many of the phenomena recorded by historians and sociologists of science have their counterparts in the biological world[281]. Indeed, these phenomena are often so characteristic and yet so complex that it is easier to denote them in biological language than to describe them in detail. For a biologically literate metascientist, a term such as 'emergence', 'extinction', 'vestige', 'niche', 'diversification', 'speciation', 'convergence', 'evolutionary drift', 'stasis', 'punctuated equilibrium', 'entrenched *bauplan*', 'developmental lock', 'environmental adaptation', 'arms race', 'co-evolutionary stable strategy' (etc.!) would often seem the most succinct way of encapsulating a particular type of episode in the history of science.

This informal list of some of the detailed *phenomenological* similarities between bio-organic and epistemic evolution is too long to be decoded and interpreted here. Some of these analogies are, perhaps, a bit forced. But they depict genuine episodes of epistemic change that are much more diverse and complex than can adequately be covered by the general term 'revolution'[282]. For example, 'punctuated equilibrium' at any level, on any time scale, may signal the 'emergence' of new theoretical concepts without implying the 'extinction' of old ones, or it may arise through the 'convergence' of existing concepts to 'adapt' to a new techno-scientific environment.

Indeed, we are easily led into interpreting epistemic change in terms of the *structural* analogies between academic science and a biological system. As we have seen, *competition* for recognition animates the research world [3.8, 9.1]. But as feminists [5.10, 6.3, 7.5] have pointed out[283], 'mutualistic' relationships – which are much more usual in biology than is often realized[284] – are frequently found in science, for example, between a mathematical technique and a physical theory, or a conceptual scheme in psychology and its sociological counterparts.

Again, scientific concepts, like many of the traits of evolving organisms, can be traced back many 'generations', almost without change. On the other hand, they undergo 'variation' through both 'mutation' and 'recombination' – that is to say, concepts that were previously never connected may be combined in a novel fashion[285]. In fact, quite unrelated conceptual 'genealogies' can form viable 'hybrids' – for example, carbon isotope dating in archaeology or economic game theory in ecology. It is sometimes argued [8.8, 8.14] that most scientific 'creativity' – like most of the bio-organic variation that fuels Darwinian evolution – can be accounted for by this mechanism[286].

There is evidently much more to the notion of 'epistemic evolution' than a vivid but vague metaphor. But there are serious practical and conceptual obstacles to developing it into a comprehensive theoretical model for scientific change. These same obstacles emerge in the attempt to use bio-organic evolution as a model for the historical trajectory of any system of *cultural* entities, such as languages, scientific theories, research communities, technological artefacts, commercial firms, social customs, political institutions, or legal precedents[287].

The first obstacle is that cultural entities do not have a hidden, atomistic *genetic* base. Modern evolutionary biology is dominated by 'neo-Darwinism' – that is, by the addition of molecular genetics to Darwin's original theory[288]. Molecular biology has triumphantly confirmed the material reality, in the form of segments of DNA, of the discrete, semi-permanent, self-replicated 'genes' hypothesized by Mendel. These genes determine all the heritable, variable, combinable etc. traits that are operated on by natural selection. Indeed, it is widely held that they are the real actors in the evolutionary drama – as, for example, in the imaginative slogan of 'the selfish gene'[289].

To preserve the analogy, then, it has been suggested[290] that a cultural entity such as a scientific theory or social custom can be similarly characterized in terms of 'memes', with similar properties to genes. A 'meme' is supposed to be an elementary epistemic entity, capable of existing, replicating and moving from mind to mind. This is obviously a theoretical construct, with no other substance than its use in this context. But it provides a graphic way of talking about changing systems of ideas, their diffusion through society, and their effects – good and bad – on human thought and behaviour[291]. One might say, for example, that Darwin's idea of *evolution by natural selection* is a 'meme' that demonstrably improves the cognitive 'fitness' of a great many scientific paradigms far outside organic biology.

It is doubtful, however, whether evolutionary models of cultural change are strengthened by this usage. In metascientific discourse, a 'meme' is almost always synonymous with an 'idea' or a 'concept', and is best interpreted in that sense. It is misleading to project properties of atomicity, permanence and precise operational efficacy on cognitive objects that are intrinsically vague, mutable and intangible, or to risk the sweep of Occam's razor by postulating a hidden world of gene-like entities [8.12] where cultural evolution 'really' takes place.

What is more, there is no proof that the BVSR mechanism absolutely requires entities with these properties. Darwin's original theory was perfectly convincing without Mendelian genetics, and is still valid for systems such as natural languages which cannot be chopped up into discrete 'atoms' – and where, incidentally, the segmentation of time into 'generations'[292] would also be purely arbitrary.

Indeed, one of the major difficulties in setting up an evolutionary model of scientific change is that the action does not take place entirely in an abstract domain of 'knowledge' [6.11]. Scientific concepts operate in and through the cognitive domains of individual researchers [8.14] and the archival domains – 'World 3' – of research communities [9.3]. For example, scientific memes are stored in libraries, and replicate by citation – a cultural practice at the interface between these two domains[293].

These domains are not just passive environments. They are directly involved in the evolutionary process. For example, the fate of a novel scientific idea is bound up with the professional career of its discoverer, with the institutional history of a subdiscipline and in many cases with the availability of experimental apparatus to test it. But individuals, institutions and instruments have other evolutionary interactions besides those directly mediated by scientific 'memes'. In fact, a quasi-economic analysis of academic science reveals a number of interconnected 'markets', each of which is effectively an evolutionary domain in this sense[294].

This type of complication is familiar in biology. Genes are not selected as such, but survive and replicate through the organisms they encode. But the neat biological duality between 'genotypes' and 'phenotypes' – or equivalently, between 'replicators' and 'interactors' – does not really hold in the scientific world. This is a source of much diversity in the literature[295]. Some authors have designated individual researchers as 'interactors', or as the 'vehicles' of scientific memes: others have assigned this function, equally plausibly, to research groups or institutions. Here again, the bio-organic analogy is misleading. The genotype–phenotype duality is not, as far as we know, essential to the evolutionary process[296]. Every system has its own dynamical structure, which needs to be modelled as such in an evolutionary analysis.

These obstacles arise primarily out of the assumption that the evolution of scientific knowledge has to be structurally homologous to the evolution of living organisms[297]. But is this assumption really necessary? The neo-Darwinian model of modern evolutionary biology is only one of several bio-organic 'selectionist' systems, including the immune and neural systems of animals[298], each of which operates along different lines. A number of different computer programs, such as 'genetic algorithms' and simulations of 'artificial life', also exhibit typically 'evolutionary' features[299]. In general, it seems that any BVSR system may be expected to show these features, regardless of the nature of the evolving entities, the details of their interactions, or the precise mechanisms by which they are varied and selectively retained.

A naturalistic evolutionary account of epistemic change in science thus seems a reasonable metascientific goal. Unfortunately, there remains a very serious objection, which applies in principle to all evolutionary models of 'cultural' change[300]. The objection is that the processes of variation and selection are never 'blind' in the full Darwinian sense. Scientists do not make conceptual or factual research contributions at *random*. Scientific communities are not just harsh environments for unwitting *natural* selection of the fittest of these contributions. Scientific research is usually an exquisitely *planned* activity [9.2], where only the results are uncertain. Scientific scepticism is systematic and highly organized, more like the *artificial* selection practised by plant and animal breeders than the chancy outcome of haphazard tests[301]. As is often (though speciously) remarked, cultural evolution is not 'Darwinian' but 'Lamarckian': it operates through the inheritance of characteristics acquired by experience, rather than through uncoordinated cycles of chance and necessity[302].

There is no virtue in going into denial on this point. There is often as much rational design in the creation of scientific theories [8.10] as there is in the determination of 'facts' by observation and experimentation [5.2, 5.6]. The putative discombobulation of all scientific reasoning by quantum uncertainties in the brain is a reductionist fantasy [10.8], and the notion that 'anything goes' is a wildly romantic slogan.

In practice, most of the variant concepts that individual scientists offer for communal acceptance are skilfully crafted, and the criteria by which they are selected are carefully thought out and argued[303]. Science is notable for its reliance on 'learning' – that is, rational reconstruction of the remembered past – and on 'prediction' – that is, rational pre-construction of an imagined future. Of all cultural 'memes', those of academic science are required to be the 'clearest' and the 'most enlightening' – that is, the least 'blind'.

Nevertheless, as an epistemic institution, academic science deliberately leaves openings for uncertainty, chance, and other genuinely stochastic factors [8.10]. Barely credible conjectures are seriously entertained, disconfirmed predictions are reported with glee, serendipitous discoveries are warmly welcomed[304], and even inexplicable observations do not always go unrecorded. 'Selection' is not an authoritative yes/no decision, but an extended, open-ended process involving numerous uncoordinated acts of explicit appraisal or tacit use in practice [8.11, 9.2]. Research communities, like populations of evolving organisms, are not composed of entirely identical, like-minded individuals. At any one moment, the scientific front line is being pushed forward into a fog of conflicting research claims and of diverse assessments of their credibility. In this fog, every new claim, however brilliantly conceived or minutely investigated, is to some degree 'blind'.

The question is, then, whether the stochastic factors in epistemic change are sufficient to justify an essentially evolutionary model[305]. This same question arises in connection with technological change[306]. It is quite clear, in both cases, that 'design' plays a major role in the production of variants, and that selection is largely a 'social' process. But even if they are not perfectly 'Darwinian' systems, how far do they deviate from the Darwinian ideal type?

The key point is whether the search for the solution to a scientific or technological problem is typically spread over a more or less random collection of possibilities, including some that seem wildly implausible[307], or whether it is concentrated on just a few of the most obvious candidates. There is clearly a trade-off here between efficacy and efficiency. A horde of relatively 'blind' variants may hit on a solution that would be missed in a narrowly focussed search, even though very wasteful of resources in those variants that fail.

Academic research communities are notoriously fecund and profligate with their intellectual offspring[308]. This shows up, for example, in the small proportion even of *published* research claims that are eventually accepted as well established. This suggests that academic science is not too far from behaving as a genuine BVSR system, governed more by 'selection' than by 'design'. But as we have seen, the post-academic aspiration to 'steer' research in the service of public policy [4.9, 7.9] is pushing the system back towards 'design', with consequences that can only be guessed at until they become apparent, in retrospect, some time in the next century!

9.8 Complexity and progress

An optimistic summary of the above discussion might be that it is feasible in principle to transform the evolutionary *metaphor* into a more formal *model* of scientific change. This possibility, however unrealizable in practice, is of the greatest epistemological significance. All complex evolutionary systems have certain general properties, which go beyond particular structural and phenomenological similarities[309]. No special argument is needed, therefore, to explain how it is that scientific knowledge has these same properties.

Thus, science is minutely subdivided into a nested hierarchy of specialties, paradigms and disciplines [8.3, 8.4]. At first sight, this is surprising. One might have thought that a self-regulating complex system with a large admixture of near-random elements would be *chaotic*, both metaphorically and mathematically. Perhaps, after all, there is a supervisory intelligence at work – a faceless bureaucrat in the office of the Royal Society or of the National Science Foundation issuing instructions that keep all those proud scientific barons and their unruly knights in line as they advance against inchoate ignorance.

But intelligible orderliness does not need to have been intentionally planned[310]. A complex system often has a remarkable capacity for *self-organization*[311]. Provided that its internal connections are not too dense[312], it generates internal order as it evolves. This is a spontaneous process, whose tangible outcome may be very striking – witness, for example, the elaborate taxonomic order of the living world [6.3].

What is more, the potential for order is usually not visible before it emerges, and is often difficult to explain after it has become established [10.8]. The 'map' of scientific knowledge is usually well defined, but apparently very illogical. Perhaps this is because, in the terminology of complexity theory, it is the surface manifestation of a hidden web of 'strange attractors' whose actual rationale is very difficult to fathom.

Another characteristic property of science is that the actual trajectory of change is an *unpredictable, irreversible* historical process[313]. Even without any stochastic elements, the detailed behaviour of a genuinely complex system[314] cannot be calculated from a computable algorithm. It is just too complex. Apparently insignificant factors emerge from obscurity, grow rapidly, interact with one another and produce unforeseen effects. In principle, *fulguration*[315] – the emergence of novelty 'like a lightning bolt' – is step-wise deterministic. But the tangle of interactions cannot be unravelled afterwards, and their results undone piecemeal, even in imagination.

Take, for example, the increasing role of computers in every field of research [6.10]. They seem to make scientific reasoning more rigorous and deterministic. But every computation takes as its starting point a specific configuration of knowledge, and operates on it by an effectively opaque process into a new configuration. Traced back in retrospect, this process can be seen to draw upon a wide variety of facts and concepts, instruments and people, that 'happened' to be available at that particular moment. Its outcome would not necessarily have been the same if an identical computation had been undertaken with slightly different assumptions about boundary conditions, classification criteria, approximation limits or other arbitrary parameters. In other words, epistemic change, like biological evolution, is a non-linear *path-dependent* process[316]. The previous history of the system is an integral factor in the way it subsequently evolves.

This is a natural property of scientific knowledge. It is more generalized than the unpredictability of the results of any particular experiment, the irreversibility of a once-glimpsed insight, or the historical validity of the genealogies showing how scientific ideas begat one another in many-branched families. The intrinsic *historicity* of the scientific enterprise accords with all our individual and collective experience of it. And yet it contrasts epistemologically with the strong feeling, enshrined in the traditional Legend, that research is simply a matter of uncovering and deciphering a pre-ordained text.

Perhaps this sense of 'discovering what is there' arises out of another general evolutionary property of science – its *adaptive fit* to the requirements of its environment. Scientific knowledge is by no means arbitrary or unconstrained. The whole BVSR process ensures that it comes closer and closer to satisfying the criteria for which it is selected. If, as I argue throughout this book, research communities accept knowledge on the basis of consensus on their credibility [9.2], then what accumulates as established knowledge must seem increasingly credible and consensual. The researcher who makes a highly acceptable contribution is bound to feel that things could scarcely have been otherwise – which prompts an unconscious inference that they already existed, just waiting to be 'discovered' [8.9].

This inference is an almost unshakable psychological conviction that is damnably difficult to justify by other means [10.7]. The evolutionary model of academic science is peculiarly unhelpful on the vexed question of scientific realism, except as one component of a completely generalized 'evolutionary epistemology' [10.3]. Indeed, the theory of complex evolutionary systems does not support the notion that a hypothetico-deductive process of 'conjectures and refutations'[317] must eventually home in on the 'absolute truth'. In general, a BVSR algorithm generates *satisficing*[318], but not necessarily *optimal* solutions to any problem that it is set. That is, it searches around until it begins to produce results that satisfy the requirements of the problem up to a reasonable level of approximation. It may accidentally hit upon a 'better' solution by a serendipitous discovery[319], or by a conjectural 'jump' to an apparently unwarranted conclusion. But it has no mechanism for taking a synoptic view where the ideal solution would stand out, in all its perfection, even though located in some other region of the search domain.

This principle is now well established in evolutionary theory. Even biological organisms with very ancient lineages are never 'perfectly' adapted to their environment[320]. In evolutionary economics, even very successful firms never have 'perfect' routines for making decisions[321]. It must surely apply to epistemic change in science, where researchers are seldom even presented with a static environment of 'do-able' problems [8.1]. The solutions that eventually emerge out of the co-evolution of conceptual 'memes', research techniques, personal careers and scientific institutions are often remarkably elegant and reliable [7.3], but there is no way of knowing whether or not they are optimal.

Nevertheless, there is an overwhelming conviction that science makes *progress*[322]. How could this be doubted? Ceaseless epistemic change is innate to academic science [9.6]. Sometimes, especially in the human sciences, new knowledge simply accumulates, without significantly improving on what was known before[323]. In general, however, the knowledge base of every discipline is continually being constructed and reconstructed[324] [8.13, 9.3]. In the end, even the most famous original contributions – Einstein's papers on relativity, Crick and Watson on the structure of DNA – sink out of direct scientific use into a purely historical cultural memory[325]. On a broad scene, in addition to the addition of myriads of particular details, there are usually major changes in the conceptual language in which knowledge is represented within a research community simply to accommodate newly revealed regularities[326].

The trouble is, as evolutionary biologists have long realized[327], that the concept of progress in an evolutionary system is tautologous. How should it be measured? Surely in terms of increasing adaptive fitness of the evolving entities. But fitness is ultimately defined in terms of survivability. So an evolutionary process that ensures the 'survival of the fittest' cannot fail to indicate progress. QED. In the case of science, the fact that scientific knowledge is increasingly reliable, increasingly self-consistent, covers a wider range of phenomena, etc. is simply due to the fact that reliability, consistency, universality, etc. are the qualities for which research claims are selected and reproduced[328]. If those are your criteria for scientific progress, then you may indeed insist that academic science satisfies them abundantly – but don't imagine that when you say that 'science is making progress in revealing the nature of things' (etc.) you are really saying anything more than that science is as science does.

Some biologists have argued, however, that bio-organic evolution has an intrinsic direction. They insist, for example, that it inevitably generates increasingly diverse and complex organisms[329]. Whether or not this is true in general[330] it certainly looks like that in science. But perhaps this is just a sign of the self-organizing capabilities of the system, which produces a radiating pattern of evolutionary branches rather than a single line of 'descent'[331]. And how much should we allow for co-evolving technologies that facilitate the realization and testing of increasingly elaborate and complex scientific entities[332]? The question whether scientific progress has a 'natural' direction is either trivial, or much more obscure than it sometimes seems[333].

Science policy [4.9], of course, is based upon the presumption that the advance of knowledge can be 'steered'. That is to say, it is assumed that the direction of progress can be influenced by a systematic bias in the population of variant research claims – typically, through the operation of social, political, economic, institutional and personal interests in the formulation of research projects [7.9, 8.7].

But the evolutionary model of academic science is fundamentally *selectionist*. In the end, it requires that the research results that count as scientific knowledge should have had to satisfy the acceptance criteria of a 'sceptical' research community. In principle, what comes out of the whole process should be determined solely by these criteria, rather than by whatever may have been driving and shaping the original research claims. Certainly, this bias should not affect the reliability, consistency, objectivity and other more general epistemic qualities of the knowledge thus produced[334].

In practice, however, if research is deliberately not undertaken on a particular topic, then we are likely to remain somewhat ignorant about that topic [7.6]. What is more, the variation and selection processes in academic research are not really distinct. 'Discovery' and 'justification' are interdependent phases of an unbroken cycle of action [8.11]. This cycle is driven by conscious design. And quite apart from the general cultural atmosphere favouring pragmatic, instrumental projects and choices[335], the whole action is increasingly permeated with socio-economic purpose. These factors are not included in the basic BVSR model of epistemic change, and have no biological analogues, yet they obviously influence the direction and quality of scientific progress.

The selection criteria implicit in the notion of 'justification' also influence scientific progress. Paradoxically, post-academic science could become so obsessed with accountability, performance monitoring, contractual scrutiny and other forms of 'quality control'[336] that it sacrifices the quality of the procedures themselves to their sheer quantity. Mode 2 research[337] does not promote the establishment of groups of practitioners in stable positions of intellectual authority. In the absence of human reference groups, assessment procedures may be automated. Quality control is then made to rely on surrogate indicators of performance, whose legitimacy may well be questioned on scientific grounds.

In other words, Mode 2 downplays the role of systematic intellectual criticism, which is the key to the validity of academic science. In contexts of application, practical utility must eventually be effective as a selection mechanism, even if only in pragmatic terms [7.3]. But in fundamental research, organized scepticism is the only real protection against the embodiment of serious errors in the knowledge that is produced. Perhaps a higher level of cognitive insecurity is a price that will have to be paid as post-academic science becomes more entangled with 'trans-epistemic' issues, involving societal, environmental and humanistic values.

The big question about progress, of course, is how the overall sweep of scientific change correlates with changes in the larger world. That there are powerful linkages in both directions is perfectly evident[338]. Ever since Bacon, it has been a matter of faith that scientific progress is closely connected with socio-economic progress, both as a cause and as an effect. But that has a lot to do with what we mean by socio-economic progress – and leads on into a much wider field of thought and argument than we can hope to explore here.

The metascientific project of producing a systematic evolutionary model of scientific change is just too ambitious. It requires too many simplifications and approximations. Considered in detail, the diverse practices of the scientific culture do not reduce naturally to 'blind' processes of 'variation', 'selection' and 'retention'. The elaboration of the BVSR algorithm to include intentional features such as 'learning' and 'design' soon gets out of hand, and leads away from the human realities of a system which is in many ways 'creationist' in its operations[339].

Nevertheless, 'evolutionary reasoning' is a vital tool of metascientific analysis. I have only touched on some of the insights it stimulates[340]. In particular, it enables us to make good sense of many paradoxical aspects of academic science – its anarchic order, its wasteful efficacy, its committed neutrality, its futuristic historicity. These insights are particularly valuable when we come to consider the transition to a post-academic regime.

Take, for example, the way in which science policy is often treated as an exercise in *management*, or *agriculture* – the deployment and resourcing of productive entities according to a rational plan. Evolutionary reasoning would suggest rather that it should be considered an application of *ecology* – the creation and maintenance of an environment in which a hundred exotic, hybrid flowers will bloom naturally and produce desirable fruits[341]. We would then begin to ask whether post-academic science still retains the selectionist practices that would make such an environment epistemically productive and sound. This is a mode of metascientific thought that should be given much more weight than it is currently allowed.

10

What, then, can we believe?

10.1 Understanding and explanation

Science produces *knowledge*. This is something more than codified *information*. As we have seen, the notion of 'knowledge without a knower' [9.3] cannot be taken literally. The myriads of facts and theories in the scientific archives have been shaped by the requirements of interpersonal communication. They have to be *meaningful*: they have to be capable of being *understood*.

This meaning may only apply in a very esoteric context. The necessary understanding may be limited to a tiny, highly specialized research community. The production process may have involved an opaque conglomeration of automatic instrumentation, computation and symbolic manipulation. Nevertheless, the norms and practices of academic science require that the nature of this process and its final products should have been communicated to and consciously accepted by human minds. As I have repeatedly stressed, the *epistemology* of science is inseparable from our natural faculty of *cognition*.

But scientific communications must not only be *comprehensible*: they also, typically, *enable comprehension*. For reasons that we have discussed at length, they relate directly or indirectly to shared aspects of the life experiences of those who utter and receive them. In a word, they are messages that we send to each other about the 'world' that we seem to have in common. They thus help us to *understand* that world.

'Understanding' is a complex process, of which we know less about the parts than about the whole. Once again, we are up against a primitive notion, a brittle 'natural kind', that shatters under the impact of analysis. But for communicating individuals to understand each other, they must somehow establish common mental structures [5.12, 8.14] into which each newly shared item can be fitted[1]. When, therefore, we say that an individual 'understands' a non-cognitive entity, we imply that she has established an internal mental structure representing that entity – in particular, a mental structure that might well be held by another individual with whom she desires to communicate on the same subject. Thus, if I say that I 'understand carburettors', you take it that I have a sufficiently distinct 'mental model' of a carburettor that I think that I can discourse meaningfully with other persons who are similarly learned on this mysterious topic.

The 'maps' – that is, the 'models', 'theories', 'formulae', 'classification schemes', 'factual data sets', etc. – by which scientists represent various aspects of their shared experiences are thus ideal instruments for this purpose. Sometimes, as in the human sciences, even the most elegant explanatory models are really very rough and ready[2]. At their best, however, scientific 'maps' [6.4, 6.5] are self-consistent, empirically reliable, effectively consensual, and well adapted for mental manipulation. In other words, once it is grasped, scientific knowledge can often provide us with such an attractive way of 'understanding' things that it seems perverse to look further.

This type of understanding is particularly compelling when it constitutes an *explanation* of some previously unaccountable feature of the world. Indeed, science is frequently characterized as a search for 'explanations'. Scientists celebrate explanatory victories – planetary motion, earthquakes, chemical reactions, pulsing blood, biological reproduction, etc. – and are spurred on by explanatory challenges – high-temperature superconductivity, the origins of language, monetary inflation, etc. It sometimes seems as if only explanatory modes of understanding are to be considered truly scientific.

In practice, however, scientific explanations take many different forms[3]. Sometimes, as with earthquakes, they are *causal*, in that a specific *effect* is traced back to a specific *cause* [6.4, 10.5]. Sometimes, as with chemical reactions, they are *reductive* [10.8] – that is, complex manifestations of the behaviour of simpler deeper structures[4]. Sometimes, as with the movements of the planets, they are hard-wired *mechanically* into the system[5]. In biology, as with the circulation of the blood, they are typically *functional*. In the human sciences, as with inflation, they are often about the unintended consequences[6] of rational action. They are also entertained with various degrees of credibility. Biological reproduction is *obviously* explained by the chemical replicability of DNA. The spread of malaria was *surprisingly*[7] explained by mosquitoes. The existence of the solar system is *speculatively* explained by a whole raft of theoretical calculations. Neuroses are *purportedly* explained by childhood traumas. And so on.

The only common element seems to be the linkage of a known empirical phenomenon into a wider network of accepted – or at least potentially acceptable – 'facts' and concepts. But the production of such linkages is the main business of research. A disconnected 'fact', an anomalous phenomenon or a singular entity has no secure place in 'World 3' [9.3]. Somehow, it has to be joined up consistently – ideally by a feedback loop from consequences to causes – with what we already know, or suspect, or uncertainly infer[8]. When we 'explain' it we locate it on one of our scientific 'maps' and incorporate it therein.

'Explaining things' is just science in action. An 'explanation' is not a special form of scientific understanding. After the event, it is no more than what we see if we zoom into a particular region of our 'map', taking a particular phenomenon as the centre of focus and ignoring the lack of definition as we approach the edges of the image. Good 'explanations' certainly require 'good science'[9]. But like 'discoveries', 'solutions to problems' and 'answers to questions', they refer to the processes and contexts where knowledge is produced[10], not to the ultimate meaning of that knowledge.

Thus, the cliché version of science as 'the search for understanding' is true almost by definition. But it is not adequate as a recipe for action. We may well exhort would-be scientists to go out and try to 'understand' the world. But this is not a mission that can be accomplished directly by even the most dedicated and perceptive individuals on their own. It is only by their personal commitment to a very elaborate and demanding social institution, whose practices may at times seem directed towards quite other goals, that scientists are enabled to achieve collectively the powerful but peculiar mode of understanding that we call 'scientific'.

In previous chapters we have tried to discover what sort of 'understanding' this is. We have looked at various characteristics of the research claims collected in the scientific archives [9.3] and shown how closely they are related to the social norms and epistemic practices of research communities. In this concluding chapter, I shall try to place this body of 'knowledge', this way of 'understanding', in a larger frame. What might people mean, for example, when they say – or deny – that scientific knowledge is 'true', or that they 'believe' in it, or that it tells us what the world is 'really' like? In other words, we round off our study by circling back to the grand epistemological issues raised in the very first chapter, using what we have laboriously learnt about the *nature* of science to gather fruit from some of the hardiest perennials in the metaphysic gardens of philosophy.

10.2 Life-world knowledge

Naturalism locates science in the *life-world* [6.11]. Edmund Husserl's technical term *Lebenswelt*[11] (meaning literally 'the world of life') is useful because it avoids the many epistemological implications of more familiar terms, such as 'ordinary', 'everyday', 'common sense', 'mundane' etc. In effect, scientific knowledge is amongst the many things human beings know about the world in which they live[12]. The 'natural attitude'[13] is to take for granted a whole body of knowledge of which a great part is usually not considered to be 'scientific' [5.10]. But that does not necessarily mean that this *life-world knowledge* is just a context or frame[14] for something quite different called SCIENCE, or alternatively that scientific knowledge is just life-world knowledge carried to a higher power[15]. For the moment, we can only say that life-world knowledge constitutes a distinctive epistemic domain[16], with characteristic features that determine its relationship with the domain of scientific knowledge.

   The most striking feature of life-world knowledge is, of course, that it is not striking at all! Like the air we breathe, it is pervasive and unobtrusive, yet extraordinarily difficult to take hold of and examine. Even if we begin to think that it is just 'the dream that dreams us'[17] we have no choice but to go on dreaming it, for it is impregnable to global doubt[18]. It makes no sense to suppose the life-world to be comprehensively different from what it is for us, except perhaps in the eyes of a hypothetical Other Being able to correct our illusions[19]. Indeed, it is unanalysable in principle[20], for it is often the only feasible stopping point in an infinite recess of analyses of analyses[21].

   Life-world knowledge appears particularly ineffable because it is largely *tacit*[22], and evinced primarily through action [5.8, 5.9, 6.6, 8.10]. As competent persons, we just 'know' a great many of the things about the world that enable us to do what needs to be done – get up in the morning, make breakfast, go to work, greet our colleagues, etc. etc. That is not to say that such behaviour is so habitual that it excludes conscious thought or reasoned argument. On the contrary, normal human action is expected to be 'rational'[23] – that is, accompanied, shaped, or at least explicable by mental operations. These operations may include very complex chains of reasoning. For example, in everyday life we are continually making inferences about hidden circumstances, such as 'there must be a road-block ahead'; ' she is unhappy because she has had a row with George' or 'that bank will soon go bust'. Nevertheless, although such inferences are often convincing and reliable, they are usually reached by unformulated, non-demonstrable modes of argument[24]. Thus, they rely heavily on the testimony of others rather than on our own direct observations[25]. *Practical reasoning* [5.7, 6.8], which is essential to life-world knowledge, is not 'irrational' but it does not require logical proof.

How has life-world knowledge come into being? This is a question to which we shall return later [10.3, 10.4]. But conscious life-world knowledge, having once evolved amongst articulate humans, is regenerated in the mind of each individual in the course of personal development. Throughout childhood, from the earliest infancy to maturity, personal sensorimotor experience is interwoven with social perspectives acquired through language[26]. On the one hand, the life-world is directly accessible to vision, hearing, etc., and is responsive to actions such as speech and touch[27]. On the other hand, information about its contents, properties, values, capabilities, limitations etc. is continually being amassed indirectly, along with other language skills, through informal or didactic discourse with trusted adults[28]. The two components are inseparable. Natural language [5.12, 6.6] and the world are learned together in a single process[29].

Thus, life-world entities are simultaneously particular and general. They are particular in that we directly experience them as specific and concrete – *my* hand, *our* house, *your* mother, *this* village, *that* dog, etc. At the same time, however, they are general, in that they are implicitly classified by the words we use to talk about them – my hand has the properties of all 'hands', that dog is a 'dog', like all dogs, and so on. In using these words, we assume that they have the same meanings for all competent speakers of our mother tongue. In spite of its subjective diversity and plasticity in our private thoughts, the public life-world is intersubjectively unique and resistant to doubt. This shared knowledge both enables and is enabled by interpersonal communication in a natural language[30]. Indeed, much life-world knowledge is so readily translated from language to language that it is a communal resource for the whole human species[31].

The heterogeneity of 'what everybody knows' is often minimized, both by metascientists who would rather that life-world knowledge didn't exist at all[32] and by computer wizards who try desperately to simulate it. Nevertheless, this heterogeneity is essential to its role in integrating quite ordinary thought and talk about the human condition[33]. Life-world knowledge includes numerous biological 'natural kinds' [6.3], as well as the names of parts of the body, geographical features – oh, all the innumerable objects of direct experience to which we customarily give names [34]. Typically, these are the first words that must be memorized in learning a foreign language.

But the basic 'vocabulary' required for meaningful communication includes adjectives and prepositions, as well as nouns. Life-world entities are implicitly divided into categories, such as by colour or shape, and structured by mutual relationships such as being 'inside' or 'distant from' one another [8.14]. Whether or not these are 'universal' to all humanity[35], they are the same within large linguistic communities and can be made intelligible between them.

This basic vocabulary includes *verbs*. The life-world is not just a 'world picture'[36], providing a 'background'[37], or 'public space'[38], for interesting natural and social events: in part it *is* those events. This is not just metaphysical mystery-mongering. Take, for example, the words *throw* and *catch*. To make sense of these words, we must obviously know that we inhabit what a geometer would describe as a three-dimensional spatio-temporal frame of reference which is invariant from the point of view of both throwers and catchers[39]. But this life-world knowledge must also include what physicists now call gravity – i.e. that the apple would *fall* if left *unsupported*, that we could not even *move* it if it were the size of an elephant, and so on[40]. It also has physiological elements, in that we are able to rehearse mentally the different bodily actions of *lifting, chucking, hurling, tossing* (etc., etc.) and of *grasping, gripping, clutching, seizing* (etc., etc.). The 'tacit' knowledge embodied in such actions is shared publicly through the many very ordinary words we have for describing them.

What is more, any person from anywhere – a New Guinea tribesman, for example – watching a film about, say, Californian co-eds, would know at once that these were *people*. He would not need to be told that they inhabited, perceived and experienced the same physical world as he did, that they had inner thoughts that influenced their behaviour, that they could partially convey these thoughts to one another in order to coordinate their actions[41], and so on. Life-world knowledge thus incorporates perception, cognition, intentionality, intersubjectivity, interpersonal cooperation and many other familiar features of personal and communal living. These are not apprehended as problematic socio-psychological 'phenomena' but as 'trivial truths' so central to our experience from infancy that they require no explanation at all[42].

The life-world is not only a vital *ingredient* of social life: it *includes* social life[43]. It is a 'many-personed' space[44], structured by behavioural conventions and institutions through which we learn to navigate as through the streets of a city[45]. The rituals and symbolic representations of a temple or bank are not less 'taken for granted' and compelling [10.7] than its marble steps and echoing corridors, its sinister guardians and occasional handouts. A blessing or a cheque is as much a life-world entity as a stone or a sparrow. The social role of 'being an aunt' is normally just as unquestioned as the physiological one of 'being a mother'. Indeed, human kinship relations are often considered so 'natural' and unequivocal that they are used as models for shared representations of relationships between other life-world entities, such as biological species[46].

In the life-world, then, everyone is as competent a psychologist or social scientist[47] as they are a geometer, physicist, naturalist or anatomist. This is not to say that 'folk psychology'[48] and 'lay sociology'[49] [5.10, 5.12, 6.10, 8.14] are necessarily more valid than other ideas about social life. In practice, they can be very reliable locally over short periods of time. But (like our everyday understanding of geodesy, gravity, visual perception, etc.) they do not operate satisfactorily outside their customary contexts[50]. Indeed, part of our life-world knowledge of social entities like families and banks is that (like the multitudinous mortal denizens of forests, prairies, deserts and oceans) they are impermanent, locally specific and bewilderingly diverse.

Indeed, although people perceive their knowledge of social life as immutable, history shows that it changes slowly but radically over time[51]. 'Common sense' about family life is not the same now as it was in 1950[52]. High-street banks were not 'everyday' institutions in the Middle Ages. Temples as we now know them probably emerged less than 10,000 years ago. But the same applies to life-world knowledge of more tangible natural and cultural objects. What every aboriginal hunter-gatherer knew about 'familiar' plants and animals is scarcely now remembered by anyone. Vitamins are as much 'taken for granted' now as were formerly the herbal remedies of our grandmothers. Automobiles and electronic calculators have replaced horse-drawn vehicles and mental arithmetic in 'ordinary' life. And so on, *ad infinitum*. . . .

10.3 The epistemology of the life-world

Ad infinitum indeed. The life-world is endless in its multiplicity, diversity and heterogeneity. Its contents cannot be defined or classified [6.2], for that would mean that we no longer 'take them for granted'. For this reason, a frontier between the life-world and 'the scientific world' cannot be fenced or policed [10.6]. In any case, innumerable entities have dual nationality. The mere act of examining the epistemological credentials of an everyday item of knowledge automatically gives it an entry visa into 'science'. Indeed, one of the maxims of the scientific culture is that any life-world entity is capable, in principle, of satisfying such scrutiny. What we call scientific curiosity [2.7, 8.9] is precisely the impulse to drag some 'ordinary' object or idea across the boundary into the scientific world for detailed definition, classification, analysis and (all too often) 'correction'.

Nevertheless, it is a fantasy [6.6, 10.8] to suppose that this process could be applied simultaneously to every element in a typical chain of scientific reasoning. All scientific knowledge is riddled with unscrutinized life-world 'impurities', which can never be entirely flushed out[53]. At the same time, novel 'facts', concepts and artefacts that were once strictly scientific – e.g. clock time[54], dinosaurs[55], monetary inflation, vitamin supplements, etc. – lose their original 'scientificity' and are incorporated into 'everybody's' knowledge of the life-world [10.6]. Indeed, most attempts to chart 'public understanding of science' have foundered on the impossibility of differentiating between 'scientific' and other forms of general knowledge over a broad swathe of epistemic territory [9.5]. Whom do we follow: the pedantic palaeontologist who snorts at elementary public misconceptions about dinosaurs, or the cynical economist who holds that the views of most of his professional colleagues on inflation are no better warranted than those of the average punter? The more expertly such issues are analysed, the less clear they become.

Nevertheless, there is universal agreement that life-world knowledge normally lacks certain features which are typical of scientific knowledge[56]. As we have seen, it is very difficult to specify those features directly. Our whole argument, however, is that the peculiar characteristics of scientific knowledge stem from the collective operation of the academic ethos. 'Common sense' has many 'unscientific' characteristics simply because it does not have to observe the CUDOS norms.

Thus, life-world knowledge, although widely shared, is largely tacit, since its mental models are not required to be explicitly 'communal'; it is typically local, incoherent and frequently inconsistent, because it does not have to be 'universal'; it is practical, pragmatic and often deeply prejudiced, because it is not constrained to be 'disinterested'; its collective maxims combine heuristically into 'the conventional wisdom', which has no place for originality, or scepticism. One might add that it facilitates and recognizes action, rather than contributions to understanding, and that its most celebrated exponents are generalists, who decry specialization.

Now this puts the Legend into a spot of bother. It claims to be building a firm foundation for belief in science. But scientific knowledge cannot be purged of life-world knowledge. Therefore the credibility of life-world knowledge is a vital factor in the credibility of science[57]. But the Legend associates credibility with logical consistency, conceptual coherence, observational consensus, empirical reproducibility, resistance to critical testing, etc., etc. – i.e. with epistemic properties that are characteristic of science. These are properties that life-world knowledge does not normally possess. Therefore, the extreme epistemological claims of the Legend are bound to fail.

This is the general form of an argument that we have already encountered at various points – for example, in discussing the roles of personal testimony [5.7], intersubjectivity [5.10], visual representations [5.12] and natural language [6.6] in scientific reasoning. To proceed any further, we have to come to a decision about the epistemological status of life-world knowledge itself, independently of its scientific connections or putative scientific corrigibility. In other words, we must find some ground on which 'common sense' can stand securely on its own two feet, and not have to rely on the hope that it will eventually be made safe and underpinned by 'science'.

The *phenomenological* solution to this problem is to 'bracket it out'[58]. We simply adopt the 'natural ontological attitude'[59] [5.10], which is that many features of the life-world simply *are* as we experience them personally and try to describe them to each other. Although that need not deter us from trying to understand them better, we should not be dismayed if we fail to 'reduce' them to anything else. In effect, this attitude is consistent with the normal scientific strategy of formulating 'do-able' research projects [8.2] by freezing their inexplicable features into the boundary conditions and axioms.

The trouble is that as scientists we are well aware that much of the life-world knowledge that we take for granted in our research lacks scientific verification [5.8]. Mistakes, illusions, misconceptions, secrets, lies, contradictions, *non-sequiturs* and so on are normal features of 'everyday' life. We want to be sure that our results are not compromised by some error or omission from this source. Since it is never feasible to check every point of weakness that we might think of, we unconsciously classify them in terms of their likelihood and possible impact on our work. This is usually an automatic, unreflective cognitive process. Nevertheless, in scientific practice – as in all forms of practical reasoning – we do not really take all life-world knowledge to be *absolutely* reliable. Like the scientific knowledge that we apply more consciously [8.11], it is tacitly weighed up for 'credibility' as it is put to use.

But by what criteria can we assess the credibility of a particular item of life-world knowledge, other than by those of science? This is not, as is so often supposed, a matter of 'rationality' [6.8]. When necessary, common sense can be quite as well reasoned as theoretical physics, and a lot more credible. The real problem is the credibility of *empirical* knowledge, especially where this is eventually based on personal testimony [5.7]. The formal scientific strategy [5.8] is to 'collectivize' this knowledge by deliberately and systematically seeking confirmatory evidence from a multiplicity of independent witnesses. We instinctively adopt the same strategy, albeit subliminally and unsystematically, for much of our empirical knowledge of the life-world. What absolutely everybody seems to believe, we also believe.

Strictly speaking, however, the knowledge accredited by this strategy is bounded by the range of consensual witnesses. Our knowledge of the social aspects of the life-world is necessarily subject to this restriction. What we feel that we know for sure about people and institutions is limited in practice to what is accepted without question by fellow members of our community, which may be quite small. This is a real problem, to which we shall shortly return [10.4].

But we do also have individual shares, both as participants and as informants, in a substantial body of universally attested human experience. From this we gain unassailable knowledge of such basic 'facts' as that we are active, cognitively competent beings who inhabit with others the same three-dimensional world of material objects subject to gravity, death, taxes, etc. Here again, intersubjective consensus is the best policy – essentially because it is the only game in town. The alternative strategy of grounding life-world credibility on personal intuition – e.g., Descartes' *cogito ergo sum*, 'I think therefore I am' – is no more than the individualistic aspect of this same socio-psychological process [8.14].

For those aspects of the life-world studied in the natural sciences, the argument from universal consensus is normally so compelling that it scarcely needs further strengthening. But it can be reinforced by *evolutionary epistemology*[60] – that is, by systematic reference to the biological evolution of *cognition* [5.9]. In essence, this reminds us that the life-world is the environment in which, as organisms, we have to survive and reproduce. Our life-world knowledge is not merely stored mentally; it is coded organically into our behaviour, genetic make-up and bodily form[61]. In all these modes, it has been shaped by adaptive selection from the innumerable random variations and vicissitudes of the historical experience of our whole lineage, back to the beginnings of life on earth[62]. What we thus 'know', as individuals, about our life-world environment must, therefore, be very reliable, for it is a literally vital element in our 'fitness' for survival. If that philoprogenitive ancestor we worship had been seriously mistaken about, say, the intrinsic geometry of his jungle home, then that hungry tiger would have had him, for sure!

This argument is directly naturalistic[63] and does not involve any elements of intentional design. In this respect, it differs in principle from the evolutionary metaphor for scientific change[64], to which it is otherwise closely connected [9.7]. From a high metascientific viewpoint the biological and cultural modes of epistemic evolution merge conceptually into a single, continuous, *selectionist,* or *adaptationist* process[65]. In a certain sense, a colony of amoeba getting to know the chemical taste of a food is using just the same method of variation and selection as a community of cosmologists getting to know about the origins of the universe. But this move towards 'universal Darwinism' blurs a highly significant distinction between biological and cultural selection, and introduces metaphysical implications which are outside the scope of a naturalistic account of science.

Nevertheless, our modern scientific understanding of non-cultural epistemic evolution is entirely relevant to our theme. It shows us how our basic knowledge of the life-world is both justified and shaped by the way that the 'innate' cognitive capabilities of *homo sapiens* have evolved in response to the demands of that world[66]. For example, it provides a rationale for Immanuel Kant's characterization of (Euclidean) space and (Newtonian) time as 'synthetic, *a priori* categories', given to us by intuition rather than by the conscious exercise of reason[67]. The life-world is not necessarily, 'in itself', spaced and timed, but this way of representing its structure to ourselves and to each other has proved so indispensable for survival that it is now 'hard-wired' into our cognitive operating systems. The same goes for other Kantian categories, such as 'substance' and 'causation', which have come to seem *a priori* because they are so well adapted to the requirements of ordinary interpersonal communication[68].

From the perspective of the Legend the most disputable of these intuitive principles is probably 'induction'. But as we have seen, this is a sound epistemic strategy [8.10] with a simple evolutionary rationale[69]. Even for an amoeba, a propensity to act as if events that have already occurred several times will continue to do so is a natural adaptive response to the regularities of its life-world. Human beings owe much of their success as organisms to the further evolution of more complex cognitive capabilities, such as recognizing patterns, defining similarity classes, constructing 'maps' and mental models, and transforming these socially, through communication, into intersubjective representations. In other words, various modes of practical reasoning which are fundamental to science emerged originally as cerebral tools for coping with the hominid life-world[70].

Evolutionary arguments thus support the empirical observation that human cognition has a uniform substructure, right across the species. This applies not only to direct neuro-physiological traits such as the perception and differentiation of colours[71], but to our more abstract 'powers of thought'[72]. The cultural differences lovingly studied by anthropologists tend to conceal the degree to which all people have to be intellectually and emotionally alike to perform all the elementary interpersonal operations of social life – communicating, collaborating, mating, caring and so on[73].

This domain of human psychological uniformity, like the domain of human anatomical and physiological uniformity, must surely be quite extensive [10.2], although its actual scope is entirely a matter for conjecture. Other sentient beings – e.g. dolphins or extraterrestrials – would presumably 'know' things differently[74], but science fiction 'thought experiments' in which we imagine trying to communicate with them [5.9] produce nonsensical results unless we assume a solid substratum of shared perception and rationality.

Evolutionary epistemology thus affirms that 'knowing' is an integral part of 'living'[75], not just an add-on accessory, retrofitted when Adam and Eve ate that apple. It also indicates that our elementary representations of the life-world must correspond well enough with that world to have been naturally selected and re-selected in innumerable close encounters with it. In effect, the entities about which we have knowledge have been actively involved in the making of that knowledge. This participation in its very own construction gives much of our basic life-world knowledge exceptional epistemic authority – an *authenticity* that puts it beyond all reasonable doubt.

But this argument is obviously circular. It does not suffice to *prove* that this sort of knowledge is necessarily true. Just to know, for example, that our knowledge of life-world entities (including human beings) is being constructed requires that we already know a lot about human beings and their cognitive capabilities. In that light, all our understanding of the life-world, and of ourselves in it – including all of science – might be considered no better than a self-referential, bootstrap operation[76], ultimately unsustained. What evolutionary epistemology does demonstrate, however, is that this operation is self-consistent, socially coherent, and inclusive of both the natural and humanistic aspects of life[77]. This global 'hermeneutic circle'[78] is a self-organizing 'feedback loop'[79], a Buddhist wheel from which even death does not free us.

What is more, although our elementary representations of the life-world are surely authentic, they cannot be supposed perfect. As we have said, the outcome of the 'blind variation, selective retention' algorithm of evolution is not necessarily optimal, or even unique: it simply 'satisfices' the selective conditions of the problem on which it operates [9.8]. Thus, 'flat-earth' geography satisficed all the requirements of neolithic life, despite the roundness of our planet.

Life-world problems, moreover, occur in very different contexts, bounded by a great variety of conditions. The basic Kantian categories are cognitive devices by which we are adapted for biological survival in many different environments: they cannot be entirely truthful to all of them. Take the principle of induction. When a tiger is sighted, a snap decision based on fragmentary circumstantial evidence is the life-saving strategy: when gathering mushrooms, a well-informed toxicological inference after meticulous inspection is much safer. Inductive reasoning is a powerful instrument, but it has to be very flexible to fit us to all the irregularities of the life-world.

In any case, even as scientists we need different 'maps' to represent different aspects of the world [6.5]. The intuitive knowledge required to cope with a changing natural environment is selected for prediction and control, not for scientific understanding[80]. This applies even to our inborn capacity for reason and our latent propensities for mathematics.

These, too, evolved by 'natural' rather than 'rational' selection, and do not uniquely mirror some pre-existing properties of the life-world. Indeed, one of the paradoxes of 'cultural Darwinism' is that some of its most favoured products, such as the scientific solutions to technological problems, although selected for their 'rationality', can turn out to be so ill adapted to the life-world that they threaten the survival of our whole species[81].

10.4 Cultural contexts

But look at life-world knowledge from the social end. One thing we do know is that we are each of us born into, and immersed throughout life in, a particular human *culture*. What we directly experience or are told at second hand about most things in the life-world comes to us through the filter of a particular natural language, already entangled in a web of particular meanings [10.3]. Some of this knowledge is specifically individual – I am I, this is *my hand*, she is *my mother*, etc. – and some, as we have seen, is humanly universal. But a great part of it is shared only with other members of a particular human group. To belong to a culture requires active knowledge of a variety of *social* entities, such as personal roles, representational codes, symbolic objects, organized collectives and other *public institutions* characteristic of that culture. Whatever similarities these may turn out to have with corresponding entities in other cultures, their differences from culture to culture are just as noteworthy[82].

Respectful recognition of significantly different human cultures on this planet is not just politically correct or spiritually humble[83]. It is a prerequisite for any general understanding of those aspects of the life-world studied in the human sciences[84]. Any generalization about a particular type of social entity cannot be considered knowledge – even life-world knowledge – without an indication of its cultural scope. In the end, whether we are political scientists or hard-headed political activists, we have to be able to say 'thus it is amongst Europeans, but not amongst Africans', or 'ethnic Serbs and Croats do, in fact, run their villages in the same way.'

At first sight, this seems to raise purely methodological difficulties, to be met, for example, by systematic empirical cross-cultural comparisons. But taken to extremes it poses a serious threat to all the human sciences, if not to the whole scientific enterprise. The point is very simple. As we have seen, academic science relies on the exercise of intersubjectivity

[5.10], directed towards the achievement of consensus [9.2]. In the human sciences, this is achieved through *empathy* – shared understanding of personal and social experience [5.11]. But if the social life-world is subdivided into distinct cultures which are intelligible only to their respective members, this shared understanding must be similarly restricted[85]. Thus, if only Serbs can empathize with Serbs, and only Croats with Croats, then it is impossible to make sense of a statement about human behaviour in both cultures, even by way of comparison or contrast. Indeed, this limitation would apply not only to would-be scientific statements but to quite ordinary life-world statements, such as a newspaper report that Serbs and Croats all speak essentially the same language.

This argument is obviously very damaging to the credibility of all general discourse about human behaviour. It is not wrong in principle, as shown, for example, in the perennial debate about whether 'insiders' or 'outsiders' make the more reliable witnesses of social action [5.3, 5.7]. But its apparently devastating conclusion rests upon a number of assumptions that are only partially true.

In the first place, it assumes uncritically[86] that a 'form of life' is *holistic* and *hermetic* – that each culture has a complete answer to every question, and thus surrounds itself with a picket fence of incontrovertible life-world propositions barring every path into meaningful knowledge of any alien culture. But life-world knowledge is not a coherent epistemic structure like a formalized science[87][10.3]. From within its customary frame, all alien ideas may seem contradictory, fanciful, unrealistic, and so on[88]. But this frame is a collection of disconnected statements about what 'is', not a network of ideas that can ward off any intellectual probe that threatens to penetrate the gaps and glitches between 'self-evident truths'.

In most cultures, of course, life-world knowledge is partially structured by organized belief systems [10.5]. Practical sciences, transcendental religions and other established epistemic institutions tend to discourage awkward questions[89]. But such systems are never anything like as closed and comprehensive as they profess. This applies even to modern academic science, which breaks with traditional cultural norms by elevating the asking and answering of hypothetical questions into a positive virtue. In the life-world, moreover, just as in science, serendipity tears open the veil around what we think we know. This arouses curiosity – a propensity for sensing and trying to grasp unfamiliar meanings – which is another adaptive psychological trait shaped by biological evolution.

Another unrealistic assumption is that cultures are well defined, and quite distinct from one another – like, say, the nation states of political geography. In other words, each of us is supposed to have a specific 'culture', just as we are each supposed by international law to have a specific nationality. Whether or not this was ever the case, it is certainly false nowadays[90]. It is not just that most modern societies are multicultural mosaics: it is that most people are multicultural hybrids. I am an *elderly upper middle-class English male academic* of *Jewish* parentage, brought up in *New Zealand*. Each of these terms locates me in a community whose culture I am deemed to share. But these various communities are neither mutually exclusive nor coextensive. Am I so untypical in the number and variety of my cultural connections? All our experience of social life indicates that 'cultures', like scientific paradigms [8.4–8.5], nest inside one another in a vast hierarchy that extends from the individual person to the whole of humanity[91]. They partially overlap, are not homogeneous internally [92], and change radically even within a single lifetime[93]. Life-world social entities and their meanings – for example, natural languages – are partially shared, or merge across twilight zones[94]. The supposition that they are distinct, invariant and sharply delimited from one another is not sustainable.

In fact, we continually demonstrate that cultures are permeable and mutually intelligible by *translating* representations of their entities from one to another. Within every natural language there are resources for understanding other natural languages [6.6, 6.8], including its own past[95]. What is more, when two linguistic cultures come into contact, intermediate, hybrid languages – 'pidgins' and 'creoles' – inevitably appear in the 'trading zone' where people have to work together. This occurs even between scientific disciplines, such as between physicists and engineers, or between theoreticians and experimentalists in the same discipline[96]. People who are fluent in several languages disconfirm the notion that the same person cannot partake of two distinct 'forms of life'[97]. Similarly, legal systems have established procedures for comparing judgements between different jurisdictions[98].

This mutual interpretability may well be based primarily on the substratum of universally shared life-world entities [10.2. 10.3] – material objects, organic species, the necessities of human biology, kinship relations, the Kantian categories of practical reason and so on – that underlies all languages and all cultures[99]. But the global market in translations of novels and dramatic scripts – even of poetry – amply demonstrates that very complex social and psychological meanings can also be transmitted more or less satisfactorily across cultural boundaries.

Of course, literary translation is an extremely subtle art that must never pretend to mirror-like fidelity. But that applies equally to the initial literary challenge of representing the writer's personal slice of the human condition to fellow members of that culture. All that we ask of life-world representations and translations is that they should satisfice the requirements of their contexts. They cannot be expected to satisfy universal 'scientific' criteria of unambiguous communication – criteria which are, as we have seen, so strict that they are unattainable in practice.

Thus, epistemic naturalism does not support the notion that life-world knowledge is subdivided culturally into incommensurate domains. Indeed, this notion is incoherent, for it would exclude the possibility of obtaining any empirical evidence from which it might be inferred. Taken literally, it would make every man an island, so enclosed in his own particular 'form of life' as to be deaf to the tolling of any distant bell[100]. That describes the malady of autism [5.11], not the normal state of an articulate human being.

Let me insist that this is not an attack on the general principle that all life-world knowledge is culturally shaped and that where it applies specifically to social life it is socially 'structurated' [8.13]. Indeed, that principle is the very core of the argument of this book. Anthropologically speaking, all human ideas, emotions and institutions – the elements of 'human nature' as we know it – are cultural artefacts[101]. In general, these are far from universal in shape or function. But they are rough-hewn out of the natural materials of our common biological and physical circumstances – including 'society' itself[102]. These circumstances have regularities that generate a certain number of widely shared epistemic entities – cultural artefacts that are sufficiently universal to make possible the intersubjectivity and voluntary consensus on which science itself is based. Without this broad social base in life-world knowledge, there could be no scientific knowledge.

In confuting cultural incommensurability I am not downplaying the diversity of cultural contexts and the variety of meanings that people construct for the entities that constitute their social and personal lives. I am not disputing, for example, that gender differences [7.5] of life-world cognition and perception arising out of strongly institutionalized differences in the social roles of men and women, now and in the past, may well have a significant influence on what is sought and accepted as scientific knowledge[103]. Life-world knowledge with supposedly strong scientific backing, such as the perception of risk, can turn out to depend almost capriciously on its social context[104]. Classification schemes[105], representations of space, time[106] and number[107], and other Kantian 'common-sense' categories are social institutions[108]. They are encoded in natural languages which are not fixed over time, and may vary noticeably from culture to culture[109]. Indeed, cultural variations in 'what is found persuasive' in everyday reasoning are scarcely more extreme than disciplinary variations [8.11] in the 'logic of justification' in academic science[110].

Of course, tidy minds find cultural diversity messy and 'illogical'. But it is not an obstacle to science – not even to the human sciences – that ought to be eliminated quickly in the name of rationality and progress. On the contrary it is one of the glories of our life-world, a natural phenomenon to be celebrated, sustained and explored[111]. If that presents certain methodological or even epistemological difficulties[112], so what? As scientists, we can never possibly know *everything*, but we can at least get to know *something* about such difficult but fascinating matters.

10.5 Sciences, religions, and other belief systems

One of the 'trivial truths' [10.2] about the life-world is that human behaviour is *intentional*. We humans frequently undertake *actions*, that is, behaviour guided by conscious *thoughts*. The modes of thought thus put into operation are called *beliefs*. Indeed, it is just this consonance between thought and action [8.10] that distinguishes beliefs from other mental entities[113]. To be rational at all one must have beliefs and be ready to act on them – however misconceived they may be. And beliefs are, of course, the very stuff of scientific knowledge[114]. Indeed, the significance of *causality* [6.4] in scientific theorizing is probably no more than that it projects metaphorically on to non-sentient entities and events the human capacity for *agency* whereby we manifest our beliefs[115].

In principle, beliefs are capable of being rehearsed mentally, and expressed meaningfully[116]. But they are not just 'sentences stored in the head', waiting to be uttered to the next opinion pollster or saloon-bar acquaintance[117]. Indeed they usually only take shape as they are explicitly formulated, which often makes them difficult to state coherently in words. Yet they are not entirely elusive, haphazard or inconsequential[118], for they provide the actor with an intelligible narrative concerning what she does, or might possibly do. They are the structures in 'World 2' – the realm of our individual inner thoughts [8.14] – that link us dynamically with the life-world.

All life-world knowledge is by definition 'believable', in that it is normally 'taken for granted' when we decide to do anything. When, for example, I walk down the street to a letter box, and post a letter, I am acting on my beliefs in the material solidity of the pavement, the spatial receptivity of the letter box, the social reliability of the postal service, and so on and so on. These beliefs are so well founded that I am scarcely conscious of them. I just carry them around in my head during the ordinary course of life and only reflect on them when there is an earthquake or a postal strike. Sometimes they are stored as propositions – 'this letter box is cleared daily at 4 p.m.' – but very often they constitute mental models, 'maps' or other non-verbal internal representations[119], which I manipulate hypothetically [8.14] as I imagine my action – 'let me see, should I turn left or right as I go out of the house?'

Indeed, the mundane beliefs that directly guide everyday action are so closely tied into that action that it is often difficult to separate the word from the deed [10.2]. They form a tightly woven substructure whose elements seem almost 'unnatural' when looked at separately[120]. But when things don't go quite according to plan, mental linkages are activated to a superstructure of more general beliefs, such as practical maxims, legal principles, religious teachings, tribal myths – or, as it may be, even scientific theories. The human mind is capable of sustaining complex *belief systems*, which are not closely coupled with on-going life-world action but which people need to obtain the straws of guidance or crumbs of comfort required to go on acting rationally in the face of the unexpected[121].

Such an over-arching belief system is, of course, a social institution[122]. It permeates the language and logic that people ordinarily use to communicate with one another. In consulting it for guidance we are seeking support from our community in terms of our shared values and meanings. Nothing can be more distinctive of a culture than its belief system. The human sciences and liberal arts celebrate the marvellous diversity of the religions, myths, traditions, sacred writings, folk sciences, supernatural entities, etc., etc., in which various peoples variously believe.

But a belief system is also, of course, highly personal. Individual members of a social group feel that their beliefs are their own, even though these, when revealed, are usually only minor variations on a common theme. Psychically, we are each of us very much what we believe in[123], generally as well as specifically, abstractly as well as concretely, for eternity as well as for now. The liberal arts and human sciences also explore this rich landscape of the mind, continually uncovering new cognitive and emotive linkages to the network of words, images, symbols and 'maps' by which people represent to themselves the meaning of things.

We must stop here, on the verge of much deeper waters. Any account of the origins, forms, functions and dynamics of belief systems must have both psychological and sociological dimensions[124]. Why is it, for example, that many widely accepted beliefs, even amongst well-informed, well-intentioned and intelligent people, are patently false or absurd[125]? Do they simply minister to hidden personal needs, or are they typically *ideologies* serving powerful societal interests? The balance between the individual and social aspects of belief systems is not at all irrelevant to our theme, but far too difficult to discuss further in this book.

But in what sense is such a system held to be 'true'? From a naturalistic point of view, there can be no doubting the sincerity of those who swear by it, and are ready to live or die for it. But most people are not zealots: they simply 'take for granted' the system of general beliefs established in their culture, accepting it loyally without thinking seriously about its validity. This is not because they are infinitely credulous, or whipped into conformity. It does not imply that 'falsity' and 'absurdity' are just social conventions[126] which have no meaning for the isolated individual. It is because the superstructure of abstract beliefs is *metaphysical*. It seems to belong to a 'world' of its own, cut off from the life-world. The gods, so to speak, are so remote, so little involved in earthly affairs, that there are very few opportunities for directly challenging their decrees.

Life-world reasoners are perfectly acquainted with the notion of credibility in mundane matters [5.7, 8.11, 9.5]. But they are seldom under any compulsion to apply this notion to their over-arching beliefs. In practice, discrepancies between beliefs and realities depend on the context. A general doctrine cannot be applied to a specific case until it has been *interpreted*. And for ordinary people in unusual circumstances, that requires the services of specialists. As Azande witch-doctors, Delphic priests, Marxist intellectuals, Freudian psychoanalysts and other professional interpreters have amply demonstrated, it is easy to distract attention away from the credibility of generalities by focussing on particulars.

Thus, when specialist advice leads to disaster, people are told, or argue for themselves, that they must have misunderstood what ought to have been done in these specific circumstances, or sacrificed this time to the wrong god, or inverted the symbolic meanings in this latest dream. They find it easier to accept such explanations than to risk losing faith in the general principles on which they think they have acted, or in the collective wisdom of the specialists whom they have had to trust[127].

As history shows, belief systems not only *belong* to communities: they also *generate* communities and social organizations, such as churches, typically led by specialist interpreters. Nevertheless, as history also shows, even the most entrenched belief system is vulnerable to external competition or internal 'heresy'. Such events, like scientific revolutions [9.6], have social, psychic and cognitive elements that interact elaborately. But they always show that the old regime was nowhere near as coherent, systematic or comprehensive as its defenders claimed.

These failings are characteristic of even would-be 'universal' religions. It is notorious that the century-long labours of whole colleges of theologians tend rather to fragment than to consolidate their respective doctrines. To the cold eye of a non-believer, every religion reveals itself as an irregular mosaic of wise teachings, natural observations, traditional practices, metaphysical dogmas, humanistic values, wishful thoughts and so on. Omnicompetence and infallibility is not a demonstrable holistic property of a belief system. It is either 'taken for granted', or it has to be affirmed by the Faithful as a specific item amongst their beliefs – e.g., 'There is no God but Allah!'

What is more, the protagonists of rival belief systems are seldom unwilling to debate their respective beliefs. But their invariable inability to achieve agreement is not evidence of total 'incommensurability' between their systems [10.4]. On the contrary, such disputes usually show that on many points they understand the issue perfectly well, but cannot agree on how it should be resolved. Every such system is riven with incongruities, heterogeneities, ambivalences, sectarian variants and other epistemic imperfections. These provide plenty of entry points for rational criticism, as well as logical minefields to hold off organized attack.

I am not here denigrating organized religion, or any other supposedly 'universal' belief system. I am just pointing out that in many respects all general belief systems are alike. What is more, this applies also to science, for that too operates as a general belief system in relation to the life-world[128].

To see this, remember that scientific knowledge is much more than a great body of specific 'facts' and 'concepts'. To count as 'scientific', these have to be mapped into a network of theories, models, classification schemes, etc. [6.4, 6.11]. In life-world practice, this network constitutes a system of general beliefs, linked back and forth to the innumerable mundane beliefs that directly guide our everyday actions. This belief system sustains and is sustained by the culture of 'modernism' that permeates contemporary life. In use, moreoover, scientific knowledge requires specialized interpretation, which is provided, of course, by organized communities of academic scientists.

Not surprisingly, scientific knowledge is normally 'taken for granted' by those who do not fully understand it, but need it to guide their actions. It is commonly presented in an orderly form [6.6], like the articles of a creed. What is more, science has developed many of the institutional features of an established church[129]. Its devotees often treat it as if it had all the other attributes of an organized religion[130]. They try to model all their thinking on its epistemic practices[131] and come to believe that it is omnicompetent[132], even though this is no more justifiable than it is for any other system of general beliefs [10.8].

In other words, 'sciences' and 'religions' are very much alike, in that they are general systems of belief from which people seek guidance in their life-world thoughts and actions. But where they overlap they usually offer quite different 'maps' of the same aspects of 'reality' – the origins of the universe, the unity and diversity of organic life, human prehistory, the nature of consciousness, personal identity, the significance of history, the social roles of individuals, and so on. These are matters on which scientific and religious beliefs notoriously disagree and often arouse violent social conflicts.

Indeed, members of different religious groups – Christians and Buddhists, Moslems and Mormons, Shintoists and Quakers – disagree profoundly on many issues. And so too do members of different research communities. Quantum cosmologists, molecular biologists, evolutionary ethologists, behavioural psychologists, functional sociologists, classical economists, cultural theorists and other academic tribes all profess allegiance to the doctrine of Unified Science. But their grand paradigms [8.6] often 'map' the same features of the world in very different ways, and offer very different guides to practical action. Thus, the relatively recent notion that 'science' and 'religion' are so different as to be incompatible[133] plays down the diversity and discord on each side of the divide.

What then is the real distinction between science and religion? Our naturalistic approach might suggest that it is not so much in the *substance* of their respective beliefs as in the *attitude* of the believers. In particular, the scientific ethos rejects the *dogmatism* that is said to be so typical of religion. In principle, even the best-established scientific paradigms are never given absolute credence, and are always open to correction [9.2]. Science is devoted to doubt, discovery and epistemic change [9.6], whereas religion emphasizes faith, revelation and orthodoxy.

This distinction is surely valid, but very far from absolute. As we have seen, science rarely lives up to its ideals. Scientific paradigms often become socially entrenched, and are presented as if entirely beyond question. The notion that science is never dogmatic is one of its dogmas! At the same time, not all religious systems are hostile to originality and scepticism. Hinduism and Buddhism are continually open to new wisdom gained by personal enlightenment. Even a 'revealed' religion such as Judeo-Christianity or Islam, where any line of argument can be closed off by reference to a text provided by an omnipotent deity[134], can never be systematically fundamentalist. Its teachings are reshaped by Prophets and Saints. Its founder texts become the focus of creative heresy, critical debate and doctrinal re-interpretation. For example, vigorous scholastic controversy within mediaeval Christendom created a fertile intellectual seedbed for new belief systems, such as Reformation theology, Renaissance humanism and scientific naturalism[135]. The notion that 'religion' is always dogmatic is also a scientific dogma!

The key difference may well be that religion, unlike science [ch.7], is not required to be 'disinterested'. This norm excludes from science the specific personal values and social interests [7.4] that always loom so large in the thoughts accompanying life-world action. Religious beliefs, by contrast, concentrate on the ethical, aesthetic and spiritual features of the human condition[136]. To give meaning to these features, they have to be much more sensitive to cultural contexts[137] [10.4], and much less finicky about standards of credibility, than scientific theories. Religious knowledge differs profoundly from scientific knowledge because it is directed towards the achievement of different goals[138].

But here again, the demarcation line has shifted over time, and is still far from fixed or precise[139]. When particle physicists quip that their 'Holy Grail' is a Theory of Everything [6.5, 10.8], is their quest so mockingly irreligious? When molecular biologists announce that the essence of life is 'information', are they not echoing ancient theological claims? When culture theorists deconstruct ideologies, do they not feel as virtuous as they would if they were hunting witches? And conversely, is it futile to search for ultimate meanings in Big Bang cosmology or the Gaia hypothesis?

To go much further on this issue would need as elaborate an analysis of 'religion' as we are trying to give of 'science'. Academic science is only one of the plurality of institutions and sub-cultures that co-exist uneasily in modern society. It is differentiated from these other institutions by a number of peculiar social practices, such as those summed up in the Mertonian norms. As a consequence, scientific knowledge is somewhat different from the other belief systems with which it competes for life-world influence. Science is a positive, aggressive, proselytizing institution, but its jurisdiction over human action is limited by its own norms. In many areas of knowledge it can confidently expect to supplant its rivals, but there are many fields of action where it can only complement them – and then not always for the best[140].

For this reason, the programme of *relativism* [1.2, 1.4, 3.7, 10.7] espoused by the 'sociologists of scientific knowledge'[141] [1.3] cannot succeed in being as 'strong' as they would like[142]. As we have seen, the scientific culture supports a distinctive system of beliefs, but this is not privileged above, sharply differentiated from, or completely incommensurable with, the belief systems current in other cultural contexts[143]. In a pluralistic society, these systems overlap and contradict one another at many points. That is to say, there are many occasions when we face sharply conflicting opinions as to the reliability of a particular item of knowledge, or the wisdom of a particular course of action. In the end, what we decide to be 'the truth of the matter' is bound to be relative to the system of beliefs that we happen to hold most firmly.

Metascientists and other epistomological analysts are obviously very interested in the considerations that enter into such decisions[144]. This research is facilitated by an open-minded stance where conflicting belief systems and categorial frameworks are taken at their own valuation, or at least at a fair estimate of their credibility in the mind of the person making the decision. The history, sociology and philosophy of science would be meaningless without such an attitude towards the superseded science of the past, or to the views of the various parties in a scientific controversy[145].

But a generous methodological stance of 'cognitive relativism' cannot be inflated into an independent belief system in its own right. Although the streetwise maxim that 'everybody is right in their own way' is unassailable in principle, and in appropriate circumstances much to be commended, it is futile as a general guide to action. There are many variants of 'relativism'[146], but the 'stronger' they become the nearer they get to the black hole of total scepticism[147], from whence no philosophical traveller returns. It may well be instructive to say that the truth about planetary motion was genuinely different for Ptolemy, Copernicus, Newton and Einstein respectively, but that would not justify setting off on a space mission as if one had no reliable knowledge of the workings of gravity. *Systematic* scepticism [*8.10, 9.1*] is a nihilistic metaphysical doctrine, which denounces the rationality of any form of belief. It is quite the opposite of *organized* scepticism, which assesses rationally the credibility of the beliefs that we must have in order to exist at all[148]. As Sir Thomas Browne put it, 350 years ago[149]: 'The Sceptics that affirmed they knew nothing, even in that opinion confuted themselves, and thought that they knew more than all the world beside.'

10.6 Science and common sense

Science must not be taken entirely at its own valuation. Its nature cannot be properly expressed in its own terms. The 'epistemic context' of academic science is clearly just as rich and diverse as its 'social context'. Let us now look at it against that background.

The relationship between science and its life-world counterpart – 'common sense' – is particularly complex [*10.3*]. They are sometimes presented as adversaries, at other times as allies. They are often contrasted – as in Wittgenstein's metaphor of an ancient haphazard city with newly built orderly suburbs[150] – and yet turn out to be intricately mingled[151]. We presume that we know the difference in principle, and could, when necessary, separate the precious metal of scientific knowledge from the worthless ore of mere opinion. But then in doing this we find that we have to rely on our everyday grasp of 'the way things are', and exercise the powers of reasoning that we applaud as 'common sense'[152]. Thus, although scientific knowledge has many very distinctive features[153], it cannot be formally constituted as a separate, self-contained epistemic domain[154]. Even its grandest theoretical paradigms are inferred from and rooted in down-to-earth empirical 'facts' and always have to incorporate uncritically a great deal of 'taken for granted' life-world knowledge[155].

Scientific knowledge is only a small component of what is known[156].

And yet science is not just a systematic enlargement of 'folk science' [5.12], or 'common sense writ large'[157] [10.3]. On the contrary, it is in its conformity to the *small* print of common sense that science is distinctive. Even its most empirical components – field observations, numerical data, experimental results, classifications of natural entities, etc. – have to be carefully selected and elaborately reshaped to be made acceptable to a research community. Taken one by one, the cognitive norms that have to be satisfied – accuracy, specificity, reproducibility, generality, coherence, consistency, rigour, and so on – are all perfectly commonsensical: but they are seldom applied simultaneously outside science[158]. Indeed, any collection of 'facts' conforming outwardly to these norms is commonly called 'scientific', and given credence on that basis alone[159].

But 'facts' [ch.5] really only acquire scientific significance through their connections with 'theories' [ch.6]. Scientists represent various aspects of the world by means of elaborate networks of laws, concepts, models, classification schemes, formulae and other mental 'maps' [6.4, 6.10]. In their most sophisticated form, these are symbolic constructs in abstract 'scientific domains' [6.11] that are often strikingly unlike the life-world domains they represent. Nevertheless, in combination with the 'facts', instruments and research practices that support them, they are organized into the nested hierarchies of paradigms [8.4] that give scientific meanings to actions at every level down to the most mundane. The ultimate epistemic significance of science is as a general system of *beliefs* of this kind[160] [10.5].

As we have seen, scientific beliefs often conflict with other beliefs about the life-world. This can occur at the most mundane level. Systematic investigation on scientific lines often produces empirical evidence confuting a generally held opinion on some practical matter, such as that roses are best pruned in the spring rather than the autumn, or that it is necessary to pay business executives enormous salaries to get good work out of them. Much of what we call 'common sense' is associated with dubious maxims that can be disconfirmed and rejected – or, alternatively, confirmed and strengthened – on the basis of phenomenological theories inferred from more rigorous analysis. At this level, 'scientific rationality' [6.8] – very often in technological form – operates as an adjunct to ordinary modes of practical reasoning in revising or extending 'common sense'[161].

Frequently, however, a conflict between mundane beliefs is referred upwards to higher authorities in their respective belief systems [10.5].

The conventional rationale of executive pay scales, for example, is part of a system of very general popular beliefs about the economic efficiency of market competition. What counts as 'common sense' cannot be separated from commonly held belief systems of much wider scope[162], even when these are in direct conflict with scientific beliefs of similar scope. The assertion that a particular belief is 'unnatural', or 'contrary to good sense' then becomes a typical rhetorical resource in a conflict where much more is thought to be at stake than the rationality of this specific idea. That, surely, is how people reacted to Copernicus and Darwin, whose theories were derided as 'nonsense' primarily because they were at variance with the religious doctrines of the day.

A stand-off between 'science' and 'common sense' becomes more understandable as one moves upwards from the life-world into the higher reaches of scientific theory. As we have seen, the scientific domain contain some very strange creatures [8.12, 9.5]. It is not so much that many hypothesized entities are very abstract[163], and remain 'invisible' long after their existence has been established beyond reasonable doubt. Indeed, theoretical physicists are like pure mathematicians, in that they are often interested in the hypothetical behaviour of entirely imaginary objects, such as parallel universes, or particles travelling faster than light, whose actual existence is not being seriously proposed at all.

The trouble is that that some of the entities that scientists do hold to be part of the natural order have such peculiar properties[164]. It offends against ordinary reason to be expected to believe in quantum objects like quarks, or relativistic objects like black holes, whose behaviour is so patently counter-intuitive, even logically self-contradictory. Many people – scientists amongst them – insist on a showdown. These mental monsters, they argue, must either be re-defined in commonsensical terms, or else assigned to a limbo of 'unnatural', at best provisional, scientific notions whose only merit is that they have useful predictive capabilities.

But as we have seen, that is not what usually occurs. In the crucible of scientific practice, hypothetical entities either melt into oblivion [9.2], or crystallize into firm belief. In due course, they become part of the mental furniture of a research community[165], and are simply 'taken for granted' in the private thoughts and public discourse that accompany scientific action[166]. This local 'common sense' diffuses outwards, along with the more tangible objects and technological artefacts produced by research, until – perhaps a century later – it has become part of what 'everybody knows'. That is what eventually happened to the 'irreligious' concepts of

Copernicus and Darwin, and is slowly happening to the 'illogical' concepts of Einstein and Bohr.

In other words, 'common sense' is not a fixed epistemic context for science. It is not a standard mode of reasoning based on a static body of knowledge. It changes over time – just like science[167] [9.6]. For example, much of what passed for 'common sense' in the Middle Ages would still be perfectly acceptable now – but much would now be considered untenable. It is not our task here to analyse such changes, which are only partly due to the rise of modern science and technology. It is quite clear that scientific, religious and 'common-sense' beliefs are the products of distinct social institutions and have distinct cultural functions [10.4]. But they are strongly linked in thought and action, and define each other as they evolve together[168] [9.8]. Thus, when we contrast a scientific belief with 'common sense' we are indicating that we think we could define it precisely and give some coherent account of why it should be relied on, rather than just 'taking it for granted' as an undeniable truth[169]. In practice, that may make little difference to what is believed – and yet it may make all the difference in our attitude to what we and our fellow citizens think and do[170].

10.7 Realism

Much metascientific discourse is about whether scientific entities are *real*. In everyday usage, this merely implies that we believe unreservedly in something and act accordingly. But this apparently simple term[171] is now so encrusted with philosophical fossils that even its dictionary definition begs many questions.

In a naturalistic perspective, however, scientific realism signifies belief in *public invariants*. It implies that some of the features on the mental 'maps' on which we individually base our actions are *cognitively objective*[172]. That is to say [7.9], they are not only shared in the thoughts of other people, but are also not affected by our (or anybody's) actions, perceptions or thoughts[173]. Typically, these features interweave into a stable, richly textured context for our personal existence. This is what we and fellow mortals call 'the external world'[174]. The 'reality' of this world is not something that we infer about it: it is an inseparable characteristic of our developing knowledge of it, as persons, as communities, and as an evolving species of conscious beings[175].

Now here is a paradox. Why should we even question the reality of scientific knowledge? Isn't the whole purpose of science to explore and map 'the real world'? Isn't the scientific 'method' just a way of ensuring that this map represents all those features, and only those features, on which everybody agrees? For nine long chapters I have been invoking this quasi-geographical metaphor to make just this point – and now I am casting the whole enterprise into doubt. For most people, reliable, well-tested scientific knowledge is the acme of 'reality'. This attitude may be naive, but what can be wrong with it? What is bugging those pettifogging philosophers and their anti-science allies?

The problem is the norm of scepticism. This seems to have an anti-realist slant. Although it does not license comprehensive, all-embracing, metaphysical doubt, it does require scientists to look critically at every new hypothesis, and accept the possibility that even their most settled beliefs may turn out to be unsound. This can be psychologically devastating[176]; witness the astonishment of the geologists when they discovered that the continents themselves were not invariant features of the global map. And yet working scientists go to great lengths to explore the consequences of quite wild conjectures [8.10, 9.5, 10.5]. Indeed, one of the fascinations of the scientific life is the extraordinary ideas that apparently sober scientists sometimes arrive at personally and treat as real, even though their colleagues do not find them in the least bit credible[177]. What is more, most of the entities studied in the human sciences are clearly 'constructs', often with quite arbitrary characteristics [8.13]. Is a 'research community', for example, a real 'public invariant', or just a name for an evanescent association of individuals?

From a scientific point of view, 'reality' and 'objectivity' are correlated with 'credibility'. As we have seen [8.10], this is a variable quantity that lies between zero and unity, but can never attain either of its limits. There are thus many ambiguities and ambivalences in scientific practice that are difficult to reconcile with the absolute certainty – yea or nay – of the quintessential realist[178]. Can they be resolved in principle, or is this a deep fault line in the logic of scientific discovery?

The Legend favours 'strong' scientific realism, but is very divided on how to justify it[179]. With one hand it still encourages the traditional endeavours of those seeking a formal proof that science will ultimately yield the 'truth'[180] – that is, will give an account of the world as it 'really' is. As we have seen, this is a fruitless enterprise that was scuppered just 250 years ago by David Hume. With the other hand it brushes aside all objections of principle and points to the practical *reliability* of scientific knowledge [7.3, 7.9] – 'science must be for real because aircraft fly', etc. This pragmatism[181], often reduced puritanically to technological *instrumentalism*[182], is so deeply engrained in the scientific culture that it is the 'natural attitude' there. But although it amply confirms our prior belief that there is a physical 'reality' which we all share[183], it fails to deal with genuine anxieties about what we think we know about it[184].

By contrast, *constructivism* [8.13] is anti-realist. In its most extreme form, it argues that scientific knowledge is the product of intentional human activity[185], and is therefore essentially a 'fabrication'. But as we have seen, research results are as much 'discovered' as 'made'. Overzealous constructivists systematically underplay the refractory nature of the material out of which scientists have to fashion credible research claims, and ignore the processes by which these have to be shown to be publicly invariant before they are accepted. What are we to make of a 'sociology of scientific reality' that divorces itself thus from institutionalized scientific practice?

What is more to the point, scientific knowledge is so intermingled with 'common sense' [10.6] that it cannot be supposed to be 'unreal' without supposing the same of the whole life-world. But that would negate our 'taken for granted' belief that other people are in the same world as ourselves [5.10]. This belief is not only our basic paradigm of 'reality'[186]; it is also an essential presupposition of rationality[187]. Intersubjectivity is vital to the notion of a 'social construct'[188]. Doctrinaire constructivists who deny this are busy sawing off the very branch on which they are trying to build their nest!

As we have seen [10.3, 10.6], 'common-sense' realism extends to the social aspects of the life-world[189]. Considered as public invariants, governments and their taxes are as real and cognitively objective as mountains and snow. A social institution is a coherent entity, an assembly of 'routines' whose intersubjective reality is collectively maintained by the similar representations that we have of it in our individual thoughts and conversations[190]. In other words, it is one of the facts of life about which we talk together, and thus a suitable subject for scientific study.

Indeed, it is impossible in practice for the human sciences to adhere strictly to the positivist dictum that they should limit their accounts of 'social reality' to its material manifestations [5.4, 5.12]. They cannot count heads, or make videos of behaviour, whilst ignoring the institutions that give meaning to all social action[191]. What is more, most of the material objects that figure in our representations of the life-world are closely connected with social institutions. Thus, for example, a stone axe is not just a lump of rock: it is 'in reality' a meaningful artefact that cannot be separated conceptually from its mode of use[192].

Nevertheless, as I hinted above, the reality of social institutions is compromised by their *reflexivity* [7.7]. That is to say, they are created and held together by *symbolic representations* which are always open to questioning and re-interpretation. The potential power of a commercial firm is in the numbers on its bank statement and the legal wording of its property deeds. It is structured internally by managerial memoranda and externally by invoices. It arose out of a prospectus – and yet may end in bankruptcy. In other words, for all its personnel and materiél, it is essentially a 'social construct' [8.13]. It may have too much collective credibility to be wished away by any single individual, but it does not exist independently of human thoughts, perceptions, words, actions, values and dreams[193].

Once again, we must not step further into the sociological jungle. It is clear, nevertheless, that the reality of the social life-world does not require all social entities to be permanent and unchanging. Indeed, the birth and death of institutions, as of people, is one of the publicly invariant features of 'real' social life. What we 'take for granted' is the stability and integrity of the structure as a whole. We know very well that its individual components, although immaterial and transcendental, are not immortal. But we often treat them as such as a matter of policy – although not without qualifications or criticisms of the way that they are constructed. Thus, the naturalistic realism[194] of our beliefs about many such life-world entities might variously be described as 'transcendental realism'[195], 'policy realism'[196], 'qualified realism'[197], 'convergent realism'[198], 'critical realism'[199], or 'constructive realism'[200], rather than the metaphysical 'naive realism' invented and belaboured by philosophers.

From this standpoint, the reality of scientific entities scarcely seems problematic. It is abundantly clear that a generally realist orientation is basic to all the paradigms of rationality embodied in the scientific culture[201]. In effect, the norms of academic science require scientists to behave as if they believed in a shared external world which is sufficiently uniform that they can usefully exchange information with one another about it[202]. One might say that they do this as a matter of policy, but it becomes so ingrained by training and practice that it is completely 'taken for granted'[203]. This applies even to scholars in the human sciences – and even when they are remarking on the influence of ideological constructs, reified ideals and other unreal entities in social life!

This belief is qualified by the knowledge that even the most intensive scientific research can only produce imperfect representations of 'reality'[204]. Indeed, the quantum-mechanical representation of the world of microphysics is so systematically 'indeterminate' that it escapes all our common-sense intuitions [10.6]. But this 'reality' is common to all researchers. Thus, it makes sense for them to share, compare, revise and combine their diverse 'facts' and 'maps' into a version, or a portfolio of equivalent versions[205], on which they all agree[206].

In practice, the structuration of scientific knowledge is an elaborate process, which is often aborted and never finally completed [8.13, 9.2]. Nevertheless, along the way, it acquires increasing credibility until it becomes integral to the personal 'reality' [8.14] of those scientists who habitually base their actions on it[207]. This belief diffuses into wider research communities, extending eventually to the general public. In other words, in becoming 'established' a research claim is transformed into a general social institution, and can thus be treated as 'real' in the same naturalistic sense. Scientific knowledge joins life-world knowledge in the same spirit of 'policy', 'qualified', 'critical' or 'constructive' realism as most of the other things that we 'take for granted' in everyday life.

Science, as we have seen, produces a diversity of these agreed representations of particular aspects of 'the external world' [6.5, 6.10]. Each of these is open to revision, theoretically or empirically. In the course of time, however, they tend to fuse into a more general *schema*, which is taken to represent 'reality' on a much broader scale. As we have seen, this process is driven by the norms of communalism and universalism, which impel scientists to seek linkages, to reconcile inconsistencies [8.11], and even to gloss over the actual gaps and misfits between their best-established paradigms[208]. The highest-level, over-arching paradigm [8.5] that is thus generated is what we call 'scientific reality'. Like 'social reality', this is far too large, elaborate and pervasive to be challenged or disregarded as a whole by any sane person. But it is a joint product of 'nature' and 'society', and like all scientific knowledge slowly evolves through its own inner dynamic.

This argument can be traversed in reverse order. Suppose that we look carefully at the origins of scientific beliefs, and from this analysis decide that these beliefs should be assigned to a certain category of realism – 'policy', 'qualified', 'critical' or 'constructive', as it may be. Then as we have seen [10.3, 10.6], this category must also include the common-sense beliefs whose credibility depends on scientific research or some similar evolutionary process of variation and intersubjective selection[209]. In the end, we find that even the most ordinary life-world things, such as aunts and uncles, tables and chairs, cats and dogs, are no more or less real than electrons, genes, dinosaurs or other things that we normally think of as distinctively 'scientific'. The challenge to general philosophy is not to show how scientific realism differs fundamentally from, or negates, life-world realism but to investigate the close relationships between them[210].

True believers in absolute scientific realism see post-modern scepticism as the ultimate flight from reason[211]. But they fail to understand that the post-modern critique is strictly non-metaphysical, and applies only to attempts to *represent* objects existing outside the mind. Such a representation can be perfectly 'real' as a public guide to thoughts and actions, but that does not require it to be complete, timeless or unique[212].

What is more, a scientific 'map' can always be analysed more and more precisely, and traced further and further back into wider networks of representation. Post-academic science will be enlarged and enriched by this process of deconstruction – typically towards greater generality and abstraction. But 'Mode 2' research [4.8, 7.9] will also be more deeply rooted than Mode 1 in problems arising in life-world contexts of application[213], where purely pragmatic reliability [7.3] is the principal criterion of validity. There is no reason to suppose that it will lose contact with the 'common-sense' operational realism that has always flourished at the scientific coal-face.

10.8 Unified by reduction

The principal tenet of *scientism* is that science is progressing towards producing a complete, comprehensive 'scientific world picture', which will constitute the ultimate 'reality' [10.7]. But scientific knowledge excludes many human aspects of the life-world, such as moral and aesthetic values [7.4, 10.5], which are just as real as physical data and biological traits[214]. This is therefore a metaphysical goal[215] which is not attainable even in principle[216].

Nevertheless [6.11, 8.3], in spite of its fragmentation into separate disciplines, academic science does have an inbuilt tendency towards epistemic *unification*[217]. In effect, the norms of communalism and universalism encourage research into all intersubjectively accessible aspects of the world, and the elimination of all apparent inconsistencies between the various ways in which they are represented[218]. If this resulted in a unified 'map' of 'the world according to science', it would be an immensely powerful instrument of thought, for it would allow one to reconstruct mentally a representation of any aspect of the world as perceived by any organism from any chosen point[219]. Indeed, instead of seeming just a map or picture, it would be a virtual reality, ready to be 'taken for granted' in life-world action[220].

In practice however, as we have seen at a number of points, the scientific culture is not systematically directed towards the construction of a global map of knowledge[221]. The projects, procedures, instruments, epistemic criteria and putative discoveries of specialized research communities are not coordinated in advance[222], are often incompatible[223], and are only rendered partially coherent by later social structuration and codification. What results is more like intercalated brickwork[224], a cobbled-up net[225], a fish skin of overlapped scales[226], a many-stranded cable[227], or a much-repaired clinker-built ship[228] than, say, an extremely long, error-corrected sentence about the whole cosmos[229].

In any case, even if the sciences were to get their act together to join up their 'maps', the task would be endless. The potential subject matter of research is inexhaustible[230]. Every solution breeds more problems [8.1]. Every detail that we at last understand has finer details requiring further explanation. Every local approximation is a challenge to stricter analysis in a wider frame[231]. The horizons of achieved knowledge contract as we dig deeper, but recede as we struggle towards them[232]. Scientific knowledge is always expanding too rapidly to be grasped as a whole.

Thus, the 'scientific world picture' about which people often talk is not a representation of the cosmos as directly revealed by scientific research. That would be complicated beyond comprehension [6.5]. It is a highly simplified and stylized representation of certain general features of the knowledge stored in the scientific archives [9.3]. A small 'map' of a large territory may well convey valuable, well-attested communal knowledge, but without the aid of satellite imagery it is not the outcome of direct perception.

Even the truest of maps [6.5] is necessarily a *schema*, whose intelligibility depends on the principles on which it is constructed. As we zoom out to enlarge the area of coverage, these principles become more abstract. In other words, the 'scientific world picture' is a theoretical construct, a pattern that people claim to be able to recognize in scientific knowledge, taken as a whole. Thus, many of the general features that we think we find on such a map are very likely to be ones that we have unwittingly put onto it as we selected and represented the items that it depicts.

Dame Nature does not provide us with a single, simple set of principles for classifying scientific knowledge [6.2]. There is no innate order to the great forest of paradigms [8.4] under whose canopy scientists hunt their prey. In practice, however, academic science is roughly mapped into disciplines in terms of their supposed 'complexity'. The conventional schema is of a stratified hierarchy [2.6, 8.5], with the human sciences at the top, the biological sciences below them, and the physical sciences at the base.

When people talk about the need for science to be 'unified' – as distinct from merely constituting a multiply connected network of facts and theories[233] [6.4, 6.11] – this is the type of structure that they usually have in mind[234]. Indeed, the 'layer-cake' metaphor is so much taken for granted that we usually forget its obvious imperfections. In reality [8.5], the objects and concepts of scientific research are elaborately cross-connected[235], and do not fall neatly into semi-autonomous 'levels' that can be explored independently of one another[236].

Nevertheless, despite its deficiencies, this metaphor is very attractive. This is because it points to the 'reductive' relationships that seem to hold between certain fields of research. That is to say, many disciplines are concerned with compound entities [6.9] whose components are studied separately, from a more 'fundamental' point of view, under another disciplinary heading. The chemistry of molecules is related in this way to the physics of atoms, just as the sociology of institutions is related to the psychology of individuals, and so on.

A 'systems analysis' of the natural world [6.9] can never be better than an approximation[237]. Nevertheless, 'weak' reductionism is a powerful research strategy. A great many features of the world can be explained [10.1] by separating a complex entity into simpler constituents, or by inferring a general theoretical principle that covers a number of special cases. Explanatory linkages can thus be made through the interfaces between apparently distinct fields of knowledge[238].

What is more, this strategy is often so successful that it seems as if it could be inverted into a process of synthesis. Researchers begin to think that if their analysis were sufficiently thorough it might provide an adequate explanation of every aspect of the behaviour of the more complex entities studied in the 'higher-level' discipline. In time they come to believe that all the knowledge mapped in this 'upper' discipline might eventually be 'reduced' to no more than what is known at the 'lower' level [6.10, 8.5, 8.11].

Any such 'strong' reductionist potentiality is obviously extremely speculative, even between such closely related disciplines as molecular genetics and developmental biology[239]. It is one thing to reveal the elaborate mechanisms by which particular chunks of DNA are successively activated to produce the particular proteins that interact chemically to build a particular type of organism or internal organ: it is quite another matter to show that a specific genome must inevitably give rise to such and such organisms, or organs, or organic functions, or behavioural traits, or pathological conditions. In practice, clearly, we are far, far away from any such capability.

Nevertheless, suppose even more speculatively that this potentiality for 'strong' reduction holds throughout the scientific 'cake' – that each layer could in principle be reduced to the one below it. Then by projecting all its 'higher' levels down on to the ground plane, the whole scientific world picture could be represented rigorously on a single 'map'[240]. This is the dream that activates much of the movement for the unification of science. This movement has a hidden agenda, which is nothing less than the reduction of all scientific knowledge to its most elementary forms[241]. The ultimate goal of science becomes a 'Theory of Everything'[242] [4.8, 6.5, 8.5] symbolized by a single mathematical equation: '$X = 42$': this is our universe: QED!

Never mind for the moment that this goal is obviously wildly impractical. Indeed, even if the master equation were actually discovered next week, it would have scarcely any scientific impact outside theoretical physics[243]. It would also have nothing to say about the 'values' and 'interests' that play such an important part in the human sciences [7.4, 7.7, 7.10]. But 'strong' reductive relationships between scientific paradigms are not ruled out by the Legend[244], and are therefore widely thought to be possible in principle. Indeed, personal faith in this possibility [2.6] is what inspires many academic scientists[245]. This, they would say, is what enables them to contribute to an orderly – albeit incomplete – world picture that is far from metaphysical in substance.

What is more, it is an *active* faith. The 'strong' reductionist agenda is the party manifesto of modern scientism. Only its most extreme supporters would apply it to every aspect of the life-world. But this is the doctrinal base of the advocates[246] of the dominance of physics over biology, the subordination of biology to genetic determinism, the revival of behaviourism in psychology, the enthronement of economics in the social sciences, and especially the complete eradication of sociology from the scholarly world. Here is where epistemology comes back down from the clouds to blitz the sociology of knowledge.

It would go beyond the purpose of this book to pursue such doctrines into the academic and political spheres where they are so influential[247]. Indeed, I think I have said enough already about the nature of science to indicate the many excellent reasons why 'strong' reductionism is untenable, as a general metascientific principle and at every 'level' of complexity[248]. These reasons are usually presented as essentially pragmatic, but they are actually much deeper.

It has always been obvious, for example, that it would never be feasible to fully test out a 'reduced' model of an elaborate system by a computer simulation of its behaviour under all circumstances [6.10]. Engineers know all too well[249] that there is no substitute in practice for 'higher-level' knowledge of the performance of the artefact as a whole. It is now clear that this is not just a matter of improving the equations, obtaining more detailed data and enlarging the computer. It is impossible in principle to calculate the long-term behaviour of a non-linear system, such as the weather, that is mathematically *chaotic*. Even though the system may be well represented by a closed, deterministic model, an infinitesimal change in the initial conditions can produce, after a surprisingly short time, an entirely different outcome[250]. In the end, intrinsic *quantum* indeterminacy must ultimately confound all reductionist aspirations in this direction.

But there is a deeper objection to 'strong' reductionism. This objection applies even to theoretical systems that are microscopically deterministic and not chaotic – for example, a 'soup' of molecules modelled out of atoms interacting according to finite rules of chemical bonding. If such a system is sufficiently large and long-lived then it is likely that it will *evolve*[251]. As we have seen [9.8], the historical unfolding of an *evolutionary* process is irredeemably unpredictable[252]. In essence, this is because the state of the system as a whole is sensitive to the influence of innumerable, incidental, past and present micro-scale configurations. These configurations can be so diverse and numerous that their theoretical occurrence can only be represented probabilistically – in effect, as if the system were truly open and infinite in extent. And yet a small proportion of these 'variants' may be capable of producing very dramatic effects.

Thus, for example, the whole soup may suddenly be taken over by the accidental formation of a small set of molecules that mutually catalyse their own replication. A novel state of the whole system can thus *emerge* spontaneously by internal processes, without any external interference[253] [8.9].

The point is not just that the moment of emergence could not, in practice, have been predicted: it is that the nature of what eventually emerges could not have been foreseen, even in principle[254]. Even a hypothetical re-run of the whole evolutionary process, in actuality or on a cosmically vast computer, would not necessarily produce the same result[255]. To replicate or simulate the historical event, time itself would have to have been magically clocked back and restarted for the purpose – a divine intervention that not even a mediaeval theologian would consider seriously. The evolutionary emergence of a novel mode of order in a complex system can sometimes be understood, explained and made to seem inevitable – but only after it has already taken place. Without prior knowledge of the outcome, the universe of theoretical potentialities for this mode is so vast that it cannot be spanned rationally even by computer-assisted conjecture.

Thus, 'strong' reductionism is a fantasy. Indeed, its extreme positivism is as damaging to science as the extreme scepticism of its arch-enemy, 'strong' relativism. For example, it simply cannot cope epistemologically with one of the most elementary aspects of the natural world – the successive emergence of novel life forms through biological evolution[256]. What is more, the spontaneous emergence of novel modes of order in complex systems [8.9] is not just an intellectual deviation on the path to a final theory. It is a natural phenomenon that renders science possible. Otherwise, the universe would be just an unpatterned jumble. The assembly of primary entities into more or less distinct compound entities that can interact as wholes [6.9] is not a tiresome complication. On the contrary, it is a *simplification* of nature, and of human cognition as naturally evolved [8.14, 10.3], that actually makes scientific research feasible[257].

But this feasibility can only be exploited by undertaking research wholeheartedly at each of its many 'levels'. However successful our exercises in 'weak' reduction, we must continue to study all kinds of naturally occurring entities, from quarks through quaggas to quangos, in their own terms. This is the rationale of the conventional classification of academic disciplines [8.5] according to the presumed complexity of their research objects. This way of carving nature scientifically makes a lot of sense, provided that it is not elevated into a hierarchy of esteem, or collapsed schematically on to its supposed foundations.

10.9 Post-academic knowledge

Our investigation thus arrives at a paradoxical conclusion. Academic science, the spearhead of modernism, is *pre-modern* in its cultural practices [4.11]: and yet it turns out to be *post-modern* in its epistemology.

Contrary to the Legend, science is not a uniquely privileged way of understanding things, superior to all others. It is not based on firmer or deeper foundations than any other mode of human cognition. Scientific knowledge is not a universal 'metanarrative' from which one might eventually expect to be able to deduce a reliable answer to every meaningful question about the world. It is not objective but reflexive: the interaction between the knower and what is to be known is an essential element of the knowledge. And like any other human product, it is not value-free, but permeated with social interests.

In describing these features as 'post-modern', I am, of course, going with the contemporary intellectual flow. But terms such as 'modernism' and 'post-modernism' are very ill defined, and apply differently in different sectors of life[258]. Most scientists only know of them as slogans, uttered wholesale by the partisans of the most diverse fashions and fads. I believe they do have a serious core of meaning[259], but would not pretend to be able to define it. Indeed, one of the defining characteristics of post-modernism is that it cannot be defined! It is not an '-ism', but a trenchant critique of all '-isms'.

What I am really saying is that the post-modern critique effectively demolishes the Legend, along with the more general philosophical structures that support it. That does not mean that we should go to the opposite extreme of a purely anarchic or existential philosophy where everything is 'socially relative', and 'anything goes'. 'Negativism' is not the only alternative to undue 'positivism'! But it does mean that science needs a new *affirmative* philosophy more in keeping with its actual capabilities.

What I have outlined in this book is a naturalistic epistemology that is consistent with the way that scientific knowledge is actually produced and used. In effect, we have inverted the metascientific tradition. Instead of trying to justify scientific practice from a prior set of idealized philosophical principles, we have derived a more realistic account of its cognitive methods and values out of an analysis of the social institution where they operate. By first building up a sociological model of science, we have been able to explicate its real philosophy.

What *is* this new philosophy, then? Now that I have come to the end of the book, surely I should try to express it in a few well-chosen words, or summarize it in a brief list of 'regulative principles', or 'epistemic norms'? No, that is precisely what I cannot do – for that list would end up just as long and elaborate as this whole book!

What would be the significance, for example, of the bald assertion that scientific knowledge is always *uncertain*? One first thinks this proposition merely refers to the technical problems of eliminating observational error. But then one remembers that instruments and experiments are threaded through with models and theories which can never be perfectly validated. What is more, these are largely derived from the scientific and technological literature, which aims at credibility but does not pretend to be entirely trustworthy. And yet, scientific knowledge is often very reliable, because it has had to survive stringent practical tests and communal criticism. And so on. In other words, this apparently straightforward epistemic proposition is meaningless unless it is linked to all the other epistemic, material and social elements of the research system.

Notice, moreover, that this proposition cannot be made more meaningful by formulating it more precisely, for that would involve cutting these vital linkages. It is clearly impossible to give a formal definition of scientific uncertainty that would cover with equal precision the whole range of observational, instrumental, experimental, inferential, model-making, theorizing, rhetorical, critical (etc., etc.) practices with which it is intimately connected. That is why, throughout this work, I have had to resist my training as a scientist to try to say *exactly* what I mean. The intrinsic interconnectedness of the whole system can only be grasped by leaving its component elements somewhat vague.

Scientific knowledge does indeed have a number of distinctive epistemic features, but these cannot be characterized purely epistemically. They can only be understood in terms of the distinctive human circumstances – individual and collective, local or more general, practical or speculative – in which this knowledge is typically produced and used. The essence of the new philosophy is in its open-minded, open-ended approach to a complex form of life where cognitive, social and material processes are inextricably commingled. It simply cannot be encapsulated in a formula.

This approach in no way devalues the scientific enterprise or its products. I recognize, of course, that many scientists believe firmly in the

Legend, try to live and work by its precepts, and sincerely propagate its teachings. In spite of its formal deficiencies, it remains a powerful belief system [10.5], motivating much beneficial human action. But its internal inconsistencies and external disharmonies are beginning to show through, and even as a vocational ideology it is becoming a liability. It can no longer provide science with a credible defence against sympathetic, well-informed criticism, and thus lays it open to devastation by less scrupulous, more ignorant opponents.

The best way now to defend the research culture is not to try to patch up the Legend but to depict science honestly as a systematic, rational, human activity performed by ordinary people on common-sense lines. Instead of trying to deduce familiar scientific working methods such as 'observation', 'theory', 'discovery', 'induction', 'model', 'explanation', etc. from more abstract principles, it is better to show how essentially *reasonable* these methods are, and how well adapted they can often be to the logic of the situations where they are employed. This approach also encourages both researchers and their 'customers' to estimate much more realistically the quality of the goods they are exchanging – the actual strengths and weaknesses of the knowledge that society gets from science. By renouncing all transcendental pretensions to authority, and presenting science as an epistemic institution trading publicly in credibility and criticism, we establish a stable place for it in our culture.

Notice, moreover, that our epistemological analysis – especially in this chapter – has been focussed almost entirely on academic science, as traditionally practised. But we originally chose this mode of knowledge production as the 'exemplar', or 'ideal type' of 'science' in general [2.1, 4.1]. In the first instance, therefore, this analysis ought to apply to the knowledge produced by the whole range of institutions and cultures that we call 'scientific'. The extent to which these actually follow the same epistemic practices, even though their sociological norms are somewhat different, is obviously a matter for detailed investigation. But from a broad epistemological perspective – in relation to 'common sense', for example [10.6] – all modern science ought to look very much the same.

Certainly, the transition to post-academic science is making subtle changes in the *form* of knowledge that is being produced. Some of the 'post-modern' features of Mode 2 research are much-needed corrections to the excesses of 'scientism' [10.8]. Others are welcome antidotes to the decontextualized rationalism [6.8] that has long plagued the philosophy of science. Other features, again, help rescue the scientific imagination from entrenched specialization [8.8]. In its post-academic mode, science can no longer evade all social responsibility by pretending that the production of universally valid, value-neutral knowledge is its only goal and only achievement [7.4]. And localized pragmatism largely compensates for the dilution of theoretical standards of scientific validity with 'trans-epistemic' considerations [7.10].

On the down side, the moral integrity of science has become more debatable. Post-academic science is not directly guarded, institutionally or ideologically, against societal interests[260]. Governments and industrial corporations now have considerable power over both the initiation of research projects [8.7] and the publication of their results [5.13]. In consequence, it has become more difficult to enlist science as a non-partisan force against obscurantism, social exploitation or folly.

In the long run, moreover, the post-academic drive to 'rationalize' the research process [8.2] may damp down its creativity. Bureaucratic 'modernism' presumes that research can be directed by policy. But policy prejudice against 'thinking the unthinkable' aborts the emergence of the unimaginable [9.8, 10.8]. The evolutionary dynamism of academic science fits it well for its established role in a pluralistic open society [3.1]. Globalized post-academic science may be less epistemically adventurous, in keeping perhaps with a globalized post-industrial society that may be less pluralistic and open than we like to suppose.

At the coal face, however, science is not noticeably 'going postmodern'[261]. The transition from academic to post-academic research is too recent to have affected its operational philosophy[262]. Indeed, in spite of the post-modern critique of its underlying assumptions, this philosophy remains defiantly affirmative. There is still little evidence that working scientists are infected with the philosophical scepticism, sociological constructivism, political cynicism, ethical nihilism and historical incommensurabilism projected on to them by some of their wilder critics.

To put it simply: post-academic scientists still formulate and try to solve practical and conceptual problems on the basis of their shared belief in an intelligibly regular, not disjoint, world outside themselves. They still go on theorizing, and testing their theories by observation and experiment. They still try as best they can to eliminate personal bias from their own findings and are extremely canny in their acceptance of the claims of others. To that extent at least, we, the public at large, have just as good grounds as we ever did for believing (or doubting!) the amazing things that 'science' tells us about the world in which we live.

Endnotes

Preface

1 Polanyi 1958
2 Popper 1935 (1959)
3 Ziman 1960, 1981
4 Ziman 1968
5 Ziman 1978
6 Ziman 1984
7 Ziman 1994
8 Ziman 1995

21 Callon 1995
22 Popper 1963
23 Kuhn 1962
24 Thagard 1993
25 Fuller 1988
26 Chalmers 1982
27 Fine 1991
28 Schmaus, Segerstrale & Jessuph
   1992

Chapter 1

1 Holton 1993
2 Wolpert 1992
3 Kitcher 1993
4 Böhme 1992
5 Slezak 1989
6 Collins & Pinch 1993
7 Callon 1995
8 Ziman 1996a
9 Oakeshott 1933
10 Collins 1982
11 Shils 1982
12 Elkana 1981; Böhme 1992
13 Popper 1959
14 Giere 1988; Hooker 1995
15 Fuller 1988
16 Lucas 1986
17 Callebaut 1993, 1995
18 Campbell 1974
19 Ezrahi 1990
20 Hooker 1995

Chapter 2

1 Harré 1986; Atran 1990
2 Goffman 1974; de Mey 1982
3 Shapin & Schaffer 1985
4 Ezrahi 1990
5 Bodmer 1985
6 Ziman 1994
7 Vincenti 1990
8 Salomon 1973
9 Ziman 1981a, 1995
10 Popper 1963; Laudun 1977
11 Ziman 1994
12 Bernal 1939
13 Spiegel-Rösing & Price 1977;
   Cozzens et al. 1990
14 Ziman 1991 (1995)
15 Smit 1995
16 Irvine, Martin & Isard 1990
17 Beck 1986; Ezrahi 1990; Böhme
   1992; Restivo 1994
18 Wolpert 1992; Gross & Levitt 1994

Chapter 2 cont.

19 Latour 1987; Collins & Pinch 1993
20 OECD 1980
21 NAS 1972
22 Rosenberg & Nelson 1994
23 Rothschild 1972
24 Ziman 1994
25 Mitchell 1986
26 Midgley 1992
27 Cohen & Stewart 1994
28 Harré 1986
29 Rosenberg & Nelson 1994
30 CSS 1989
31 Atiyah 1995
32 Shils 1982
33 Bernal 1939
34 Harris 1993
35 Restivo 1994
36 Cooper 1952; Polanyi 1958
37 Odlyzko 1995
38 Salomon 1973
39 Lindblom 1990
40 Elkana 1981; Fuller 1988; Traweek
    1988; Mukerji 1989; Restivo 1994;
    Pickering 1995
41 Barnes 1995
42 Collins English Dictionary
43 ABRC 1987
44 Irvine, Martin & Isard 1990
45 Ben-David 1971
46 Ziman 1981 (1974); Carlson &
    Martin-Rovet 1995
47 Ziman 1987
48 Lodge 1975; Martin-Rovet &
    Carlson 1995
49 Rose & Ziman 1964
50 ABRC 1987

Chapter 3

1 Polanyi 1958
2 Hesse 1980; Harré 1986; Coady 1992
3 Cole 1992; Hooker 1995
4 Knorr-Cetina & Mulkay 1983;
   Callon 1994,
5 Campbell 1983
6 Latour & Woolgar 1979; Knorr-
   Cetina 1981

7 Latour 1987
8 Bourdieu 1975; Mukerji 1989
9 Sklair 1993
10 Restivo 1994; Hooker 1995; Ziman
    1996b
11 Mann 1983
12 Harré 1986; Turner 1990
13 Hooker 1995
14 Hjerrn & Porter 1983
15 Polanyi 1969,
16 Toulmin 1972; Shapin & Schaffer
    1985; Ezrahi 1990; Porter 1995
17 Dews 1986; Bernstein 1991
18 Merton 1942 (1973); Ezrahi 1990
19 Price 1963
20 Ziman 1968; Meadows 1980;
    Bazerman 1988
21 Merton 1938
22 Shapin & Schaffer 1985; Shapin 1994
23 Ben-David 1974
24 Morrell & Thackray 1981
25 Crawford 1984
26 Shapin & Schaffer 1985; Segerstrale
    1991
27 Campbell 1979
28 Hagstrom 1965; Whitley 1984;
    Becher 1989
29 Merton 1973
30 Waddington 1948; Kantorovich 1993
31 Gouldner 1970; Mulkay 1979
32 MacIntyre 1981
33 Campbell 1977, 1979
34 Bourdieu 1975; Gieryn 1983;
    Barnes 1985; Arbib & Hesse 1986;
    Traweek 1988; Smith 1990
35 Barnes 1995; Hooker 1995
36 Merton 1973; Harré 1986
37 Mulkay 1979
38 Traweek 1988; Restivo 1994
39 Ravetz 1971
40 Henderson 1990; Callon 1994
41 Mulkay 1979
42 Ziman 1968
43 Taylor 1985
44 Latour & Woolgar 1979; Knorr-
    Cetina 1981
45 Ziman 1969 (1981)
46 Myers 1991a; Sinding 1996
47 Hagstrom 1965

Chapter 3 cont.

48 Traweek 1988
49 Hagstrom 1965
50 Hagstrom 1965
51 Ziman 1968, 1978
52 Nowotny & Felt 1997
53 Merton 1973
54 Merton 1973
55 Keller 1988; Haraway 1991
56 Hagstrom 1965
57 Böhme 1992
58 Merton 1973
59 Hagstrom 1965
60 Bernal 1939; de Jouvenel 1961;
    Campbell 1979
61 Mitroff 1974
62 Goffman 1974
63 Merton 1973
64 Ziman 1968; Mulkay 1991
65 Hagstrom 1965
66 Merton 1973
67 Harré 1986; Markus 1987
68 Harré 1986; Hull 1988; Shapin
    1994
69 Ziman 1971 (1981)
70 Traweek 1988
71 Hagstrom 1965
72 Kuhn 1962
73 Ziman 1987, 1995
74 Hagstrom 1965
75 Merton 1973
76 Hull 1988
77 Brannigan 1981
78 Merton 1973
79 Merton 1973
80 Ziman 1968; Polanyi 1969
81 de Jouvenel 1961
82 Polanyi 1969
83 Ezrahi 1990
84 Hagstrom 1965
85 Hagstrom 1965; Galison 1987
86 Foucault 1969
87 Hagstrom 1965; Merton 1973
88 Ziman 1995
89 Hull 1988
90 Mukerji 1989
91 Hagstrom 1965
92 Merton 1973
93 Merton 1973

94 Zuckerman 1977
95 Merton 1973
96 Latour & Woolgar 1979; Knorr-
    Cetina 1981; Hull 1988; Restivo 1994
97 Merton 1973
98 Mulkay 1979
99 Bourdieu 1975
100 Schaffner 1994
101 Lemaine, *et al.* 1976; Toulmin 1990;
    Weingart 1995
102 Blume 1988
103 Toulmin 1972; Chubin 1976
104 Gieryn 1983; Fuller 1988
105 Hagstrom 1965
106 Whitley 1984; Becher 1989
107 Chubin 1976
108 Hill 1995
109 Ziman 1987
110 Ziman 1987, 1995
111 Price 1961
112 Price 1963
113 Bernal 1939; Studer & Chubin 1980;
    Ziman 1987
114 Hagstrom 1965
115 Mukerji 1989
116 Hill 1995
117 Merton 1973; Hagstrom 1965
118 Morrell & Thackray 1981
119 Ziman 1995
120 Ziman 1987
121 Caplow & McGee 1958
122 Hill 1995
123 Bourdieu 1984
124 Caplow & McGee 1958; Hagstrom
    1965
125 Bernal 1939
126 Ziman 1995
127 Crawford 1984
128 Mansfield 1991
129 Turner 1990
130 Morrell & Thackray 1981
131 Byerly & Pielke 1995
132 Mukerji 1989
133 Rip 1994
134 Turner 1990
135 Weber 1918
136 Restivo 1994
137 Restivo 1994
138 Wolpert 1992

Chapter 3 cont.

139 Ziman, Sieghart & Humphrey 1986
140 Bernal 1939; Hill 1995
141 Studer & Chubin 1980; Ziman 1987
142 Ziman 1980
143 Ziman 1995
144 Bodmer 1985
145 Galison 1997
146 Ziman 1984
147 Elzinga 1986

Chapter 4

1   Popper 1963; Naess 1972; Hesse 1980
2   Ziman 1996b
3   Shapin & Schaffer 1985; Shapin 1994
4   Restivo 1994
5   Ben-David 1974
6   Hooker 1995
7   Gibbons *et al.* 1994
8   Arbib & Hesse 1986; Giere 1988;
    Kantorovich 1993; Kitcher 1993;
    Hooker 1995
9   Fuller 1988
10 Arbib & Hesse 1986; Giere 1988;
    Callebaut 1993; Cohen & Stewart
    1994; Hooker 1995
11 Ziman 1968, 1978
12 Ziman 1960 (1981)
13 Cozzens *et al.* 1990; Gibbons *et al.*
    1994; Ziman 1994
14 Feyerabend 1975
15 Restivo 1994
16 Ziman 1987; Mukerji 1989; Cozzens
    et al. 1990; Gibbons *et al.* 1994;
    Ziman 1994; Nowotny & Felt 1997
17 Galison 1987; Knorr-Cetina 1995
18 Etzkowitz 1992
19 McCormmach 1982
20 Cooper 1952
21 Watson 1968
22 Weingart 1989
23 Cozzens *et al.* 1990; Ziman 1994
24 Nowotny & Felt 1997
25 Restivo 1994; Hooker 1995
26 Ziman 1995
27 Cozzens *et al.* 1990; Ziman 1994
28 Gross & Levitt 1994

29 Fuchs 1996
30 Gibbons *et al.* 1994
31 Collins English Dictionary
32 Hooker 1995
33 Gibbons *et al.* 1994
34 Weingart 1989
35 Bernal 1939
36 Galison 1997
37 Price 1986
38 Ziman 1983, 1984, 1995
39 Weinberg 1967; Galison 1997
40 Galison 1997
41 Mukerji 1989
42 Price 1986; Hicks & Katz 1996;
    Galison 1997
43 Etzkowitz 1983; Weingart 1989;
    Mukerji 1989; Etzkowitz 1992
44 Galison 1997
45 Price 1961
46 Lemaine *et al.* 1976
47 Price 1961
48 Ziman 1994
49 Ziman 1985, 1995
50 Böhme *et al.* 1983; Jagtenberg 1983
51 Gibbons *et al.* 1994
52 Nowotny & Felt 1997
53 Salomon 1973
54 Midgley 1989
55 Ziman 1995
56 Graham 1979; Ziman 1980
57 Greenberg 1969
58 Salomon 1973
59 Rip 1994
60 Mukerji 1989
61 Mukerji 1989; Rip 1994
62 Merton 1973
63 Latour & Woolgar 1979; Knorr-
    Cetina 1981; Hull 1988; Restivo 1994
64 Etzkowitz 1992
65 Ziman 1991 (1995)
66 Mukerji 1989
67 Ravetz 1971; Galison 1997
68 Ziman 1987
69 David 1995
70 Rosenberg & Nelson 1994;
    Nowotny & Felt 1997
71 Weingart 1989; Foray 1997; Meyer-
    Krahmer 1997
72 Latour & Woolgar 1979; Ziman 1994

Chapter 4 cont.

73 Ziman 1984
74 Salomon 1973
75 Galison 1987, 1997; Knorr-Cetina 1995
76 Price 1979
77 Elzinga 1986
78 Gibbons *et al.* 1994
79 Gibbons *et al.* 1994
80 Turner 1990
81 Weingart 1989; Rip 1994
82 Galison 1997
83 Weber 1918 (1948)
84 Ziman 1994
85 Toulmin 1990

Chapter 5

1 Michalos 1980
2 Galison 1997
3 Restivo 1994
4 Ziman 1981; Galison 1997
5 Campbell 1983
6 Giere (Callebaut 1993)
7 Cassirer 1956; Hesse 1980; Hooker 1995
8 Hooker 1995
9 Kantorovich 1993
10 Naess 1972; Campbell (Callebaut 1993); Yearley 1994
11 Polanyi 1958; Putnam 1991; Shapin 1994
12 Coady 1992
13 Merton 1942
14 Schmaus, Segerstrale & Jessuph 1992
15 Oakeshott 1933
16 Millikan 1984; Arbib & Hesse 1986; Boyd 1991
17 Mayr 1988
18 Fleck 1935; Beveridge 1950; Popper 1972; de Mey 1982; Secord 1990
19 Churchland 1990
20 Fleck 1935; Barber & Fox 1958; Kantorovich 1993
21 Oxford English Dictionary
22 Nowotny & Felt 1997
23 Pasteur 1854
24 Harré 1986

25 Oakeshott 1933; Mead 1938; Goonatilake 1984
26 Shapin & Schaffer 1985
27 Harré 1979; Hesse 1980; Campbell 1982, 1983
28 Merton 1973; Mukerji 1989
29 Gouldner 1979; Restivo 1994; Geertz 1995
30 Midgley 1992a
31 Keller 1988; Haraway 1991; Dupré 1993
32 Mulkay 1979; Knorr-Cetina 1981a
33 Markus 1987
34 Galison 1997
35 Campbell 1921
36 Jaki 1966
37 Rouse 1987
38 Luce & Narens 1987
39 Stamp 1981; Porter 1992
40 Ezrahi 1990
41 Campbell 1984a
42 Ziman 1978; Harré 1979
43 Porter 1995
44 Ziman 1983 (1995); Traweek 1988; Galison 1997
45 Hooker 1995
46 Shapin & Schaffer 1985; Rouse 1987
47 Campbell 1977
48 Nagel 1979
49 Polanyi 1958, 1969; Mukerji 1989
50 Galison 1997
51 Mukerji 1989; Ihde 1991
52 Millikan 1984
53 Galison 1997
54 Callebaut 1993
55 Hacking 1983
56 Collins English Dictionary
57 Galison 1987
58 Wolpert 1992
59 Rouse 1987; Mukerji 1989
60 Fleck 1935; Cantore 1977; Franklin 1986; Traweek 1988; Galison 1997
61 Campbell 1979; Harré 1979; Campbell 1984a; Plott 1986; Rouse 1987; Secord 1990
62 Crease 1992
63 Fleck 1935; Ravetz 1971; Shapin & Schaffer 1985; Mukerji 1989; Rose 1992

Chapter 5 cont.

64 Mead 1938; Kitcher 1993; Shapin 1994
65 Winch 1958
66 Kantorovich 1993
67 Barfield 1965
68 Shapin & Schaffer 1985; Shapin 1994
69 Rudwick 1985
70 Coady 1992
71 Coady 1992; Burnyeat 1993; Shapin 1994
72 Magyar 1981; Harré 1986; Porter 1995
73 Giddens 1990
74 Shapin 1994; Knorr-Cetina 1995; Whitbeck 1995
75 Traweek 1988
76 Collins & Pinch 1993
77 Ezrahi 1990
78 Franklin 1986
79 Medawar 1967; Ziman 1968
80 Shapin & Schaffer 1985
81 Shapin & Schaffer 1985
82 Ziman 1968, 1978; Ravetz 1971; Porter 1995
83 Hull 1988
84 Franklin 1986
85 Root-Bernstein 1989
86 Popper 1963; Collins 1975; Mulkay 1979; Shapin & Schaffer 1985
87 Mayr 1988; Dupré 1993
88 Porter 1995
89 Mead 1938; Star 1985; Mukerji 1989
90 Campbell 1983
91 Gilbert & Mulkay 1984
92 Knorr-Cetina 1981, 1981a; Traweek 1988; Mukerji 1989; Collins & Pinch 1993; Porter 1995
93 Ravetz 1971; Campbell 1983
94 Polanyi 1958; Markus 1987; Galison 1997
95 Rouse 1987
96 Mulkay 1991
97 Mukerji 1989
98 Ziman 1978; Collins 1985; Collins & Pinch 1993
99 Campbell 1983; Mulkay 1985, 1991; Franklin 1986; Collins & Pinch 1993

100 Bazerman 1988
101 Kantorovich 1993
102 Callebaut 1993; Hooker 1995
103 Fuller 1988
104 Fuller 1987; Slezak 1989
105 Ziman 1968,
106 Polanyi 1958
107 de Mey 1982
108 Hill 1995
109 Weber 1918 (1948); Ziman 1987
110 Campbell 1979
111 Giere 1988
112 Cosmides & Tooby 1992
113 Boden 1988
114 Lorenz 1941; Riedl 1984; Donald 1991; Plotkin 1994
115 Campbell 1974; Munz 1993; Edelman 1992; Dennett 1995
116 Bradie 1990; Callebaut 1993; Hooker 1995
117 Vollmer 1984
118 Harré 1986
119 Collins English Dictionary
120 Mead 1938; Varela *et al.* 1991
121 Ryle 1949; de Mey 1982
122 Campbell 1977
123 Ezrahi 1990
124 Polanyi 1958
125 Ihde 1991
126 Brain 1951; Diettrich 1995
127 Polanyi 1958; Merleau-Ponty 1962; Tricker 1965; Ihde 1991
128 Polanyi 1958; Mukerji 1989
129 Gregory 1973; de Mey 1982
130 Harraway 1991
131 Hanson 1958
132 Churchland 1990
133 Campbell 1921; Levi 1989; Burnyeat 1993
134 Kuhn 1984
135 Coady 1992
136 Ziman 1978
137 Nicod 1923; Popper 1945; Polanyi 1958; Naess 1972; Remmling 1973; Capaldi 1975; Campbell 1977
138 Mead 1938; Bernstein 1991
139 Pivcevic 1970
140 Schütz 1967; Schütz & Luckmann 1974

Chapter 5 cont.

141 Hesse 1980; Campbell 1984a; Gilbert & Mulkay 1984; Kuhn 1984; Vollmer 1984; Taylor 1985; Arbib & Hesse 1986; Harré 1986; Markus 1987; Longino 1990; Edelman 1992; Kitcher 1992; Restivo 1994; Shapin 1994; Hill 1995; Hooker 1995

142 Fine 1991

143 Denbigh 1975

144 Geertz 1983

145 Pivcevic 1970; Harré 1983; Churchland 1990; Coady 1992

146 Millikan 1984; Taylor 1985; Doyal & Harris 1986; Edelman 1992

147 Polanyi 1958

148 Millikan 1984; Williams 1985; Munz 1993

149 Ziman 1968, 1978, 1984

150 Rouse 1987

151 Collins English Dictionary

152 Boden 1988

153 Pivcevic 1970; Gouldner 1970; Williams 1985

154 Nagel 1961; Lear 1985; Taylor 1985

155 Dennett 1969; Weizenbaum 1976; Cosmides & Tooby 1992

156 Baron-Cohen 1990

157 Berlin 1979

158 Geertz 1973; Mead 1976

159 Turkle 1982

160 Shweder 1992

161 Cassirer 1956

162 Collingwood 1946; Klemke, Hollinger & Kline 1980

163 Winch 1958

164 Ziman 1978; Hesse 1980

165 Michalos 1980; Fuller 1988

166 Blume 1988

167 Hesse 1980

168 Campbell 1983

169 Doyal & Harris 1986

170 Fuller 1988

171 Keat & Urry 1975; Putnam 1978; Knorr-Cetina 1981, 1981a; Markus 1987; Rouse 1987; Knorr-Cetina (Callebaut 1993)

172 Ravetz 1971; Wynne 1995; Watson-Verran & Turnbull 1995

173 Ivins 1953

174 Gilbert & Mulkay 1984; Hooker 1995; Galison 1997

175 Ziman 1978; Bazerman 1988; Henderson 1991

176 Fleck 1935

177 Geertz 1973; Shapin & Schaffer 1985

178 Medawar 1964; Ziman 1968; Cole 1992

179 Markus 1987

180 Porter 1995

181 Mead 1938; Harré 1979; Millikan 1984; Barnes 1995

182 Toulmin 1972; Campbell 1983; Davidson 1984; Harré 1986; Schmaus, Segerstrale & Jessuph 1992; Daston 1992

183 Gibbons *et al.* 1994

184 Mead 1938

185 Arbib & Hesse 1986

186 Geertz 1973

187 Vygotsky 1962; de Mey 1982; Hill 1995

188 Cole & Scribner 1974; de Mey 1982; Arbib & Hesse 1986; Dews 1986; Donald 1991; Cosmides & Tooby 1992

189 Winch 1958; Toulmin 1972; Johnson-Laird 1983; Taylor 1985; Haraway 1991

190 Bazerman 1984, 1988

191 Whitley 1984; Ziman 1987, 1995; Becher 1989

192 Ezrahi 1990

193 Ziman 1968

194 Schaffner 1994; Weingart 1995

195 Taubes 1996

196 Latour & Woolgar 1979; Knorr-Cetina 1981

197 David 1995

198 Galison 1997

199 Gibbons *et al.* 1994; Nowotny & Felt 1997

200 Merton 1973

201 Foray 1997

202 Restivo 1994

Chapter 6

1 Barrow 1991; Diettrich 1995

2 George 1938; Furnham 1988

Chapter 6 cont.

3 Naess 1972; Kantorovich 1993
4 Medawar 1967
5 Barnes 1981
6 Haraway 1991
7 Bloor 1976; Margolis 1991
8 Collins English Dictionary
9 *ibid.*
10 Kuhn 1962
11 Feyerabend 1975; Fuller 1988; Pickering 1995
12 Galison 1997
13 Cohen & Stewart 1994
14 Cohen & Stewart 1994
15 Grene 1961; Hesse 1974
16 Rosenmayer 1983
17 Toulmin 1990
18 Knorr-Cetina 1981a
19 Hesse 1974
20 Campbell 1977; Hofstadter 1985; Donald 1991
21 Whorf 1941; Feyerabend 1975; Aitchison 1996
22 Edelman 1992
23 Cohen & Stewart 1994
24 Porter 1995
25 Garfinkel 1967
26 Atran 1990
27 Hesse 1974
28 Polanyi 1969
29 Ziman 1978; Hall 1995
30 Knorr-Cetina & Amann 1990; Crease 1993
31 Galison 1997
32 Denbigh 1975; Hesse 1980; Harré 1986
33 Galison 1997
34 Polanyi 1958; Gombrich 1960; de Mey 1982; Churchland 1990; Edelman 1992; Cohen & Stewart 1994
35 Varela *et al.* 1991
36 Whorf 1941; Lyons 1981; Høeg 1993
37 Durkheim 1903; Douglas 1973; Barnes 1981; Douglas 1987
38 Donald 1991
39 Cole & Scribner 1974; Varela *et al.* 1991; Shepard 1992
40 Forge & Mayer 1970
41 Aitchison 1996
42 Barkow, Cosmides & Tooby 1992
43 McLuhan 1962; Carroll 1985
44 Boden 1988
45 Ziman 1978
46 Arbib & Hesse 1986
47 Phillips 1981
48 Campbell 1977; Johnson-Laird 1983; Harré 1986; Atran 1990; Hull (Callebaut 1993); Kitcher 1993; Rose 1979
49 Churchland 1990
50 Mayr 1988; Atran 1990
51 Hofstadter 1985; Smith & Torrey 1996
52 Harré 1979; Atran 1990; Thagard 1993
53 Cohen & Stewart 1994
54 Mayr 1982; Hull 1988; Yoon 1993
55 Mayr 1988
56 Edelman 1992; Dupré 1993
57 Douglas 1987
58 Steen 1988
59 Millikan 1984
60 Harré 1986
61 Star & Griesemer 1989; Turnbull 1995
62 Mayr 1988
63 Durkheim & Maus 1903
64 Mayr 1988; Hull 1988
65 Dennett 1995
66 Toulmin 1953; Polanyi 1958; Kuhn 1962; Ziman 1978; Millikan 1984; Aronson 1984; Rudwick 1985; Harré 1986; Giere 1988; Mukerji 1989; Turnbull 1989
67 Rudwick 1976
68 Ezrahi 1990
69 Harré 1986; Turnbull 1989
70 de Mey 1982
71 Boden 1990
72 Campbell 1921
73 Toulmin 1953; Ziman 1978
74 Kantorovich 1993
75 Hesse 1974, 1980; Ziman 1978; Glymour 1980; de Mey 1982; Harré 1986; Fine 1991; Leydesdorff 1991; Hooker 1995; Galison 1997

Chapter 6 cont.
76 Rouse 1987
77 Barnes 1977
78 Ivins 1953; Mukerji 1989
79 Ziman 1978
80 Nagel 1986
81 Turnbull 1989
82 Ziman 1978
83 Haraway 1991
84 Mukerji 1989
85 Barnes 1995
86 Jarvie 1972
87 Ziman 1978; Turnbull 1989
88 Campbell (Callebaut 1993); Cohen & Stewart 1994
89 Toulmin 1953; Ziman 1978
90 Lindblom 1990
91 Gilbert & Mulkay 1984; Turnbull 1989
92 Rose 1997
93 Turnbull 1989
94 Hesse 1980
95 Toulmin 1990
96 Gilbert 1976; Harré 1986; Giere 1988, 1990; Hull 1988
97 Toulmin 1972
98 Jarvie 1972
99 Carroll 1893; Hull 1988; Campbell (Callebaut 1993)
100 Cohen & Stewart 1994
101 Toulmin 1953
102 Körner 1966; Munz 1993
103 Giere 1988
104 Diettrich 1995
105 Cole 1992
106 Hofstadter 1985
107 Donald 1991
108 Suppé 1974
109 Foucault 1969
110 Popper 1959; 1963
111 Körner 1970
112 Hesse 1963, 1974
113 Edelman 1992
114 Porter 1995
115 Millikan 1984; Harré 1986; Cupitt 1990
116 Popper 1963
117 Jaki 1966; Barrow 1991
118 Toulmin 1972

119 Harré 1986
120 de Mey 1982
121 Harré 1986
122 Richards (Callebaut 1993)
123 Lloyd 1979; Kuhn 1984
124 Ziman 1978
125 Taylor 1985; Munz 1993
126 Fuller 1988
127 Rescher 1988; Kitcher 1993
128 Kern, Mirels & Hinshaw 1983; Hull 1988
129 Ryle 1949; Harris 1990
130 Lyons 1970
131 Piaget 1954; Toulmin 1972
132 Thom 1975
133 Campbell 1977
134 Oldfield & Marshall 1968
135 Giere 1988
136 Kantorovich 1993; Munz 1993
137 Galison 1997
138 Bradie 1990; Diettrich 1995
139 Wolpert 1992
140 Restivo 1990; Kantorovich 1993
141 Slaughter 1982
142 Boden 1988; Hooker 1995
143 Churchland 1990
144 Putnam 1975
145 Toulmin 1972; Ziman 1978
146 Ziman 1978
147 Kantorovich 1993
148 Barrow 1991
149 Steen 1988; Cohen & Stewart 1994
150 Gell-Mann 1992
151 Giere 1988
152 Watson 1990
153 Bloor 1976; Restivo 1983; Rescher 1988; Diettrich 1995; Eglash 1997
154 Jaki 1966
155 Harré 1960; Duncan & Weston-Smith 1977
156 Cohen & Stewart 1994
157 Ziman 1978
158 Wigner 1969; Mayr 1982
159 Suppé 1974
160 Roberts 1974
161 Coleman 1964
162 Toulmin 1972; Putnam 1978; Hull 1988; Collins 1993; Hooker 1995

Chapter 6 cont.

163 Toulmin 1990; Cohen & Stewart 1994
164 Barnes & Edge 1982
165 Galison 1997
166 Hooker 1995
167 Toulmin 1980, 1990; Kantorovich 1993; Kitcher 1993; Restivo 1994; Hooker 1995
168 Lindblom 1990
169 Campbell 1983
170 Kantorovich 1993; Shapere (Callebaut 1993)
171 Feyerabend 1975
172 Watkins 1970
173 Bernstein 1991; Cohen & Stewart 1994
174 Kitcher 1993
175 Nagashima 1973
176 Bernstein 1991
177 Aitchison 1996
178 Ziman 1978
179 Geertz 1973
180 Jarvie 1970
181 Donald 1991
182 Geertz 1973
183 Ziman 1978; Bernstein 1991
184 Galison 1997
185 Toulmin 1990; Yearley 1994
186 Arbib & Hesse 1986; Thagard 1993
187 Collins English Dictionary
188 Collins English Dictionary
189 Checkland 1981
190 Fuller 1988
191 Harré 1960
192 Aronson 1984
193 Cohen & Stewart 1994
194 Lehman 1977; Stevens 1988
195 Nagel 1961; Garfinkel 1991
196 Root-Bernstein 1989
197 Harré 1960
198 Hesse 1963
199 Harré 1960; Ziman 1965 (1981); Hofstadter 1985; Shinn 1987; Mukerji 1989
200 Nagel 1961
201 Pickering 1984
202 Naess 1972

203 Giere 1988
204 Bailey 1992; Hall 1995
205 Burian (Callebaut 1993)
206 Weizenbaum 1976
207 Hooker 1995
208 Yonezawa 1993
209 Oreskes *et al.* 1994
210 Galison 1997
211 Ziman 1978; Stevens 1988
212 Zahler & Sussmann 1977; Marmo & Vitale 1980,
213 Cohen & Stewart 1994
214 Ziman 1965 (1981); Colby 1981; Slezak 1989
215 Handy 1964; Inbar & Stoll 1972; Binmore 1977; Ziman 1978
216 Naess 1972
217 Chote 1992; Evans; 1997
218 Hesse 1980; de Mey 1982; Arbib & Hesse 1986; Harré 1986; Holton 1986; Atran 1990; Rose 1997
219 Davidson 1984; Taylor 1985; Cohen & Stewart 1994
220 Schon 1963; Rorty 1986; Arbib & Hesse 1986
221 Fleck 1936; Hesse 1980; Arbib & Hesse 1986
222 Holton 1973; Trenn 1974; Kitcher 1993
223 Kantorovich 1993
224 Keller 1988; Haraway 1991
225 Rose 1992
226 Mulkay 1979; Pickering 1984; Thagard 1993
227 Ziman 1978
228 Thom 1975; Restivo 1990; Gell-Mann 1992; Cohen & Stewart 1994.
229 Beveridge 1950; Hesse 1974; Lemaine *et al.* 1977; Galison 1997
230 Shinn 1987
231 Mukerji 1989
232 Giere 1988
233 Ziman 1984 (1995)
234 Mukerji 1989
235 Solomon 1992; Cohen & Stewart 1994; Hallam & Malcolm 1994
236 Harré 1979; Jagtenberg 1983; Arbib & Hesse 1986

Chapter 6 cont.
237 Mukerji 1989
238 Giere 1988
239 Pivcevic 1970; Schütz & Luckman 1974; Kantorovich 1993
240 Böhme 1992; Wolpert 1992;
241 Cohen & Stewart 1994
242 Körner 1966; Toulmin 1980
243 Vollmer 1984
244 Millikan 1984
245 Atran 1990; Solomon 1992
246 Böhme 1992
247 Zhao 1989
248 Hesse 1980
249 Mead 1938
250 Ihde 1990
251 Mukerji 1989
252 Rescher 1988
253 Markus 1987
254 Hallam & Malcolm 1994; Barnes 1995
255 Restivo 1994
256 Jarvie 1972
257 Gouldner 1970; Geertz 1973
258 Bomer 1985
259 Gibbons *et al.* 1994
260 Galison 1997

Chapter 7

1 Merton 1973
2 Holton 1992
3 Ziman 1968; Markus 1987
4 Fleck 1935
5 Segerstrale 1991
6 Laudun 1977
7 OTA 1986
8 Rouse 1987
9 Ezrahi 1990
10 Hesse 1980; Arbib & Hesse 1986; Putnam 1991
11 Collins English Dictionary
12 Churchland 1990; Hooker 1995
13 Harris 1990; Böhme 1992; Diettrich 1995
14 Lindblom 1990
15 Ziman 1978
16 Mukerji 1989

17 Hacking 1983
18 Shapin & Schaffer 1985
19 Laudun 1990
20 Hull 1988; Mukerji 1989
21 Merton 1973; Knorr-Cetina 1981
22 Ziman 1991 (1995)
23 Cole 1992
24 Latour 1987; Amsterdamska 1990
25 Hagstrom 1965; Hull 1988; Knorr-Cetina 1995
26 Bourdieu 1975, 1984
27 Broad & Wade 1983
28 Crawford 1984
29 Greenberg 1969
30 Hagstrom 1965
31 Gross & Levitt 1994
32 Mokyr 1990
33 Kuhn 1962
34 Ziman 1978; Giere 1988
35 Merton 1973
36 Jagtenberg 1983; Schmaus, Segerstrale & Jessuph 1992
37 Sismondo 1993
38 Isambert 1985
39 Bourdieu 1975; Restivo 1994; Rip 1994
40 Porter 1995
41 Barnes 1977, 1995
42 Waddington 1948
43 Bernal 1939
44 Mukerji 1989
45 Segerstrale 1991
46 Shapin & Schaffer 1985
47 Bird 1996; Etzkowitz 1996
48 Zuckerman 1977
49 Mukerji 1989
50 Collingridge & Reeve 1986
51 Restivo 1994
52 Latour & Woolgar 1979; Rudwick 1985; Dupré 1993
53 Longino (Callebaut 1993)
54 Gibbons *et al.* 1994; Restivo 1994
55 Longino 1990
56 Laudun 1990
57 Segerstrale 1992
58 Kolata 1986b
59 Rorty 1996
60 Ziman 1981 (1995)

Chapter 7 cont.

61 Bourdieu 1975; Beck 1986; Ezrahi 1990; Fisher 1990
62 Capaldi 1975
63 Hessen 1931
64 Forman 1971
65 Toulmin 1990
66 Klausner 1983; Porter 1995
67 Lindblom & Cohen 1979
68 Douglas 1987
69 Harré 1979
70 Taylor 1985; Giddens 1990; Dupre 1993
71 Weiss 1983
72 Giddens 1990
73 Klausner 1983
74 Capaldi 1975; Hesse 1980; Klausner 1983; Ezrahi 1990
75 Popper 1945; Dews 1986
76 Holzner *et al.* 1983; Ezrahi 1990
77 Orlans 1976; Rosenmayer 1983; Midgley 1992a
78 Capaldi 1975; Smith, 1997
79 Dews 1986
80 Hesse 1980
81 Klausner 1983
82 Ezrahi 1990
83 Keynes 1936
84 Weiss 1983
85 Slezak 1989
86 Rosenmayer 1983; Douglas 1987
87 Chandler 1977
88 Ezrahi 1990; Toulmin 1990
89 Weingart 1989
90 Swinnerton-Dyer 1995
91 Rosenberg & Nelson 1994
92 Campbell 1982; Rouse 1987; Jasanoff *et al.* 1995
93 Rose, H. 1997
94 Gibbons *et al.* 1994
95 Midgley 1989
96 Ravetz 1971
97 Ziman 1991 (1995)
98 Brooks 1979
99 Ziman 1996b
100 Beck 1986; Rouse 1987; Böhme 1992
101 Nowotny & Felt 1997
102 Merton 1973

103 Vincenti 1990
104 Mukerji 1989; Ezrahi 1990
105 Ziman 1978
106 Beck 1986
107 Lindblom 1990
108 Mukerji 1989
109 Brooks 1979
110 Hooker 1995
111 Smith & Torrey 1996
112 Ziman 1995
113 Gouldner 1970, 1979; Giddens 1987; Böhme 1992
114 Porter 1995
115 Mukerji 1989; Ezrahi 1990; Porter 1995
116 Gouldner 1970
117 Beck 1986
118 Hill 1995
119 Weinberg 1972; Hagendijk 1990
120 Elzinga 1986; Ezrahi 1990
121 Segerstrale 1991
122 Gibbons *et al.* 1994
123 Markus 1987; Segerstrale 1991
124 Hill 1995
125 Rescher 1980; Schäfer *et al.* 1983; Lindblom 1990; Restivo 1994

Chapter 8

1 Hagstrom 1965
2 Zuckerman 1979
3 Campbell 1974
4 Ziman 1981; Galison 1987; Traweek 1988; Knorr-Cetina 1995
5 Polanyi 1969
6 Duncan & Weston-Smith 1977
7 Laudun 1977
8 Naess 1972
9 Polanyi 1969; Rose 1992
10 Campbell 1983
11 Collingwood 1946
12 Medawar 1967
13 Merton 1973; Segerstrale 1992; Nickles (Callebaut 1993)
14 Hagstrom 1965; Langley *et al.* 1987; Lindblom 1990
15 Kitcher 1993
16 Ziman 1981

Chapter 8 cont.

17 Kantorovich 1993
18 Kuhn 1962
19 Longino 1990
20 Kant 1783 (Rescher 1978)
21 Zuckerman 1979
22 Keller 1988
23 Lemaine 1980; Jagtenberg 1983
24 Star 1983
25 Boden 1990
26 Latour 1987
27 Galison 1997
28 Traweek 1988; Galison 1997
29 Ziman 1981 (1995)
30 Ziman 1980
31 Campbell 1974; Knorr-Cetina 1981,
    1981a; Pickering 1984; Rouse 1987;
    Mukerji 1989; Kantorovich 1993;
    Restivo 1994
32 Ravetz 1971; Toulmin 1972;
    Lemaine *et al.* 1977; Fuller 1988;
    Mokyr 1990; Pickering 1995
33 Swinnerton-Dyer 1995
34 Knorr-Cetina 1981
35 Beveridge 1950; Brooks 1979;
    Lemaine 1980
36 Knorr-Cetina 1983
37 Brooks 1979; Ziman 1981 (1995);
    Schäfer *et al.* 1983
38 Ziman 1981 (1995)
39 Restivo 1994
40 Cohen & Stewart 1994
41 Weinberg 1967; 1995
42 Mukerji 1989
43 Ziman 1982, 1994, 1995
44 Ziman 1981 (1995); Jagtenberg
    1983
45 Bernal 1939
46 Jagtenberg 1983; Shrum 1984;
    Sutton 1984; Weingart 1989
47 Hagstrom 1965; Mukerji 1989
48 Galison 1997
49 Hagstrom 1965
50 Etzkowitz 1992
51 Turner 1990
52 Bernal 1939; Hagstrom 1965
53 Katz & Hartnett 1976
54 Mukerji 1989
55 Lemaine 1980

56 Chubin 1976
57 Price 1963; Ziman 1987
58 Zuckerman 1977; Galison 1997;
    Wagner-Döbler 1997
59 Traweek 1988; Galison 1997
60 Giere 1988; Midgley 1989; Meyer-
    Krahmer 1997
61 de Mey 1982
62 Lemaine 1980
63 Shrum 1984; Markus 1987
64 Hooker 1995
65 Hagstrom 1965
66 Zuckerman 1979
67 Polanyi 1958
68 Sklair 1973
69 Ziman 1985, 1995
70 Gökalp 1990
71 Ziman 1985, 1995
72 Galison 1997
73 Laudun 1977; Rudwick 1985
74 Kuhn 1962; Lakatos 1978
75 Latour & Woolgar 1979
76 Ziman 1987, 1995
77 Whitley 1984; Becher 1989
78 Hagstrom 1965; Fuller 1988
79 Toulmin 1972
80 Goonatilake 1984; Blume 1988
81 Ziman 1987, 1995
82 Goonatilake 1984
83 Edelman 1992
84 Kuhn 1962
85 Hacking 1983
86 Lakatos & Musgrave 1970;
    Churchland 1990
87 de Mey 1982; Jagtenberg 1983
88 Harré 1986; Cohen & Stewart 1994
89 Thagard 1993
90 Kuhn 1962; Ravetz 1971
91 Toulmin 1972; Restivo 1994; Hill
    1995
92 Galison 1997
93 Kitcher 1993
94 Fleck 1935
95 Toulmin 1972
96 Körner 1970, 1974
97 Fleck 1935
98 Douglas 1987
99 Geertz 1973
100 Laudun 1990

Chapter 8 cont.

101 Gouldner 1970; Hill 1995
102 Hallam 1973
103 Loasby 1989; Hodgson 1993;
     Saviotti 1996
104 Kuhn 1962
105 Laudun 1990
106 Pickering 1984
107 Körner 1970; Shils 1982; Bernstein
     1991
108 Galbraith 1987
109 Mulkay 1979
110 Tilley 1980; Ziman 1978
111 de Mey 1982
112 Mulkay 1979
113 Naess 1972; Gilbert 1976
114 Judson 1979
115 Masterman 1970
116 Searle 1995
117 de Mey 1982
118 Root-Bernstein 1989
119 Cohen & Stewart 1994
120 Lemaine *et al.* 1977; Lakatos 1978;
     Toulmin 1980; Giere 1988
121 Ziman 1968
122 Hull 1988
123 Schäfer *et al.* 1983
124 Schäfer *et al.* 1983
125 de Mey 1982
126 Kuhn 1962
127 Kitcher 1993
128 Kuhn 1962
129 Barnes & Edge 1982; Markus 1987
130 Bernal 1939
131 Lemaine 1980; de Mey 1982
132 Lemaine 1980; Hill 1995
133 Beveridge 1950; Ravetz 1971;
     Campbell 1977; Ziman 1987
134 Tilley 1980
135 Latour & Woolgar 1979
136 Mukerji 1989
137 Topitsch 1980; Hudson & Jacot
     1986
138 Rescher 1978
139 Kuhn 1962
140 Jagtenberg 1983
141 Rosenmayer 1983
142 Galison 1987
143 Schäfer *et al.* 1983; Jagtenberg 1983

144 Kuhn 1962
145 Schäfer *et al.* 1983
146 Schäfer *et al.* 1983
147 Markus 1987
148 Gibbons *et al.* 1994
149 Radnitzky 1968; Jagtenberg 1983
150 Restivo 1994; David 1995
151 Brooks 1979; Ziman 1994
152 Lindblom 1990
153 Restivo 1994
154 Restivo 1994
155 Jagtenberg 1983; Varma &
     Worthington 1995
156 Campbell 1974
157 Ziman 1996a
158 Rescher 1978; Restivo 1994
159 Studer & Chubin 1980; Schäfer *et al.*
     1983
160 Galison 1987; Knorr-Cetina 1995
161 Shrum 1984; Sutton 1984
162 Restivo 1994; Odlyzko 1995
163 Mukerji 1989
164 Brooks 1979
165 Longino 1990
166 Gibbons *et al.* 1994
167 Jagtenberg 1983; Shrum 1984;
     Sutton 1984
168 de Mey 1982
169 Galison 1987; Knorr-Cetina 1995;
     Galison 1997
170 Latour 1987
171 Latour & Woolgar 1979; Restivo
     1994; Rip 1994
172 Studer & Chubin 1980; Schäfer *et al.*
     1983
173 Tibbetts & Johnson 1985
174 Hesse 1980; de Mey 1984; Rescher
     1988; Harris 1990; Turnbull 1994
175 Schäfer *et al.* 1983; Markus 1987
176 Topitsch 1980
177 Schäfer *et al.* 1983
178 Blume 1988
179 Weingart 1995
180 Gibbons *et al.* 1994
181 Knorr-Cetina 1983; Hagendijk 1990
182 Gibbons *et al.* 1994
183 Popper 1963; Ziman 1985
184 Mukerji 1989
185 Root-Bernstein 1989

Chapter 8 cont.

186 Gibbons *et al.* 1994
187 Galison 1997
188 Gibbons *et al.* 1994
189 Gibbons *et al.* 1994
190 Lemaine *et al.* 1976; Brooks 1979; Hooker 1995; Nowotny & Felt 1997
191 Toulmin 1990
192 Gibbons *et al.* 1994
193 Boden 1990
194 Stoneham 1995
195 Kuhn 1962
196 Lemaine *et al.* 1977; Hooker 1995
197 Cantore 1977
198 Popper 1959
199 Kuhn 1962
200 Ziman 1972 (1981)
201 Ravetz 1971
202 Kantorovich 1993
203 Brannigan 1981
204 Collins English Dictionary
205 Medawar 1967; Brannigan 1981; Kantorovich 1993
206 Kantorovich 1993; Galison 1997
207 Brannigan 1981
208 Langley *et al.* 1987
209 Galison 1997
210 Odlyzko 1995
211 Caws 1969; Kantorovich 1993
212 Root-Bernstein 1989
213 Zuckerman 1979
214 Mulkay 1979
215 Fleck 1935; Giere 1988
216 Langley *et al.* 1987
217 Brannigan 1981; Kantorovich 1993
218 Mukerji 1989
219 Slezak 1989; Kantorovich 1993
220 Brannigan 1981; Traweek 1988
221 Kantorovich 1993
222 Root-Bernstein 1989
223 Merton 1973; Campbell 1974
224 Brannigan 1981
225 Hagstrom 1965
226 Root-Bernstein 1989
227 Roberts 1989
228 Kantorovich 1993
229 Dean 1977
230 Beveridge 1950
231 Zuckerman 1979

232 Causey 1968
233 Campbell 1974
234 Ziman 1994
235 Kantorovich 1993
236 Knorr-Cetina 1981; Pickering 1984; Rouse 1987; Kantorovich 1993
237 Knorr-Cetina 1981a
238 Duncan & Weston-Smith 1977
239 Hesse 1980; Elkana 1981; Rouse 1987; Root-Bernstein 1989; Kantorovich 1993
240 Huzsagh & Infante 1989
241 Harré 1986
242 Beveridge 1950
243 Giere 1988
244 Ehrenberg & Bound 1993
245 Boyd 1991; Giere (Callebaut 1993)
246 Farris & Revlin 1989; Slezak 1989; Hooker 1995
247 Barnes & Edge 1982
248 Munz 1993; Kantorovich 1993
249 Latour & Woolgar 1979
250 Collingwood 1946
251 Gieryn 1982
252 Kern, Mirels & Hinshaw 1983; Hull 1988; Tweney (Callebaut 1993)
253 Pivcevic 1970
254 Thagard 1993
255 Nagel 1961
256 Popper 1959, 1963
257 Naess 1968; Lakatos 1978; Rescher 1988; Midgley 1992
258 Kantorovich 1993
259 Capaldi 1975; Hooker 1995
260 Pirsig 1974
261 Rescher 1988
262 Ziman 1968
263 Ziman 1981, 1995
264 Harré 1986; Fuller 1988
265 Klemke, Hollinger & Kline 1980
266 Beveridge 1950
267 Linstone & Turoff 1975
268 Tricker 1965; Hesse 1974; Howson & Urbach 1989, 1991; Kantorovich 1993; Kitcher 1993
269 Howson & Urbach 1989
270 Tricker 1965
271 Langley *et al.* 1987
272 Tricker 1965

Chapter 8 cont.

273 Hesse 1974; Good 1977; Franklin 1986; Howson & Urbach 1989; Kantorovich 1993; Kitcher 1993

274 Whewell 1840; Hesse 1974

275 Giere 1988; Howson & Urbach 1989; Giere (Callebaut 1993)

276 Glymour 1980

277 Fleck 1935

278 Howson & Urbach 1991

279 Polkinghorne 1985

280 Marshall 1990

281 Popper 1959, 1963

282 Hesse 1974; Howson & Urbach 1989

283 Ziman 1984

284 Popper 1963

285 Brush 1989

286 Popper 1959, 1963

287 Magee 1973; Lakatos 1978

288 Popper 1963

289 Campbell 1982; Rouse 1987; Lindblom 1990

290 Basalla 1988; Mokyr 1990

291 Mayr 1988

292 Hull 1988

293 Howson & Urbach 1989; Segerstrale 1992

294 Shapin & Schaffer 1985

295 Duhem 1906 (1962); Quine 1953; Naess 1972; Hooker 1995

296 Coady 1992

297 Campbell 1977; Galison 1987

298 Lakatos 1970; Kern, Mirels & Hinshaw 1983

299 Root-Bernstein 1989

300 Galison 1997

301 Cohen & Stewart 1994

302 Traweek 1988; Galison 1997

303 Nagel 1979

304 Naess 1972; Mulkay 1979; Dupré 1993; Hooker 1995

305 Collins English Dictionary

306 Shapin 1994

307 Campbell (Callebaut 1993); Wimsatt (Callebaut 1993)

308 Cassirer 1956; Polanyi 1958, 1969; Chandrasekhar 1979; Giere 1988; Carlson & Gorman 1990; Sismondo 1993

309 Coady 1992

310 Toulmin 1953, 1990; Naess 1972; Hesse 1974, 1980; Putnam 1991; Shapere (Callebaut 1993); Yearley 1994; Hooker 1995

311 Feyerabend 1975; Restivo 1994

312 Kantorovich 1993

313 Kendall 1975; de Mey 1984

314 Churchland 1990; Ehrenberg & Bound 1993

315 Root-Bernstein 1989

316 Beveridge 1950

317 Anderson 1990

318 Barnes & Edge 1982

319 Midgley 1992

320 Bridgman 1991

321 Schmaus, Segerstrale & Jessuph 1992

322 Thagard 1993

323 Lindblom 1990

324 Rosenmayer 1983

325 Segerstrale 1992

326 Vollmer 1984

327 Mead 1938

328 Kitcher 1993

329 Nagel 1979

330 Ezrahi 1990; Hull (Callebaut 1993)

331 Ziman 1978

332 Isambert 1985

333 Gould 1981; Segerstrale 1992; Kantorovich 1993

334 Bhaskar 1975; Steinberger 1993

335 Pickering 1984; Galison 1997

336 Klemke, Hollinger & Kline 1980; Vollmer 1984

337 Cohen & Stewart 1994

338 Latour & Woolgar 1979; Harré 1986, 1990; Rouse 1987

339 Latour 1987

340 Solomon 1992

341 Myers 1990

342 Medawar 1967; Haraway 1991

343 Cosmides & Tooby 1992

344 de Mey 1983

345 Collins English Dictionary and Thesaurus

346 Mukerji 1989

347 Sismondo 1993

348 Pivcevic 1970

Chapter 8 cont.

349 Shapin & Schaffer 1985
350 Pickering 1984
351 Solomon 1994
352 Rorty 1986; Roth & Barrett 1990
353 Latour & Woolgar 1979; Rouse 1987
354 Giere 1993
355 Giere 1988; Isambert 1985
356 Franklin 1986
357 Weingart 1995
358 Lindblom 1990
359 Ziman 1960 (1981), 1968; Barber & Hirsch 1962; Kuhn 1962; Hagstrom 1965
360 Traweek 1988
361 Sismondo 1993; Hooker 1995
362 Restivo 1994
363 Sismondo 1993
364 Cole 1992
365 Hesse 1980
366 Ziman 1978
367 Hooker 1995
368 Isambert 1985; Ezrahi 1990
369 Suttie 1935
370 Sismondo 1993
371 Pickering 1984
372 Myers 1990
373 Jagtenberg 1983
374 Lindblom 1990
375 Varela *et al.* 1991; Sismondo 1993
376 Neurath 1932; Quine 1953; Hesse 1974; Gieryn 1982
377 Knorr-Cetina 1995
378 Doyal & Harris 1986; Hagendijk 1990; Sismondo 1993
379 Hennion 1989
380 Toulmin 1972; Bourdieu 1975; Markus 1987; Myers 1990
381 Price 1963
382 Lemaine 1980
383 Fleck 1935; Kantorovich 1993
384 Mitroff 1974
385 Caws 1969; Boden 1990, 1990a
386 Thagard 1993
387 Harris 1993
388 Lemaine *et al.* 1977; Hudson & Jacot 1986; Kantorovich 1993
389 Langley *et al.* 1987
390 Gieryn & Hirsh 1983

391 Dupré 1993
392 Polanyi 1958
393 Lyons 1981
394 Dennett 1969
395 Atran 1990; Churchland 1990; Edelman 1992
396 Toulmin 1972
397 Ziman 1978
398 Ivins 1953
399 Kosslyn 1990
400 Harré 1975, 1986
401 Ihde 1991
402 Ziman 1978; Galison 1997
403 Ziman 1978; Carlson & Gorman 1990
404 Miller 1984; Gilbert & Mulkay 1984; Tweney 1993 (Callebaut 1993)
405 Polanyi 1958; Campbell 1974; Dennett 1991; Hooker 1995
406 Piaget & Inhelder 1963; Monod 1970; Miller 1985; Shattuck 1985; Shinn 1987
407 Piaget 1954
408 Donald 1991
409 Sinclair-de-Zwart 1969; Downs & Stea 1973
410 Denbigh 1975
411 Boden 1988, 1990; Hooker 1995
412 Johnson-Laird 1983, 1988
413 Ziman 1978; Lakoff 1987
414 Oeser 1984; Giere 1988
415 Hesse 1980; Vollmer 1984; Wuketits 1984a; Edelman 1992; Munz 1993; Plotkin 1994; Cziko 1995; Hooker 1995
416 Luria 1973
417 Ziman 1978
418 Johnson-Laird 1988; Mukerji 1989
419 de Mey 1982; Kitcher 1993
420 Midgley 1992
421 Galison 1997
422 Polanyi 1958; Cohen & Stewart 1994
423 Toulmin 1972; Gilbert 1976; Callebaut 1993
424 Jagtenberg 1983
425 Arbib & Hesse 1986; Langley *et al.* 1987; Thagard 1993
426 de Mey 1982
427 Johnson-Laird 1983
428 Midgley 1992

Chapter 8 cont.

429 Bunge 1962; Wilder 1967; Ziman 1978
430 Millikan 1984
431 Toulmin 1953; Polanyi 1958; Ravetz 1971

Chapter 9

1 Merton 1942 (1973)
2 Naess 1968
3 Reid 1970
4 Ziman 1968
5 Tilley 1980
6 Medawar 1967
7 Merton 1973; Zuckerman 1979
8 Mulkay 1991
9 Sindermann 1982; Kiang 1995
10 Ziman 1968, 1978
11 Mahoney 1979; Hull 1988
12 Sindermann 1982
13 Popper 1963
14 Mukerji 1989
15 Böhme 1975; Kantorovich 1993; Hooker 1995
16 Campbell 1984; Rudwick 1985; Isambert 1985; Mukerji 1989
17 Mulkay 1985
18 David 1995
19 Hagstrom 1965
20 Mulkay 1979; Gilbert & Mulkay 1984; Schaffner 1994
21 Tilley 1980; Galison 1987; Knorr-Cetina 1995
22 Markus 1987
23 Longino (Callebaut 1993)
24 Gilbert & Mulkay 1984; Mulkay 1985; Hull 1988; Mulkay 1991
25 Collingwood 1946; Bernstein 1991; Longino (Callebaut 1993)
26 Shapin & Schaffer 1985
27 Mitroff 1974; Latour 1987; Collins & Pinch 1993
28 Habermas 1971; Hesse 1980; Topitsch 1980; Arbib & Hesse 1986; Bernstein 1991; Restivo 1994
29 Mahoney 1979; Hull 1988
30 Collins 1975; Chubin 1976; Isambert 1985; Rose 1992
31 Curtis 1994
32 Ravetz 1971
33 Mukerji 1989
34 Beveridge 1950; Bourdieu 1975
35 Segerstrale 1991
36 Shapin & Schaffer 1985
37 Toulmin 1972
38 Kitcher 1993
39 Hagstrom 1965; Hooker 1995
40 Beck 1986; Hull 1988; Kitcher 1993
41 Shapin & Schaffer 1985
42 Jaki 1966; Mitroff 1974; Bernstein 1991
43 Knorr-Cetina 1995
44 Tilley 1980
45 Feyerabend 1975; Sindermann 1982; Lindblom 1990
46 Hagstrom 1965
47 Sindermann 1982
48 Latour 1987
49 Collins & Pinch 1993
50 Latour & Woolgar 1979
51 Cole 1996
52 Galison 1987
53 Kolata 1986
54 Morell 1993
55 Nicoll & Brown 1994
56 Maddox 1995
57 Kitcher 1993
58 Rudwick 1985
59 Pinch 1986
60 Mulkay 1985
61 Collins & Pinch 1993
62 Segerstrale 1991
63 Ziman 1978
64 Mukerji 1989
65 Pinch 1986
66 Midgley 1989
67 Latour 1987
68 Hull (Callebaut 1993)
69 Campbell 1921; Shapere (Callebaut 1993)
70 Hooker 1995
71 Lemaine *et al.* 1977
72 Lloyd 1979
73 Fahnestock 1989
74 Ziman 1968
75 Myers 1991
76 Fleck 1935

Chapter 9 cont.

77 Feyerabend 1975; Campbell (Callebaut 1993)
78 Ziman 1968, 1978; Rouse 1987; Bazerman 1988
79 Donald 1991
80 Latour & Woolgar 1979
81 Law & Williams 1982; Mulkay 1985
82 Magee 1973
83 Knorr-Cetina 1981
84 Rudwick 1985
85 Laudun 1990
86 Thouless 1971
87 Hull 1988
88 Latour 1987
89 Jagtenberg 1983
90 Bazerman 1981
91 Hull 1988; Midgley 1989; Hull (Callebaut 1993)
92 Popper 1959, 1963
93 Schäfer *et al.* 1983; Restivo 1994
94 Lindblom 1990
95 Fleck 1935; Campbell 1982
96 Harré 1986
97 Giere 1988
98 Latour 1987
99 Engelhardt & Caplan 1987; Segerstrale 1987
100 Hagendijk 1990
101 Evans 1997
102 Ziman 1968, 1978; Goldstein & Goldstein 1978; Callebaut 1993
103 Hooker 1995
104 Bourdieu 1977
105 Ravetz 1971; Farley & Geison 1974
106 Traweek 1988
107 Collins & Pinch 1993
108 Hagstrom 1965
109 Lakatos 1978
110 Ziman 1970 (1981)
111 MacRoberts & MacRoberts 1984
112 Lakatos 1978; Mulkay 1979
113 Kantorovitch 1993
114 Mulkay 1979, 1991; Pickering 1984; Rouse 1987; Cole 1992
115 Hagstrom 1965; Ravetz 1971
116 Shapin 1994
117 Hagstrom 1965; Toulmin 1972; Mulkay 1979, 1985; Knorr-Cetina (Callebaut 1993), 1995; Porter 1995

118 Kitcher 1993
119 Hull (Callebaut 1993)
120 Galison 1987
121 Hesse 1980; Dolman & Bodewitz 1985
122 Hallam 1973
123 Ruse 1996
124 Markus 1987
125 Knorr-Cetina 1995
126 Toulmin 1972; Fuller 1988; Mulkay 1991
127 Mayr 1982, 1988; Callebaut 1993
128 Ziman 1978
129 Geertz 1973
130 Markus 1987
131 Knorr 1978
132 Laudun 1990
133 Ravetz 1971
134 Lindblom 1990
135 Knorr 1978
136 Knorr 1975, 1978
137 Habermas 1971; Dews 1986; Rescher 1988; Restivo 1994; Shapin 1994; Barnes 1995
138 Harré 1983
139 Schaffner 1994
140 Ryle 1949
141 Restivo 1994
142 Blakeslee 1994
143 Hagstrom 1965
144 Bazerman 1984, 1988
145 Dolman & Bodewitz 1985
146 Markus 1987; Mukerji 1989
147 Ziman 1982 (1995)
148 Markus 1987
149 Ziman 1982 (1995)
150 Shinn & Whitley 1985
151 Mukerji 1989
152 Bunders & Whitley 1985
153 Cronin 1984; Latour 1987
154 Shapin & Schaffer 1985
155 Ravetz 1971
156 Hull 1988; Myers 1990
157 Garfield 1979
158 Cronin 1984
159 Garfield 1979
160 Fleck 1935; Markus 1987
161 Ziman 1960
162 Mulkay 1979; Myers 1991a

Chapter 9 cont.

163 Ziman 1968
164 Latour & Woolgar 1979
165 de Mey 1984; Oeser 1984;
     Sokolowski 1988; Rose 1992
166 Bolter 1991; Donald 1991
167 Greenfield; Reich & Olver 1966;
     Goody 1977; Donald 1991
168 Hooker 1995
169 Geertz 1973, 1983
170 Popper 1972
171 Frege 1892
172 Popper 1972
173 Magee 1973; Harré 1986
174 Popper 1972; Jarvie 1972; Magee
     1973
175 Edelman 1992; Hill 1995
176 Frege 1892; Popper 1972
177 Magee 1973
178 Fleck 1935
179 Ziman 1978
180 Fleck 1935; Oeser 1984; Hooker 1995
181 Ziman 1978
182 Harré 1983
183 Pivcevic 1970; Hooker 1995
184 Midgley 1989, 1992
185 Popper 1963, 1972; Gould 1981
186 Segerstrale 1990
187 Laudun 1990
188 Fleck 1935; Segerstrale 1990
189 Beveridge 1950
190 Lakatos 1970, 1978
191 Weinstein 1981
192 Baltimore 1992
193 Ziman 1978
194 Schmaus 1981
195 Evans 1984
196 Root-Bernstein 1989
197 Hagstrom 1965; Ziman 1970;
     Merton 1973; Sindermann 1982;
     Franklin 1986; Hull 1988; Schmaus
     1991
198 Restivo 1994
199 Culliton 1988
200 Segerstrale 1990
201 Pinch 1993
202 Schmaus, Segerstrale & Jessuph 1992
203 Brannigan 1981
204 Nagel 1979

205 Lakatos 1978; Harré 1986; Munz
     1993; Hooker 1995
206 Capaldi 1975
207 Hesse 1974; Campbell 1983
208 Cohen & Stewart 1994
209 Menard 1971
210 Booth 1988
211 Pinch 1979; Hardin 1981; Collins &
     Pinch 1982; Schmaus, Segerstrale &
     Jessuph 1992
212 Gieryn 1983; Toulmin 1990
213 Gökalp 1990; Coady 1992
214 Kantorovich 1993
215 Nagel 1961; Millikan 1984; Star
     1985; Baltimore 1992
216 Mokyr 1990
217 Kitcher 1993
218 Mahoney 1979
219 Hull 1988
220 Zuckerman 1979; Lemaine 1980
221 Searle 1995
222 Popper 1963; Campbell 1977;
     Millikan 1984; Harré 1986; Dupré
     1993; Diettrich 1995
223 Popper 1972
224 Porter 1995
225 Wolpert 1992
226 Fleck 1935; Gouldner 1970
227 Ezrahi 1990
228 Lucas 1995
229 Furnham 1988
230 Popper 1959, 1972
231 Bourdieu 1975
232 Campbell 1977
233 Kuhn 1978
234 Collins & Pinch 1982
235 Naess 1972
236 Coady 1992; Segerstrale 1992
237 Fleck 1935; Goffman 1974;
     Toulmin 1980
238 Nowotny & Rose 1979; Regis 1991;
     Midgley 1992a
239 Restivo 1994
240 Randall 1975; Kitcher 1993
241 Langmuir 1953
242 Merton 1973; Klemke, Hollinger &
     Kline 1980
243 Bloor 1976, 1981; Knorr-Cetina &
     Mulkay 1983; Pinch 1993

Chapter 9 cont.

244 Shils 1982
245 Kitcher 1993
246 Popper 1963
247 Knorr-Cetina 1995
248 Polanyi 1958
249 Chubin 1976; Ziman 1974 (1981)
250 Ziman 1985
251 Pickering 1984; Restivo 1994
252 Mokyr 1990
253 Hull 1988
254 Goonatilake 1984
255 Kuhn 1962
256 Hallam 1973
257 Galison 1977
258 Toulmin 1972; Capaldi 1975;
     Bourdieu 1975; Capaldi 1975;
     Rudwick 1985
259 Lindblom 1990
260 Shapere (Callebaut 1993)
261 Hanson 1958; Kuhn 1962
262 Hesse 1980
263 Franklin 1986
264 Gilbert & Mulkay 1984; Hull 1988
265 Campbell 1977; Harré 1986
266 Kuhn 1962; 1984
267 Midgley 1989
268 Davidson 1984
269 Putnam 1981; Fuller 1988; Galison
     1997
270 Fleck 1935; Feyerabend 1975;
     Toulmin 1972
271 Naess 1972; Toulmin 1980;
     Kantorovich 1993; Longino
     (Callebaut 1993
272 Hesse 1980
273 Lemaine 1980
274 Lemaine *et al.* 1977; Rorty 1986
275 Fleck 1935
276 Toulmin 1972; Campbell 1974
277 Campbell 1960, 1974; 1977; Ravetz
     1971; Popper 1972; Toulmin 1972;
     Rouse 1987; Hull 1988; Root-
     Bernstein 1989; Giere 1990;
     Kantorovich 1993; Hooker 1995
278 Campbell 1960
279 Munz 1993; Plotkin 1994
280 Popper 1959, 1963
281 Ziman 2000

282 Root-Bernstein 1989
283 Keller 1988
284 Sapp 1994
285 Galison 1997
286 Koestler 1964; Lemaine *et al.* 1977;
     Hudson & Jacot 1986; Boden 1990
287 Toulmin 1972; Ziman 1996, 2000
288 Mayr 1982
289 Dawkins 1976
290 Dawkins 1976, 1986
291 Munz 1993; Csikszentmihaly 1993
292 Toulmin 1972
293 Leinfellner 1984; Latour 1987
294 Ziman 1991 (1995)
295 Toulmin 1972; Campbell 1977;
     Hull 1988; Giere 1990; Waters
     1990; Edelman 1992; Kantorovich
     1993; Hooker 1995
296 Ziman 2000
297 Hull 1988; Hooker 1995
298 Edelman 1989, 1992; Plotkin 1994;
     Cziko 1995
299 Levy 1992; Langton 1995
300 Campbell 1984; Vollmer 1984
301 Waters 1990
302 Caws 1969
303 Nickles (Callebaut 1993); Hooker
     1995
304 Kantorovich 1993
305 Campbell 1974
306 Ziman 2000
307 Dupré 1993
308 Campbell 1977; Hull 1988
309 Ziman 2000
310 Midgley 1992
311 Hjern & Porter 1983; Hooker 1995
312 Kauffman 1993, 1995
313 Lemaine 1980; Goonatilake 1984;
     Dupré 1993
314 Cohen & Stewart 1994; Gell-Mann
     1994
315 Lorenz 1973
316 Arthur 1994
317 Popper 1959, 1963
318 Simon 1956; Campbell 1984;
     Hooker 1995
319 Kantorovich 1993
320 Wimsatt (Callebaut 1993); Plotkin
     1994

Chapter 9 cont.

321 Nelson & Winter 1982; Saviotti 1996
322 Hägerstrand 1985; Ganelius 1986;
      Elzinga 1986
323 Bourdieu 1975
324 Klausner 1983; Dolman & Bodewitz
      1985; Sinding 1996
325 Markus 1987
326 Kuhn 1984; Munz 1993
327 Ruse 1996
328 Knorr-Cetina (Callebaut 1993)
329 Bonner 1988
330 Gould 1996
331 Mulkay 1991; Midgley 1992
332 Galison 1997
333 Hull 1988; Kitcher 1993
334 Kantorovich 1993; Restivo 1994
335 Rescher 1978
336 Ziman 1994
337 Gibbons *et al.* 1994
338 Toulmin 1972
339 Waters 1990; Thagard 1993
340 Campbell 1960, 1974, 1977;
      Toulmin 1972; Hull 1988; Hooker
      1995
341 Toulmin 1972; Giere (Callebaut
      1993); Hooker 1995

Chapter 10

 1 Johnson-Laird 1983; Taylor 1985a;
     Thagard 1993
 2 Lindblom 1990
 3 Callebaut 1993
 4 Campbell 1921
 5 Hull 1988
 6 Popper 1963
 7 Popper 1963
 8 Elster (Callebaut 1993)
 9 Langley *et al.* 1987; Schmaus,
     Segerstrale & Jessuph 1992
10 Toulmin 1972
11 Husserl 1924; Schütz & Luckmann
     1973; Pivcevic 1970
12 Rescher 1988
13 Fine 1991
14 Cosmides & Tooby 1992
15 Popper 1959, 1972; Medawar 1969

16 Ihde 1991
17 Cupitt 1990
18 Shapin 1994
19 Davidson 1984
20 Ravetz 1971
21 Knorr-Cetina 1981a
22 Polanyi 1958; de Mey 1982
23 Garfinkel 1967
24 Nagel 1961
25 Coady 1992
26 Piaget 1954; Piaget & Inhelder
      1963; Barnes 1995; Hill 1995
27 Ihde 1972
28 Brown 1970; Millikan 1984; Taylor
      1985a; Hill 1995
29 Coady 1992; Burnyeat 1993
30 Taylor 1985a; Cupitt 1990
31 Cosmides & Tooby 1992
32 Atran 1990
33 Remmling 1973
34 Harré 1986
35 Atran 1990
36 Davidson 1984; Diettrich 1995
37 Dews 1986; Searle 1995
38 Taylor 1985a
39 Nicod 1923; Hooker 1995
40 Brain 1951
41 Dews 1986; Midgley 1992; O'Hear
      1996
42 Pivcevic 1970; Cosmides & Tooby
      1992; Shapin 1994
43 Goffman 1974; Bourdieu 1977;
      Arbib & Hesse 1986; Atran 1990
44 Midgley 1989
45 Johnson-Laird 1983; Churchland
      1990; Searle 1995
46 Durkheim & Maus 1903 (1963);
      Munz 1993
47 Lindblom 1990
48 Peters 1970; Watson 1975; Hardin
      1981; Churchland 1990, 1991
49 Armistead 1974; Harré 1979;
      Furnham 1988
50 Rouse 1987
51 Toulmin 1972
52 Cupitt 1990
53 Armistead 1974
54 Nowotny 1994
55 Haste 1993

Chapter 10 cont.

56 Mead 1938; Garfinkel 1967; Körner 1970; Bourdieu 1977; Mulkay 1979; Tilley 1980; Jagtenberg 1983; Furnham 1988
57 Mead 1934; Henderson 1990
58 Pivcevic 1970
59 Fine 1991
60 Campbell 1974; Oeser 1984
61 Hesse 1974; Munz 1993; Diettrich 1995
62 Monod 1970
63 Cohen & Stewart 1994
64 Bradie 1990; Callebaut 1993
65 Popper 1963; Campbell 1977, 1984, 1990; Wuketits 1984a; Munz 1993; Plotkin 1994; Cziko 1995; Dennett 1995; Hooker 1995
66 Simmel 1895; Vollmer 1984
67 Körner 1966
68 Campbell 1974, 1977; Oeser 1984; Vollmer 1984; Harré 1986; Böhme 1992; Munz 1993
69 Campbell 1977, 1990; Vollmer 1984; Kantorovich 1993; Munz 1993
70 Bradie 1990; Kitcher 1993; Hooker 1995
71 Lyons 1981; Shepard 1992
72 Cole & Scribner 1974
73 Grayling 1988; Tooby & Cosmides 1992; Coady 1992
74 Toulmin 1972; Harré 1986
75 Churchland 1990; Bechtel (Callebaut 1993)
76 Glymour 1980: Nickles (Callebaut 1993)
77 Campbell 1977; Knorr-Cetina 1983; Restivo 1994; Diettrich 1995
78 Hesse 1980; Rouse 1987
79 Giere 1988
80 Hesse 1980; Böhme 1992; Munz 1993
81 Popper 1972
82 Geertz 1973; Douglas 1987
83 Ezrahi 1990
84 Handy 1964; Taylor 1985a; Schmaus, Segerstrale & Jessuph 1992; Goonatilake 1996
85 Taylor 1985a
86 Schmaus, Segerstrale & Jessuph 1992

87 Mulkay 1991
88 Fleck 1935
89 Fleck 1935
90 Goonatilake 1984
91 Midgley 1989; Cosmides & Tooby 1992
92 Goodenough 1974
93 Laudun 1990
94 Doyal & Harris 1986
95 Hesse 1980
96 Galison 1997
97 Collins & Pinch 1982
98 Toulmin 1972
99 Körner 1974; Harré 1986; Doyal & Harris 1986
100 Davidson 1984; Bernstein 1991
101 Geertz 1973
102 Shils 1982
103 Keller 1988, 1991; Haraway 1991
104 Douglas 1986
105 Douglas 1987
106 Whorf 1941
107 Watson 1990
108 Strawson 1960
109 Toulmin 1972; Rouse 1987
110 Schmaus, Segerstrale & Jessuph 1992
111 Goonatilake 1984
112 Ravetz 1971; Longino 1990
113 Simmel 1895; Ziman 1978; Harré 1983; Heath 1981; Arbib & Hesse 1986; Doyal & Harris 1986; Harré 1986; Dennett 1987; Fuller 1988; Rescher 1988; Cohen & Stewart 1994
114 Polanyi 1958
115 Rouse 1987
116 Arbib & Hesse 1986; Rescher 1988
117 Dennett 1987; Kempson 1988
118 Argyris 1980
119 Dennett 1991
120 Garfinkel 1967
121 Midgley 1992
122 Toulmin 1972; Harré 1983; Davidson 1984; Doyal & Harris 1986; Rescher 1988; Traweek 1988; O'Hear 1996
123 Polanyi 1958; Midgley 1992
124 Shils 1982; Lindblom 1990; Kitcher 1993; Munz 1993
125 Dennett 1991; Munz 1993
126 Restivo 1994; O'Hear 1996

Chapter 10 cont.

127 Barnes 1995
128 Elkana 1981
129 Ziman 1960 (1981)
130 Bloor 1976
131 Holton 1992
132 Midgley 1992, 1992a
133 Toulmin 1990
134 Cohen & Stewart 1994
135 Ziman 1968; Toulmin 1990
136 Geertz 1973; Valée 1989
137 Midgley 1989
138 Arbib & Hesse 1986; Ezrahi 1990
139 Horton & Finnegan 1973;
      Polkinghorne 1985; Cupitt 1990
140 Argyris 1980; Rescher 1988; Stevens
      1988; Böhme 1992
141 Bloor 1976, 1981; Pickering 1995
142 Shils 1982; Hooker 1995
143 Naess 1972
144 Körner 1974
145 Fleck 1935; Kuhn 1962; Elkana
      1978
146 Körner 1974; Bernstein 1991;
      Margolis 1991; Giere 1993
147 Naess 1968; Pivcevic 1970;
      Campbell 1977; Lakatos 1978;
      Rescher 1988; Midgley 1989, 1992;
      Nickles (Callebaut 1993)
148 Isambert 1985; Latour 1987;
      Haraway 1991; Munz 1993; Restivo
      1994
149 Browne 1643 – *Religio Medici*
150 Harris 1990
151 Elkana 1981
152 Barnes 1981; Atran 1990; Giere 1993
153 Körner 1970; Oeser 1984
154 Naess 1972; Mulkay 1979
155 Atran 1990; Diettrich 1995
156 Foucault 1969
157 Popper 1959
158 Campbell 1921; Schmaus,
      Segerstrale & Jessuph 1992
159 Bourdieu 1975
160 Polanyi 1958
161 Campbell 1977
162 Atran 1990
163 Kantorovich 1993; Lucas 1995
164 Pickering 1984; Böhme 1992

165 de Mey 1982; Polkinghorne 1985;
      Harré 1986; Aronson 1990; Fine 1991
166 Mead 1938: Hill 1995
167 Lloyd 1979
168 Haste 1993; Kantorovich 1993
169 Kantorovich 1993
170 Collins & Pinch 1982; Mervis 1993
171 Shapere (Callebaut 1993)
172 Arbib & Hesse 1986; Fuller 1988;
      Daston 1992; Porter 1995
173 Strawson 1960; Pirsig 1974; Capaldi
      1975; Collins, R. 1975; Ziman 1978;
      Hesse 1980; de Mey 1982; Kuhn
      1984; Vollmer 1984; Taylor 1985a;
      Rorty 1986; Cupitt 1990; Lucas
      1990, 1995; Longino 1990; Shotter
      1990; Munz 1993; Sismondo 1993;
      Restivo 1994; Diettrich 1995;
      Hooker 1995; Searle 1995
174 Fleck 1935
175 Naess 1972; Rouse 1987; Cupitt
      1990; Diettrich 1995
176 Ziman 1978
177 Vollmer 1984; Traweek 1988
178 Shapere (Callebaut 1993)
179 Elkana 1984; Harré 1986; Giere
      1988; Laudun 1990; Kantorovich
      1993; Kitcher 1993
180 Hesse 1974; Giddens 1990
181 Hesse 1980; Pickering 1995
182 Fuller 1988; Hacking 1991; Ihde 1991
183 Aronson 1984; Rouse 1987; Boyd 1991
184 Laudun 1991; Midgley 1992
185 Rorty 1986
186 Mead 1938; Restivo 1994
187 Popper 1972; Millikan 1984; Cupitt
      1990
188 Woolgar 1983; Isambert 1985;
      Grayling 1988; Bernstein 1991;
      Searle 1995; Bloor 1996
189 Rouse 1987
190 Gouldner 1970; Giere 1988; Harré
      1990; Sismondo 1993; Barnes 1995:
      Searle 1995; Edquist & Johnson 1997
191 Campbell 1982; Restivo 1994
192 Ziman 2000
193 Gouldner 1970; Searle 1995;
      Sismondo 1993
194 Giere 1988; Hooker 1995

Chapter 10 cont.

195 Bhaskar 1975
196 Harré 1986
197 Edelman 1992
198 Rouse 1987
199 Cupitt 1990; Hooker 1995
200 Giere 1988
201 Kantorovich 1993
202 Barnes 1985; Harré 1986; Giere 1988; Midgley 1992; Kantorovich 1993; Gibbons *et al.* 1994
203 Traweek 1988; Hill 1995
204 Searle 1995
205 Elkana 1984
206 Böhme 1992
207 Polanyi 1958; Hill 1995; Galison 1997
208 Harré 1986
209 Munz 1993
210 Fine 1991; Callebaut 1993; Lucas 1995; Porter 1995
211 Gross *et al.* 1996
212 Gibbons *et al.* 1994
213 Gibbons *et al.* 1994
214 Stevens 1988; Munz 1993
215 Suppé 1974; Rouse 1987
216 Popper 1963; de Mey 1982
217 Popper 1972; Holton 1974, 1986; Trouvé 1992; Kitcher 1991, 1993; Dupré 1993; Richards (Callebaut 1993)
218 Duncan & Weston-Smith 1977; Millikan 1984; Shapere (Callebaut 1993); Turnbull 1994
219 de Mey 1982; Munz 1993
220 Midgley 1992
221 Toulmin 1972, 1980; Isambert 1985; Rouse 1987; Hooker 1995; Pickering 1995
222 Markus 1987; Dupré 1993
223 Longino 1990
224 Galison 1997
225 Hull 1988
226 Campbell 1977
227 Bernstein 1991
228 Neurath 1932
229 Cupitt 1990
230 Duncan & Weston-Smith 1977
231 Hesse 1980; Giere 1988

232 Fleck 1935; Cantore 1977; Emmeche 1991; Cohen & Stewart 1994
233 Harré 1986
234 Oppenheim & Putnam 1991
235 Geertz 1973; Hesse 1974; Dupré 1993
236 Weizsacker 1952; Polanyi 1958; Kantorovich 1993; Cohen & Stewart 1994
237 Wimsatt (Callebaut 1993)
238 Nagel 1961; Arbib & Hesse 1986; Segerstrale 1992; Dupré 1993
239 Jablonka & Lamb 1995
240 Aronson 1984
241 Secord 1990; Weingart 1995
242 Barrow 1991; Diettrich 1995
243 Barrow 1991
244 Suppé 1974; Mayr 1988
245 Midgley 1992; Böhme 1992; Kantorovich 1993
246 Gross & Levitt 1994; Gross *et al.* 1997
247 Fuller 1988; Schmaus, Segerstrale & Jessuph 1992; Segerstrale 1992; Weingart 1995
248 Cassirer 1956; Polanyi 1958; Nagel 1961; Armistead 1974; Mayr 1988; Stevens 1988; Hull 1988; Longino 1990; Barrow 1991; Midgely 1992a; Dupré 1993; Munz 1993; Parisi 1993; Rosenberg (Callebaut 1993); Wimsatt (Callebaut 1993); Cohen & Stewart 1994; Rose 1997
249 Vincenti 1990
250 Cohen & Stewart 1994
251 Kauffman 1993, 1995
252 Cassirer 1956
253 Kantorovich 1993
254 Emmeche 1991; Dupré 1993; Diettrich 1995
255 Edelman 1992
256 Maynard Smith & Szathmáry 1995
257 Cohen & Stewart 1994
258 Gibbons *et al.* 1994
259 Dews 1986; Toulmin 1990; Bernstein 1991; Rose 1992; Giere 1993
260 Salomon 1973
261 Gibbons *et al.* 1994
262 Ziman 1996b

Bibliography and author index

356   Works cited in the notes to the text are indicated as follows:
'[**6** 21, 41]' – chapter 6, notes 21 and 41

ABRC 1987. *A Strategy for the Science Base* (London: HMSO)

Aitchison, J. 1996. *The Seeds of Speech: Language Origin and Evolution* (Cambridge: Cambridge University Press) [**6** 21, 41, 177]

Amsterdamska, O. 1990. Review of Latour: *Science in Action. Science Technology and Human Values* **15**:495–504 [**7** 24]

Anderson, P. 1990. Witchcraft. *London Review of Books* (8 Nov. 1990):6–11 [**8** 317]

Arbib, M.A., & Hesse, M.B. 1986. *The Construction of Reality* (Cambridge: Cambridge University Press) [**3** 34; **4** 8, 10; **5** 16, 141, 185, 188; **6** 46, 186, 218, 220, 221, 236; **7** 10; **8** 425; **9** 28; **10** 43, 113, 116, 138, 172, 238]

Argyris, C. 1980. *Inner Contradictions of Rigorous Research* (New York NY: Academic Press) [**10** 118, 140]

Armistead, N., ed. 1974. *Reconstructing Social Psychology* (Harmondsworth: Penguin) [**10** 49, 53, 248]

Aronson, J. 1990. Experimental realism. In *Harré and his Critics*, edited by Bhaskar, R., pp. 48–63 [**10** 165]

Aronson, J.L. 1984. *A Realist Philosophy of Science* (London: Macmillan) [**6** 66, 192; **10** 240]

Arthur, W.B. 1994. *Increasing Returns and Path Dependency in the Economy* (Ann Arbor MI: University of Michigan Press) [**9** 316]

Atiyah, M. 1995. Anniversary Address by the President. *Royal Society News* **8** (Nov. 1995): Supplement [**2** 31]

Atran, S. 1990. *Cognitive Foundations of Natural History: Towards an Anthropology of Science* (Cambridge: Cambridge University Press) [**2** 1; **6** 26, 48, 50, 52, 245, 395; **10** 32, 35, 43, 152, 155, 162]

Bailey, J. 1992. First we reshape our computers, then our computers reshape us: The broader intellectual impact of parallelism. *Daedalus* **121**:67–81 [**6** 204]

Baltimore, D. 1992. *On Doing Science in the Modern World*, Tanner Lecture delivered in Cambridge, UK [**9** 192, 215]

Barber, B., & Fox, R.C. 1958. The case of the floppy-eared rabbits: An instance of serendipity gained and serendipity lost. *American Journal of Sociology* **64**:128–36 [**5** 20]

Barber, B., & Hirsch, W., eds. 1962. *The Sociology of Science* (New York NY: Free Press of Glencoe) [*8* 359]

Barkow, J. H., Cosmides, L., & Tooby, J., eds. 1992. *The Adapted Mind: Evolutionary Psychology and the Generation of Culture* (New York NY: Oxford University Press) [*6* 42]

Barnes, B. 1977. *Interests and the Growth of Knowledge* (London: Routledge & Kegan Paul) [*6* 77; *7* 41]

Barnes, B. 1981. On the conventional character of knowledge and cognition. *Philosophy of the Social Sciences* **11**:303–34 [*6* 5, 37; *10* 152]

Barnes, B. 1985. Ethnomethodology as science. *Social Studies of Science* **15**:751–62 [*3* 34; *10* 202]

Barnes, B. 1995. *The Elements of Social Theory* (London: UCL Press) [*2* 41; *3* 35; *5* 181; *6* 85, 254, *7* 41; *9* 137; *10* 26, 127]

Barnes, B., & Edge, D., eds. 1982. *Science in Context: Readings in the Sociology of Science* (Milton Keynes: Open University Press) [*6* 164; *8* 129, 247, 318]

Baron-Cohen, S. 1990. Autism: a specific cognitive disorder of mind-blindness. *International Review of Psychiatry* **2**:81–90 [*5* 156]

Barrow, J. D. 1991. *Theories of Everything: The Quest for Ultimate Explanation* (Oxford: Oxford University Press) [*6* 1, 117, 148; *10* 242, 243, 248]

Basalla, G. 1988. *The Evolution of Technology* (Cambridge: Cambridge University Press) [*8* 290]

Bazerman, C. 1981. What written knowledge does: Three examples of academic discourse. *Philosophy of the Social Sciences* **11**:361–88 [*9* 90]

Bazerman, C. 1984. Modern evolution of the experimental report in physics. *Social Studies of Science* **14**:163–96 [*5* 190; *9* 144]

Bazerman, C. 1988. *Shaping Written Knowledge: The Genre and Activity of the Experimental Article in Science* (Madison WI: University of Wisconsin Press) [*3* 20; *5* 100, 175, 190; *9* 78, 144]

Becher, T. 1989. *Academic Tribes and Territories* (Milton Keynes: Open University Press) [*3* 28; *5* 191; *8* 77]

Beck, U. 1986 (1992). *Risk Society: Towards a New Modernity*. English Translation edn (London: Sage) [*2* 17; *7* 61, 100, 106, 117; *9* 40]

Ben-David, J. 1971. *The Scientist's Role in Society* (Englewood Cliffs NJ: Prentice-Hall) [*2* 45; *3* 23; *4* 5]

Berlin, I. 1979. *Against the Current* (London: Hogarth Press) [*5* 157]

Bernal, J. D. 1939. *The Social Function of Science* (London: Routledge) [*2* 12, 33; *3* 60, 140; *4* 35; *7* 43; *8* 45, 52, 130]

Bernstein, R. J. 1991. *The New Constellation: The Ethical-Political Horizons of Modernity/Postmodernity* (Cambridge MA: Polity Press) [*3* 17; *5* 138; *6* 173, 176, 183; *8* 107; *9* 25, 28, 42; *10* 100, 146, 188, 227]

Beveridge, W. I. B. 1950. *The Art of Scientific Investigation* (New York NY: Norton) [*5* 18; *8* 35, 133, 230, 242, 266, 316; *9* 34, 189]

Bhaskar, R. 1975. *A Realist Theory of Science* (Leeds: Leeds Books Ltd) [*8* 334; *10* 195]

Binmore, K. G. 1977. Mathematics, games and society. *Journal of the Institute of Mathematics and its Applications* (Nov/Dec 1977):263–71 [*6* 215]

Bird, S. J. 1996. Assessing conflicts of interest: sources of bias. *Science and Engineering Ethics* **1**:386–8 [*7* 47]

Blakeslee, A. N. 1994. The rhetorical construction of novelty: Presenting claims in a letter form. *Science Technology & Human Values* **19**:88–100 [*9* 142]

Bloor, D. 1976. *Knowledge and Social Imagery* (London: Routledge) [*6* 7, 153; *9* 243; *10* 130, 141]

Bloor, D. 1981. The strengths of the Strong Programme. *Philosophy of the Social Sciences* **11**:199–214 [*10* 141]

Bloor, D. 1996. Idealism and the sociology of knowledge. *Social Studies of Science* **26**:839–56 [*10* 188]

Blume, S. 1988. Interdisciplinarity in the Social Sciences. (London: Science Policy Support Group)

Boden, M. A. 1988. *Computer Models of Mind* (Cambridge: Cambridge University Press) [*5* 113, 152; *6* 44, 142; *8* 411]

Boden, M. A. 1990. *The Creative Mind: Myths and Mechanisms* (New York NY: Basic Books) [*6* 71; *8* 25, 193, 385]

Bodmer, W. 1985. *The Public Understanding of Science*. (London: The Royal Society).

Böhme, G. 1975. The social function of cognitive structure: a concept of the scientific community within a theory of action. In *Determinants and Controls of Scientific Development*, edited by Knorr, K. D., Strasser, H., & Zilian, H. G. (Dordrecht: Reidel) pp. 205–25 [*9* 15]

Böhme, G. 1992. *Coping with Science* (Boulder CO: Westview) [*1* 4, 12; *2* 17; *3* 57; *6* 240, 246; *7* 13, 100, 113; *10* 68, 80, 140, 164, 206, 245]

Böhme, G., van den Daele, W., & Krohn, W. 1973. Die Finalisierung der Wissenschaft. *Zeitschrift für Soziologie* **2**:128–44 [*4* 50]

Bolter, J. D. 1991. *Writing Space: The Computer, Hypertext and the History of Writing* (Hilldale NJ: Lawrence Erlbaum) [*9* 166]

Bonner, J. T. 1988. *The Evolution of Complexity by Means of Natural Selection* (Princeton NJ: Princeton University Press) [*9* 329]

Booth, W. 1988. Voodoo science. *Science* **240**:274–7 [*9* 210]

Bourdieu, P. 1975. The specificity of a scientific field and the social conditions of the progress of reason. *Social Science Information* **14** (6):19–47 [*3* 8, 34; *7* 26, 39, 61; *8* 380; *9* 34, 231, 258, 323; *10* 159]

Bourdieu, P. 1977. *Outline of a Theory of Practice* (Cambridge: Cambridge University Press) [*9* 104; *10* 43, 56]

Bourdieu, P. 1984 (1988). *Homo Academicus*. English Translation edn (Cambridge: Polity Press) [*7* 26]

Boyd, R. 1991. On the current state of scientific realism. In *The Philosophy of Science*, edited by Boyd, R., Gasper, P., & Trout, J. D. (Cambridge MA: MIT Press) pp. 195–221 [*5* 16; *8* 245]

Bradie, M. 1990. Should epistemologists take Darwin seriously? In *Evolution, Cognition and Realism: Studies in Evolutionary Epistemology*, edited by Rescher, N. (Lanham MD: University Press of America) pp. 33–38 [*5* 116; *6* 138; *10* 64, 70]

Brain, W. R. 1951. *Mind, Perception and Science* (Oxford: Blackwell) [*5* 126; *10* 40]

Brannigan, A. 1981. *The Social Basis of Scientific Discoveries* (Cambridge: Cambridge University Press) [*8* 203, 205, 207, 217, 220, 224; *9* 203]

Bridgman, P. 1991. The operational character of scientific concepts. In *The Philosophy of Science*, edited by Boyd, R., Gasper, P., & Trout, J. D. (Cambridge MA: MIT Press) pp. 57–70 [*8* 320]

Broad, W., & Wade, N. 1983. *Betrayers of the Truth: Fraud and Deceit in the Halls of Science* (London: Century Publishing) [*7* 27]

Brooks, H. 1979. The problem of research priorities. In *Limits of Scientific Inquiry*, edited by Holton, G., & Morison, R. S. (New York NY: Norton) pp. 171–90 [*7* 98, 109; *8* 35, 37, 151, 164, 190]

Brown, R. 1970. *Psycholinguistics* (New York NY: Free Press) [*10* 28]

Brush, S.G. 1989. Prediction and theory evaluation: the case of light bending. *Science* **246**:1124–9 [*8* 285]

Bunders, J., & Whitley, R. 1985. Popularisation within the sciences: The purposes and consequences of inter-specialist communication. In *Expository Science: Forms and Functions of Popularisation*, edited by Shinn, T., & Whitley, R. (Dordrecht: Reidel) pp. 61–78 [*9* 152]

Bunge, M. 1962. *Intuition and Science* (Englewood Cliffs NJ: Prentice-Hall) [*8* 429]

Burnyeat, M.F. 1993. Review of Coady 1993. *London Review of Books* (4 Nov. 1993):29–31 [*5* 71; 133; *10* 29]

Byerly, R., & Pielke, R.A. 1995. The changing ecology of United States science. *Science* **269**:1531–2 [*3* 131]

Callebaut, W. 1993. *Taking the Naturalistic Turn: How Real Philosophy of Science is Done* (Chicago IL: University of Chicago Press) [*1* 17; *4* 10; *5* 6, 10, 54, 102, 116, 171; *6* 88, 99, 122, 170, 205; *7* 53; *8* 13, 245, 252, 275, 307, 310, 330, 404, 423; *9* 23, 25, 68, 69, 91, 102, 119, 127, 258, 260, 271, 303, 320, 320, 328, 341; *10* 3, 8, 64, 75, 76, 147, 171, 178, 210, 217, 218, 237, 248]

Callebaut, W. 1995. The future of naturalistic philosophy of science. *Ludus Vitalis* **3** (5):19–52 [*1* 17]

Callon, M. 1994. Is science a public good. *Science, Technology & Human Values* **4**:395–424 [*3* 4, 40]

Callon, M. 1995. Four models for the dynamics of science. In *Handbook of Science and Technology Studies*, edited by Jasanoff, S., Markle, G.E., Petersen, J.C., & Pinch, T. (London: Sage) pp. 29–63 [*1* 7, 21]

Campbell, D.T. 1960. Blind variation and selective retention in creative thought as in other knowledge processes. *Psychological Review* **67**:380–400 [*9* 277, 340]

Campbell, D.T. 1974. Evolutionary epistemology. In *The Philosophy of Karl Popper*, edited by Schilpp, P.A. (La Salle IL: Open Court) pp. 413–63 [*1* 18; *5* 115; *8* 3, 31, 156, 223, 233, 405; *9* 276, 277, 305, 340; *10* 60, 68]

Campbell, D.T. 1977. *Descriptive Epistemology: Psychological, Sociological, Evolutionary* (unpublished draft of William James Lectures, Harvard) [*9* 340]

Campbell, D.T. 1979. A tribal model of the social system vehicle carrying scientific knowledge. *Knowledge: Creation, Diffusion, Utilization* **1**:181–201 [*3* 27, 33, 60; *5* 61, 110]

Campbell, D.T. 1982. Experiments as arguments. *Knowledge: Creation, Diffusion, Utilization* **3**:327–37 [*5* 27; *7* 92; *8* 289; *9* 95; *10* 191]

Campbell, D.T. 1983. Science's social system of validity-enhancing collective belief change and the problems of the social sciences. In *Pluralism and Subjectivities in the Social Sciences*, edited by Fiske, D.W., & Shweder, R.A. [*3* 5; *5* 5, 27, 90, 93, 99, 168, 182; *6* 169; *8* 10; *9* 207]

Campbell, D.T. 1984. Types of evolutionary epistmology extended to a sociology of scientific validity. Paper read at International Conference on Evolutionary Epistemology, at Ghent [*9* 16, 300, 318; *10* 65]

Campbell, D.T. 1984a. Can an open society be an experimenting society? Paper read at International Symposium on the Philosophy of Karl Popper, at Madrid [*5* 41, 61, 141]

Campbell, D.T. 1990. Epistemological roles for selection theory. In *Evolution, Cognition and Realism: Studies in Evolutionary Epistemology*, edited by Rescher, N. (Lanham MD: University Press of America) [*10* 65, 69]

Campbell, N. 1921. *What is Science?* (London: Methuen) [*5* 35, 138; *6* 72, *9* 69; *10* 4, 158]

Cantore, E. 1977. *Scientific Man: The humanistic significance of science* (New York NY: Institute for Scientific Humanism) [*5* 60; *8* 197; *10* 232]

Capaldi, N. 1975. The moral limits of scientific research: an evolutionary approach. In *Determinants and Controls of Scientific Development*, edited by Knorr, K.D., Strasser, H., & Zillian, H.G. (Dordrecht: Reidel) pp. 113–41 [*7* 62, 74, 78; *8* 259; *9* 206, 258; *10* 173]

Carlson, T., & Martin-Rovet, D. 1995. The implications of scientific mobility between France and the United States. *Minera* **33**:211–50 [*2* 46]

Carlson, W., & Gorman, M.E. 1990. Understanding invention as a cognitive process: The case of Thomas Edison and early motion pictures, 1888–91. *Soc. Stud. Sci.* **20**:387–430 [*8* 308, 403]

Carroll, L. 1893. *Sylvie and Bruno (Concluded)* [*6* 99]

Carroll, N. 1985. The power of movies. *Daedalus* **114** (4):79–103 [*6* 43]

Cassirer, E. 1956. *The Problem of Knowledge* (New Haven CT: Yale University Press) [*5* 7, 161; *8* 308; *10* 248, 252]

Causey, R.L. 1968. The importance of being surprised in scientific research. *Agricultural Science Review* **6**:27–31 [*8* 232]

Caws, P. 1965. *The Philosophy of Science* (Princeton NJ: van Nostrand) [*8* 211, 385]

Caws, P. 1969. The Structure of Discovery. *Science* **166**:1375–80 [*9* 302]

Chalmers, A.F. 1982. *What is this Thing called Science?* 2nd edn (Milton Keynes: Open University Press) [*1* 26]

Chandler, A.D.J. 1977. *The Visible Hand: The Managerial Revolution in American Business* (Cambridge MA: Harvard University Press) [*7* 87]

Chandrasekhar, S. 1979. Beauty and the quest for beauty in science. *Physics Today* (July):25–30 [*8* 308]

Chote, R. 1992. Why the Chancellor is always wrong. *New Scientist* (31 Oct.):26–31 [*6* 217]

Chubin, D.E. 1976. The conceptualization of scientific specialties. *Sociological Quarterly* **17**:448–76 [*8* 56; *9* 30, 249]

Churchland, P. 1991. Eliminative materialism and the propositional attitude. In *The Philosophy of Science*, edited by Boyd, R., Gasper, P., & Trout, J.D. (Cambridge MA: MIT Press) pp. 615–30 [*10* 48]

Churchland, P.M. 1990. *A Neurocomputational Perspective: The Nature of Mind and the Structure of Science* (Cambridge MA: MIT Press) [*5* 19, 132, 145; *6* 34, 49, 143; *7* 12; *8* 86, 314, 395; *10* 45, 48, 75]

Coady, C.A.J. 1992. *Testimony: A Philosophical Study* (Oxford: Oxford University Press) [*3* 2; *5* 12, 70, 71, 135, 145; *8* 296, 309; *9* 213, 236; *10* 25, 29, 73]

Cohen, J., & Stewart, I. 1994. *The Collapse of Chaos: Discovering Simplicity in a Complex World* (Harmondsworth: Penguin) [*2* 27; *6* 13, 14, 23, 34, 53, 88, 100, 149, 156, 163, 173, 193, 213, 219, 228, 235, 241; *8* 40, 88, 119, 301, 337, 422; *9* 208, 314; *10* 63, 113, 134, 232, 236, 248, 250, 257]

Colby, K.M. 1981. Modelling a paranoid mind. *The Behavioural and Brain Sciences* **4**:515–60 [*6* 214]

Cole, M., & Scribner, S. 1974. *Culture and Thought* (New York NY: Wiley) [*5* 188; *6* 39; *10* 72]

Cole, S. 1992. *Making Science: Between Nature and Society* (Cambridge MA: Harvard University Press) [*3* 3; *5* 178; *6* 105; *7* 23; *8* 364; *9* 114]

Cole, S.A. 1996. Which came first, the fossil or the fuel? *Social Studies of Science* **26**:733–66 [*9* 51]

Coleman, J.S. 1964. *Introduction to Mathematical Sociology* (New York NY: Free Press) [*6* 161]

Collingridge, D., & Reeve, C. 1986. *Science Speaks to Power: The Role of Experts in Policymaking* (London: Pinter) [**7** 50]

Collingwood, R. G. 1946. *The Idea of History* (Oxford: Clarendon Press) [**5** 162; **8** 11, 250; **9** 25]

Collins, H., & Pinch, T. 1993. *The Golem: What Everyone Should Know about Science* (Cambridge: Cambridge University Press) [**1** 6; **2** 19; 76; **5** 98, 99; **9** 27, 49, 61, 107]

Collins, H. M. 1975. The seven sexes: a study in the sociology of a phenomenon, or the replication of experiments in physics. *Sociology* **9**:205–24 [**5** 86; **9** 30]

Collins, H. M., ed. 1982. *Sociology of Scientific Knowledge: A Source Book* (Bath: Bath University Press) [**1** 10]

Collins, H. M. 1993. Untidy minds in action. *Times Higher Education Supplement* (April 9):15 [**6** 162]

Collins, H. M., & Pinch, T. J. 1982. *Frames of Meaning: The Social Construction of Extraordinary Science* (London: Routledge & Kegan Paul) [**9** 211, 234; **10** 97, 170]

Collins, R. 1975. *Conflict Sociology* (New York NY: Academic Press) [**10** 173]

Cooper, W. 1952. *The Struggles of Albert Woods* (London: Jonathan Cape) [**2** 36]

Cosmides, L., & Tooby, J. 1992. Cognitive adaptations for social exchange. In *The Adapted Mind: Evolutionary Psychology and the Generation of Culture*, edited by Barkow, J. H., Cosmides, L., & Tooby, J. (New York NY: Oxford University Press) pp. 163–228 [**5** 112, 155; **8** 343; **10** 14, 31, 42, 91]

Cozzens, S. E., Healey, P., Rip, A., & Ziman, J. M., eds. 1990. *The Research System in Transition* (Dordrecht: Kluwer) [**2** 13; **4** 13, 16, 23, 27]

Crawford, E. 1984. *The Beginnings of the Nobel Institution: The Science Prizes, 1901–1915* (Cambridge: Cambridge University Press) [**3** 25, 127; **7** 28]

Crease, R. P. 1992. How technique is changing science. *Science* **257**:344–53 [**5** 62]

Crease, R. P. 1993. Biomedicine in the age of imaging. *Science* **261**:554–61 [**6** 30]

Cronin, B. 1984. *The Citation Process: The Role and Significance of Citations in Scientific Communication* (London: Taylor Graham) [**9** 153, 158]

Csikszentmihaly, M. 1993. *The Evolving Self: A Psychology for the Third Millennium* (New York NY: Harper Collins) [**9** 291]

CSS 1985. *Academic Tenure: Luxury or Necessity?* (London: Council for Science and Society)

Culliton, B. J. 1988. Random audit of papers proposed. *Science* **242**:657–8 [**9** 199]

Cupitt, D. 1990. *Creation out of Nothing* (London: SCM Press) [**6** 115; **10** 17, 30, 52, 139, 173, 175, 187, 199, 229]

Curtis, R. 1994. Narrative form and normative force: Baconian story-telling in popular science. *Social Studies of Science* **24**:419–62 [**9** 31]

Cziko, G. 1995. *Without Miracles: Universal Selection Theory and the Second Darwinian Revolution* (Cambridge MA: MIT Press) [**8** 415; **9** 298; **10** 65]

Daston, L. 1992. Objectivity and the escape from perspective. *Social Studies of Science* **22**:597–618 [**5** 182; **10** 172]

David, E. E. 1995. A realistic scenario for US R&D. *Bulletin of Science, Technology and Society* **15**:14–18 [**4** 69; **5** 197; **8** 150; **9** 18]

Davidson, D. 1984. *Inquiries into Truth and Interpretation* (Oxford: Clarendon Press) [**5** 182; **6** 219; **9** 268; **10** 19, 36, 100, 122]

Dawkins, R. 1976. *The Selfish Gene* (Oxford: Oxford University Press) [**9** 289, 290]

Dawkins, R. 1986. *The Blind Watchmaker* (London: Longman) [**9** 290]

de Jouvenel, B. 1961. The republic of science. In *The Logic of Scientific Knowledge* (London: Routledge & Kegan Paul) pp. 131–41 [**3** 60]

de Mey, M. 1982. *The Cognitive Paradigm* (Dordrecht: Reidel) [**2** 2; **5** 18, 107, 121, 129, 187, 188; **6** 34, 66, 75, 120, 218; **8** 61, 87, 111, 117, 125, 131, 168, 174, 344, 419, 426; **10** 22, 165, 173, 216, 219] de Mey, M. 1984. Cognitive science and science dynamics. *Social Science Information Studies* 4:97–107 [*8* 313; *9* 165]

Dean, C. 1977. Are serendipitous discoveries a part of normal science? *The Sociological Review* 25:73–86 [*8* 229]

Denbigh, K.G. 1975. *An Inventive Universe* (London: Hutchinson) [*5* 143; *6* 32; *8* 410]

Dennett, D.C. 1969. *Content and Consciousness* (London: Routledge & Kegan Paul) [*5* 155; *8* 394]

Dennett, D.C. 1987. *The Intentional Stance* (Cambridge MA: MIT Press) [*10* 113, 117]

Dennett, D.C. 1991. *Consciousness Explained* (London: Penguin) [*8* 405; *10* 119, 125]

Dennett, D.C. 1995. *Darwin's Dangerous Idea* (London: Penguin) [*5* 115; *6* 65; *10* 65]

Dews, P. 1986. *Habermas: Autonomy and Solidarity* (London: Verso) [*3* 17; *7* 75, 79; *9* 137; *10* 37, 41, 259]

Diettrich, O. 1995. A constructivist approach to the problem of induction. *Evolution & Cognition* 1:95–113 [*5* 126; *6* 1, 104, 138, 153; *7* 13; *9* 222; *10* 36, 61, 77, 155, 173, 175, 242, 254]

Dolman, H., & Bodewitz, H. 1985. Sedimentation of a scientific concept: the use of citation data. *Social Studies of Science* 15:507–23 [*9* 121, 324]

Donald, M. 1991. *Origins of the Modern Mind: Three Stages in the Evolution of Culture and Cognition* (Cambridge MA: Harvard University Press) [*5* 114; *6* 38, 107, 181; *8* 408; *9* 79, 166, 167]

Douglas, M., ed. 1973. *Rules and Meanings* (London: Penguin) [*6* 37]

Douglas, M. 1986. *Risk Acceptability according to the Social Sciences* (London: Routledge & Kegan Paul) [*10* 104]

Douglas, M. 1987. *How Institutions Think* (London: Routledge & Kegan Paul) [*6* 37; *7* 68, 86; *8* 98; *10* 82, 105]

Downs, R.M., & Stea, D., eds. 1973. *Image and Environment* (Chicago IL: Abline) [*8* 409]

Doyal, L., & Harris, R. 1986. *Empiricism, Explanation and Rationality: An Introduction to the Philosophy of the Social Sciences* (London: Routledge & Kegan Paul) [*5* 146, 169; *8* 378; *10* 94, 99, 113, 122]

Duhem, P. 1906 (1962). *The Aim and Structure of Physical Theory* (New York NY: Atheneum) [*8* 295]

Duncan, R., & Weston-Smith, M., eds. 1977. *The Encyclopaedia of Ignorance: Vol. 1 Physical Sciences; Vol. 2: Life Sciences and Earth Sciences* (Oxford: Pergamon) [*6* 155; *8* 6, 238; *10* 218, 230]

Dupré, J. 1993. *The Disorder of Things: Metaphysical Foundations of the Disunity of Life* (Cambridge MA: Harvard University Press) [*5* 31, 87; *6* 56; *7* 52, 70; *8* 304, 391; *9* 222, 307, 313; *10* 217, 222, 235, 238, 248, 254]

Durkheim, E., & Mauss, M. 1903 (1963). *Primitive Classification* (London: Routledge & Kegan Paul) [*6* 37, 63; *10* 46]

Edelman, G.M. 1989. *Neural Darwinism* (Oxford: Oxford University Press) [*9* 298]

Edelman, G.M. 1992. *Bright Air, Brilliant Fire: On the Matter of the Mind* (London: Penguin) [*5* 115, 141, 146; *6* 22, 34, 56, 113; *8* 83, 395, 415; *9* 175, 295, 298; *10* 197, 255]

Edquist, C., & Johnson, B. 1997. Institutions and organizations in systems of innovation. In *Systems of Innovation: Technologies, Institutions and Organizations*, edited by Edquist, C. (London: Pinter) pp. 41–63 [*10* 190]

Eglash, R. 1997. When math worlds collide: Intention and invention in ethnomathematics. *Science, Technology & Human Values* 22:79–97 [*6* 153]

Ehrenberg, A.S., & Bound, J.A. 1993. Predictability and prediction. *J. Royal Statistical Society* 156:167–206 [*8* 244, 314]

Elkana, Y. 1981. A programmatic attempt at an anthropology of knowledge. In *Science and Cultures*, edited by Mendelsohn, E., & Elkana, Y. (Dordrecht: Reidel) pp. 1–76 [*1* 12; *2* 40; *8* 239; *10* 128, 151]

Elkana, Y. 1984. Transformation in realist philosophy of science, from Victorian Baconianism to the present day. In *Transformation and Tradition in the Sciences*, edited by Mendelsohn, E. (Cambridge: Cambridge University Press) pp. 487–511 [*10* 179, 205]

Elzinga, A. 1986. The growth of science: romantic and technocratic images. In *Progress in Science and its Social Conditions*, edited by Ganelius, T. (Oxford: Pergamon) pp. 33–47 [*3* 147; *4* 77; *7* 120; *9* 322]

Emmeche, C. 1994. *The Garden in the Machine: The Emerging Science of Artificial Life* (Princeton NJ: Princeton University Press) [*10* 232, 254]

Engelhardt, H.T., & Caplan, A.L., eds. 1987. *Scientific Controversies: Case Studies in the Resolution and Closure of Disputes in Science and Technology* (Cambridge: Cambridge University Press) [*9* 99]

Etzkowitz, H. 1992. Individual investigators and their research groups. *Minerva* **30**:28–50 [*4* 18, 43, 64; *8* 50]

Etzkowitz, H. 1996. Conflicts of interest and commitment in academic science in the United States. *Minerva* **34**:259–77 [*7* 47]

Evans, C. 1984. Empirical truth and progress in science. *New Scientist* (26 Jan.):43–5 [*9* 195]

Evans, R. 1997. Soothsaying or science? Falsification, uncertainty and social change in macroeconomic modelling. *Social Studies of Science* **27**:395–438 [*6* 217; *9* 101]

Ezrahi, Y. 1990. *The Descent of Icarus: Science and the Transformation of Contemporary Democracy* (Cambridge MA: Harvard University Press) [*1* 19; *2* 4, 17; *3* 16, 18; *5* 40, 77, 123, 192; *6* 68; *7* 9, 61, 74, 76, 82, 88, 104; *8* 330, 368, 227; *10* 83, 138]

Fahnestock, J. 1989. Arguing in different forums: The Boring cross-over controversy. *Science, Technology & Human Values* **14**:26–42 [*9* 73]

Farley, J., & Geison, G.L. 1974. Science politics and spontaneous generation in nineteenth century France: The Pasteur Pouchet debate. *Bulletin of the History of Medicine* **48**:161–98 [*9* 105]

Farris, H., & Revlin, R. 1989. The discovery process: A counterfactual strategy. *Social Studies of Science* **19**:497–513 [*8* 246]

Feyerabend, P. 1975 (1978). Against Method: Outline of an anarchistic theory of knowledge (London: Verso) [*4* 14; *6* 11, 171; *8* 311; *9* 45, 270]

Fine, A. 1991. The natural ontological attitude. In *The Philosophy of Science*, edited by Boyd, R., Gasper, P., & Trout, J.D. (Cambridge MA: MIT Press) pp. 261–77 [*1* 27; *5* 142; *6* 75; *10* 13, 59, 165, 210]

Fisher, D. 1990. Boundary work and science: The relation between knowledge and power. In *Theories of Science in Society*, edited by Cozzens, S.E., & Gieryn, T.F. (Bloomington IN: Indiana University Press) pp. 98–119 [*7* 61]

Fleck, L. 1935 (1979). *Genesis and Development of a Scientific Fact*. English translation ed (Chicago IL: U. of Chicago Press) [*5* 18, 20, 60, 63, 176; *6* 221; *7* 4; *8* 94, 97, 215, 277, 383; *9* 95, 160, 178, 180, 226, 237, 270, 275; *10* 88, 89, 174, 232]

Foray, D. 1997. Generation and distribution of technological knowledge: Incentives, norms and institutions. In *Systems of Innovation: Technologies, Institutions and Organizations*, edited by Edquist, C. (London: Pinter) pp. 64–85 [*4* 71; *5* 201]

Forge, A. 1970. Learning to see in New Guinea. In *Socialization: The Approach from Social Anthropology*, edited by Forge, A., & Mayer, P. (London: Tavistock) [*6* 40]

Forman, P. 1971. Weimar culture, causality and quantum theory, 1918–23. *Historical Studies in the Physical Sciences* **3**:1–116 [*7* 64]

Foucault, M. 1969 (1972). *The Archaeology of Knowledge.* English translation edn (London: Tavistock) [*6* 109; *10* 156]

Franklin, A. 1986. *The Neglect of Experiment* (Cambridge: Cambridge University Press) [*5* 60, 78, 84, 99; *8* 273, 356; *9* 197, 263]

Frege, G. 1892. Ueber Sinn und Bedeutung. *Z. Phil. und phil. Kritik* **100**:25–50 [*9* 176]

Fuchs, S. 1996. The poverty of postmodernism. *Science Studies* **9** (1):58–66 [*4* 29]

Fuller, S. 1987. Interview with Marc de Mey. *Social Epistemology* **1**:85–95 [*5* 104]

Fuller, S. 1988. *Social Epistemology* (Bloomington IN: Indiana University Press) [*1* 15, 25; *2* 40; *4* 9; *5* 103, 165, 170; *6* 11, 126, 190; *8* 32, 78, 264; *9* 126, 269; *10* 113, 172, 182, 247]

Furnham, A.F. 1988. *Lay Theories: Everyday Understanding of Problems in the Social Sciences* (Oxford: Pergamon) [*6* 2; *9* 229; *10* 49, 56]

Galbraith, J.K. 1987. *A History of Economics: The Past as the Present* (London: Hamish Hamilton) [*8* 109]

Galison, P. 1987. *How Experiments End* (Chicago IL: University of Chicago Press) [*4* 17, 75, *8* 4, 142, 160, 169, 297; *9* 21, 52, 120]

Galison, P. 1997. *Image and Logic: A Material Culture of Microphysics* (Chicago IL: University of Chicago Press) [*3* 145; *4* 36, 39, 40, 42, 44, 67, 82; *5* 2, 4, 34, 44, 50, 53, 57, 60, 174, 198; *6* 12, 31, 33, 75, 137, 165, 184, 210, 229, 260; *8* 27, 28, 48, 58, 59, 72, 92, 169, 187, 206, 209, 300, 302, 335, 402, 421; *9* 257, 269, 285, 332; *10* 96, 207, 224]

Ganelius, T., ed. 1986. *Progress in Science and its Social Conditions* (Oxford: Pergamon) [*9* 322]

Garfield, E. 1979. *Citation Indexing: Its Theory and Application in Science* (New York NY: Wiley) [*9* 157]

Garfinkel, A. 1991. Reductionism. In *The Philosophy of Science*, edited by Boyd, R., Gasper, P., & Trout, J.D. (Cambridge MA: MIT Press) pp. 443–59 [*6* 195]

Garfinkel, H. 1967. *Studies in Ethnomethodology* (Englewood Cliffs NJ: Prentice-Hall) [*6* 25; *10* 23, 56, 120]

Geertz, C. 1973. *The Interpretation of Cultures* (New York NY: Basic Books) [*5* 158, 177, 186; *6* 179, 182, 257; *8* 99; *9* 129, 169; *10* 82, 101, 136, 235]

Geertz, C. 1983. *Local Knowledge: Further Essays in Interpretive Anthropology* (New York NY: Basic Books) [*5* 144, 169]

Geertz, C. 1995. *After the Fact* (Cambridge MA: Harvard University Press) [*5* 29]

Gell-Mann, M. 1994. *The Quark and the Jaguar: Adventures in the Simple and the Complex* (London: Little, Brown & Co.) [*6* 150, 228, 314]

George, W.H. 1938. *The Scientist in Action* (London: Scientific Book Club) [*6* 2]

Gibbons, M., Limoges, C., Nowotny, H., Schwartzmann, S., Scott, P., & Trow, M. 1994. *The New Production of Knowledge* (London: Sage) [*4* 7, 13; *4* 16, 30, 33, 51, 78, 79; *5* 183, 199; *6* 259; *7* 54, 94, 122; *8* 148, 166, 180, 182, 186, 188, 189, 192; *9* 337; *10* 202, 212, 213, 258, 261]

Giddens, A. 1987. *Social Theory and Modern Sociology* (Cambridge: Polity Press) [*7* 113]

Giddens, A. 1990. *The Consequences of Modernity* (Cambridge: Polity Press) [*5* 73; *7* 70; *10* 180]

Giere, R.M. 1990. Evolutionary Models of Science. In *Evolution, Cognition and Realism: Studies in Evolutionary Epistemology*, edited by Rescher, N. (Lanham MD: University Press of America) pp. 21–32 [*6* 96; *9* 277, 295]

Giere, R.N. 1988. *Explaining Science: A Cognitive Approach* (Chicago IL: University of Chicago Press) [*1* 14; *4* 8, 10; *5* 111; *6* 66, 96, 103, 135, 151, 203; *7* 34; *8* 60, 120, 215, 243, 275, 308, 355, 414; *9* 97; *10* 79, 179, 190, 194, 200, 202, 231]

Giere, R.N. 1993. Science and technology studies: Prospects for an enlightened postmodern synthesis. *Science, Technology & Human Values* 18:102–112 [*8* 354; *10* 146, 152, 259]

Gieryn, T.F. 1982. Relativist/constructivist programmes in the sociology of science: Redundance and retreat. *Social Studies of Science* 12:279–97 [*8* 251, 376]

Gieryn, T.F. 1983. Boundary-work and the demarcation of science from non-science. *American Sociological Review* 48:781–95 [*3* 34; *8* 390; *9* 212]

Gieryn, T.F., & Hirsh, R.F. 1983. Marginality and innovation in science. *Social Studies of Science* 13:87–106 [*8* 390]

Gilbert, G.N., & Mulkay, M. 1984. *Opening Pandora's Box: A Sociological Analysis of Scientists' Discourse* (Cambridge: Cambridge University Press) [*5* 91, 141, 174; *6* 91; *8* 404; *9* 20, 24, 264]

Glymour, C. 1980. *Theory and Evidence* (Princeton NJ: Princeton University Press) [*6* 75; *8* 276; *10* 76]

Goffman, E. 1974. *Frame Analysis* (Harmondsworth: Penguin) [*2* 2; *3* 62; *9* 237; *10* 43]

Gökalp, I. 1990. Turbulent reactions: Impact of new instrumentation on a borderland scientific domain. *Science, Technology & Human Values* 15:284–304 [*9* 213]

Goldstein, M., & Goldstein, I.F. 1978. *How do we Know: An Exploration of the Scientific Process* (New York NY: Plenum Press) [*9* 102]

Gombrich, E.H. 1960. *Art and Illusion* (New York NY: Pantheon Books) [*6* 34]

Good, I.J. 1977. Explicativity: a mathematical theory of explanation with statistical applications. *Proceedings of the Royal Society* 354:303–30 [*8* 273]

Goodenough, W. 1974. Review of Geertz 1973. *Science* 186:435 [*10* 92]

Goody, J. 1977. Literary criticism and the growth of knowledge. In *Culture and its Creators*, edited by Ben-David, J., & Clark, T.N. (Chicago IL: University of Chicago Press) pp. 226–43 [*9* 167]

Goonatilake, S. 1984. *Aborted Discovery: Science and Creativity in the Third World* (London: Zed Books) [*5* 25; *8* 80, 82; *9* 254, 313; *10* 90, 111]

Goonatilake, S. 1996. The geo-economic shift to Asia and the future of science and technology in the 21st century. Paper read at Visions of the Asian 'Renaissance': Science Technology and Change in Asia and the Pacific, 15–17 Jan. 1996, at Seoul [*10* 84]

Gould, S.J. 1981. *The Mismeasure of Man* (New York NY: Norton) [*8* 333; *9* 185]

Gould, S.J. 1996. *Full House* (New York NY: Harmony Books) [*9* 330]

Gouldner, A.W. 1970. *The Coming Crisis of Western Sociology* (London: Heinemann) [*3* 31; *5* 153; *6* 257; *7* 113, 116; *8* 101, 226; *10* 190, 193]

Gouldner, A.W. 1979. *The Future of Intellectuals and the Rise of the New Class* (London: Macmilan) [*5* 29; *7* 113]

Graham, L.R. 1979. Concerns about science. In *Limits of Scientific Inquiry*, edited by Holton, G., & Morison, R.S. (New York NY: Norton) pp. 1–21 [*4* 56]

Grayling, A.C. 1988. *Wittgenstein* (Oxford: Oxford University Press) [*10*, 73, 188]

Greenberg, D.S. 1969. *The Politics of American Science* (Harmondsworth: Penguin) [*4* 57; *7* 29]

Greenfield, P., Reich, L., & Olver, R. 1966. *Studies in Cognitive Growth* (New York NY: Wiley) [*9* 167]

Gregory, R.L. 1973. Paradigmatic perception. *Proceedings of the Royal Institution* 46:117–39 [*5* 129]

Grene, M. 1961. The Logic of Biology. In *The Logic of Personal Knowledge* (London: Routledge & Kegan Paul) pp. 191–206 [*6* 15]

Gross, P., & Levitt, N. 1994. *Higher Superstition: The Academic Left and its Quarrels with Science* (Baltimore MD: Johns Hopkins University Press) [*2* 18; *4* 28; *7* 31; *10* 246]

Gross, P. M., Levitt, N., & Lewis, M. W., eds. 1996. *The Flight from Reason* (New York NY: New York Academy of Sciences) [*10* 211, 246]

Habermas, J. 1971. *Knowledge and Human Interests*, translated by Shapiro, J. J. (Boston MA: Beacon Books) [*9* 28, 137]

Hacking, I. 1983. *Representing and Intervening* (Cambridge: Cambridge University Press) [*5* 55; *7* 17; *8* 85]

Hacking, I. 1991. Experimentation and scientific realism. In *The Philosophy of Science*, edited by Boyd, R., Gasper, P., & Trout, J. D. (Cambridge MA: MIT Press) pp. 247–60 [*10* 182]

Hagendijk, R. 1990. Structuration theory, constructivism and scientific change. In *Structuration Theory, Constructivism and Scientific Change*, edited by Cozzens, S. E., & Gieryn, T. F. (Bloomington IN: Indiana University Press) pp. 43–66 [*7* 119; *8* 181, 378; *9* 100]

Hägerstrand, T., ed. 1985. *The Identification of Progress in Learning* (Cambridge: Cambridge University Press) [*9* 322]

Hagstrom, W. O. 1965. *The Scientific Community* (New York NY: Basic Books) [*3* 28, 47, 49, 50, 56, 59, 65; *7* 25, 30; *8* 1, 14, 47, 49, 52, 65, 78, 225, 359; *9* 19, 39, 46, 108, 115, 143, 197]

Hall, S. S. 1995. Protein images update natural history. *Science* **267**:620–4 [*6* 29, 204]

Hallam, A. 1973. *A Revolution in the Earth Sciences* (Oxford: Clarendon Press) [*8* 102; *9* 122, 256]

Hallam, J. C. T., & Malcolm, C. A. 1994. Behaviour, perception, action and intelligence – the view from situated robotics. *Philosophical Transactions of the Royal Society, A* [*6* 235, 254]

Handy, R. 1964. *Methodology of the Behavioural Sciences* (Springfield IL: Thomas) [*6* 215; *10* 84]

Hanson, N. R. 1958. *Patterns of Discovery* (Cambridge: Cambridge University Press) [*5* 131; *9* 261]

Haraway, D. J. 1991. *Science, Cyborgs and Women: The Re-invention of Nature* (London: Free Association Books) [*3* 55; *5* 31, 130, 189; *6* 6, 83, 224, 342; *10* 103, 148]

Hardin, C. L. 1981. Table turning, parapsychology and fraud. *Social Studies of Science* **11**:249–58 [*9* 211; *10* 48]

Harré, R. 1960. *An Introduction to the Logic of the Sciences* (London: Macmillan) [*6* 155, 191, 197, 199]

Harré, R. 1975. Images of the world and societal icons. In *Determinants and Controls of Scientific Development*, edited by Knorr, K. D., Strasser, H., & Zilian, H. G. (Dordrecht: Reidel) pp. 257–82 [*8* 400]

Harré, R. 1979. *Social Being: A Theory for Social Psychology* (Oxford: Blackwell) [*5* 27, 42, 61, 181; *6* 52, 236; *7* 69; *10* 49]

Harré, R. 1983. *Personal Being: A Theory for Social Psychology* (Oxford: Blackwell) [*5* 145; *9* 138, 182; *10* 113, 122]

Harré, R. 1986. *Varities of Realism: A Rationale for the Natural Sciences* (Oxford: Blackwell) [*2* 1; *2* 28; *3* 2, 12, 36, 67, 68; *5* 24, 72, 118, 141, 182; *6* 32, 48, 66, 69, 96, 115, 119, 121, 218; *8* 88, 241, 264, 338, 400; *9* 96, 205, 222, 265; *10* 34, 68, 74, 99, 113, 165, 179, 196, 202, 208, 233]

Harré, R. 1990. Exploring the human Umwelt. In *Harré and his Critics: Essays in Honour of Rom Harré, with his Commentary on them*, edited by Bhaskar, R. pp. 297–364 [*8* 338; *10* 190]

Harris, H. 1993. Dead wood. *The Oxford Magazine* (Noughth Week, Hilary Term):4–8 [*2* 34; *8* 387]

Harris, R. 1990. The scientist as *homo loquens*. In *Harré and his Critics: Essays in Honour of Rom Harré, with his Commentary on them*, edited by Bhaskar, R. pp. 65–86 [*6* 129; *7* 13; *8* 174; *10* 10]

Haste, H. 1993. Dinosaur as metaphor. *Modern Geology* **18**:349–70 [*10* 55, 168]

Henderson, D. K. 1990. On the sociology of science and the continuing importance of epistemologically couched accounts. *Social Studies of Science* **20**:113–48 [*3* 40; *10* 57]

Henderson, K. 1991. Flexible sketches and inflexible data bases. *Science, Technology & Human Values* **16**:448–73 [*5* 175]

Hennion, A. 1989. An intermediary between production and consumption: The producer of popular music. *Science, Technology & Human Values* **14**:400–24 [*8* 370]

Hesse, M. 1963. *Models and Analogies in Science* (London: Sheed & Ward) [*6* 112, 198]

Hesse, M. 1974. *The Structure of Scientific Inference* (London: Macmillan) [*6* 15, 19, 27, 75, 112, 229; *8* 268, 273, 274, 282, 310, 376; *9* 207; *10* 61, 180, 235]

Hesse, M. 1980. *Revolutions and Reconstructions in the Philosophy of Science* (Brighton: Harvester) [*3* 2; *4* 1; *5* 7, 27, 141, 164, 167; *6* 32, 75, 94, 218, 221, 248; *7* 10, 74, 80; *8* 174, 239, 310, 365, 415; *9* 28, 121, 262, 272; *10* 78, 80, 95, 173, 181, 231]

Hessen, B. 1931. The social and economic roots of Newton's 'Principia'. In *Science at the Cross Roads* (London) [*7* 63]

Hicks, D., & Katz, J. S. 1996. Where is science going? *Science, Technology & Human Values* **21**:396–406 [*4* 42]

Hill, S. C. 1995. The formation of the identity as a scientist. *Science Studies* **8**:53–72 [*3* 140; *5* 108, 141, 187; *7* 118, 124; *8* 91, 101, 132; *9* 175; *10* 26, 28, 166, 203, 207]

Hjern, B., & Porter, D. O. 1983. Implementation structures: A new unit of administrative analysis. In *Realizing Social Science Knowledge*, edited by Holzner, B., Knorr, K. D., & Strasser, H. (Vienna: Physica-Verlag) pp. 265–80 [*3* 14; *9* 311]

Hodgson, G. M. 1993. *Economics and Evolution: Bringing Life Back into Economics* (Cambridge: Polity Press) [*8* 103]

Høeg, P. 1993. *Miss Smilla's Feeling for Snow* (London: HarperCollins) [*6* 36]

Hofstadter, D. R. 1985. *Metamagical Themas: Questing for the Essence of Mind and Pattern* (New York NY: Basic Books) [*6* 20, 51, 106, 199]

Holton, G. 1973. *Thematic Origins of Scientific Thought* (Cambridge MA: Harvard University Press) [*6* 222]

Holton, G. 1974. On being caught between Dionysians and Apollonians. *Daedalus* **103** (Summer 1974):65–82 [*10* 217]

Holton, G. 1986. *The Advancement of Science and its Burdens* (Cambridge: Cambridge University Press) [*6* 218; *10* 217]

Holton, G. 1992. How to think about the 'anti-science' phenomenon. *Public Understanding of Science* **1**:103–128 [*7* 2; *10* 131]

Holton, G. 1993. *Science and Anti-Science* (Cambridge MA: Harvard University Press) [*1* 1]

Holzner, B., Knorr, K. D., & Strasser, H., eds. 1983. *Realizing Social Science Knowledge* (Vienna: Physica-Verlag) [*7* 76]

Hooker, C. A. 1995. *Reason, Regulation and Realism: Towards a Regulatory Systems Theory of Reason and Evolutionary Epistemology* (Albany NY: SUNY Press) [*1* 14, 20; *3* 3, 10, 13, 35; *4* 6, 25, 32; *5* 7, 8, 45, 102, 116, 141, 174; *6* 75, 142, 162, 166, 167, 207; *7* 12, 110; *8* 64, 190, 196, 246, 259, 295, 304, 310, 361, 367, 405, 411, 415; *9* 15, 39, 70, 103, 168, 180, 183, 205, 277, 295, 297, 303, 311, 318, 340, 341; *10* 39, 65, 70, 142, 173, 194, 199, 221]

Horton, R., & Finnegan, R., eds. 1973. *Modes of Thought* (London: Faber) [*10* 139]

Howson, C., & Urbach, P. 1989. *Scientific Reasoning: The Bayesian Approach* (La Salle IL: Open Court) [*8* 268, 269, 273, 275, 282, 293]

Howson, C., & Urbach, P. 1991. Bayesian reasoning in science. *Nature* **350**:371–4 [*8* 268, 278]

Hudson, L., & Jacot, B. 1986. The outsider in science: A selective review of evidence, with special reference to the Nobel Prize. In *Personality, Cognition and Values*, edited by Bagley, C., & Verma, G. (London: Macmillan) pp. 3–23 [*8* 137, 388; *9* 286]

Hull, D. L. 1988. *Science as a Process: An Evolutionary Account of the Social and Conceptual Development of Science* (Chicago IL: University of Chicago Press) [*3* 68; *4* 63; *5* 83; *6* 54, 64, 96, 99, 128, 162; *7* 20, 25; *8* 122, 252, 292; *9* 11, 24, 29, 40, 87, 91, 156, 197, 219, 253, 264, 277, 295, 297, 308, 333, 340; *10* 5, 225, 248]

Husserl, E. 1924. Unpublished article on Kant [*10* 11]

Huszagh, V. A., & Infante, J. P. 1989. The hypothetical way of progress. *Nature* **338**:109 [*8* 240]

Ihde, D. 1972. Sense and sensuality. In *Facets of Eros: Phenomenological Essays*, edited by Smith, F. J., & Eng, E. (The Hague: Nijhoff) [*10* 27]

Ihde, D. 1990. *Technology and the Lifeworld* (Bloomington IN: Indiana University Press) [*6* 250]

Ihde, D. 1991. *Instrumental Realism: The Interface between Philosophy of Science and Philosophy of Technology* (Bloomington IN: Indiana University Press) [*5* 51, 125, 127; *8* 401; *10* 16, 182]

Inbar, M., & Stoll, C. S., eds. 1972. *Simulation and Gaming in Social Science* (New York NY: Free Press) [*6* 215]

Irvine, J., Martin, B. R., Isard, P. A. 1990. *Investing in the Future: An International Comparison of Governmental Funding of Academic and Related Research* (Aldershot: Edward Elgar) [*2* 16, 44]

Isambert, F. A. 1985. Un 'programme fort' en sociologie de la science. *R. Franc. sociol* **26**:485–508 [*7* 38; *8* 332, 355, 368; *9* 16, 30; *10* 148, 188, 221]

Ivins, W. M. 1953 (1982). *Prints and Visual Communication*, reprint edn (London: Routledge & Kegan Paul) [*5* 173; *6* 78; *8* 398]

Jablonka, E., & Lamb, M. J. 1995. *Epigenetic Inheritance and Evolution: the Lamarckian Dimension* (Oxford: Oxford University Press) [*10* 239]

Jagtenberg, T. 1983. *The Social Construction of Science* (Dordrecht: Reidel) [*4* 50; *6* 236; *7* 36; *8* 23, 44, 46, 87, 140, 143, 149, 155, 167, 373, 424; *9* 89; *10* 56]

Jaki, S. L. 1966. *The Relevance of Physics* (Chicago IL: Chicago University Press) [*5* 36; *6* 117, 154; *9* 42]

Jarvie, I. C. 1972. *Concepts and Society* (London: Routledge & Kegan Paul) [*6* 86, 98, 256; *9* 174]

Jasanoff, S., Markle, G. E., Petersen, J. C., & Pinch, T. J., eds. 1995. *Handbook of Science and Technology Studies* (Thousand Oaks CA: Sage) [*7* 92]

Johnson-Laird, P. 1983. *Mental Models* (Cambridge: Cambridge University Press) [*5* 189; *6* 48; *8* 412, 427; *10* 1, 45]

Johnson-Laird, P. N. 1988. *The Computer and the Mind: An Introduction to Cognitive Science* (London: Fontana) [*8* 412, 418]

Judson, H. F. 1979. *The Eighth Day of Creation* (New York NY: Simon & Schuster) [*8* 114]

Kantorovich, A. 1993. *Scientific Discovery: Logic and Tinkering* (Albany NY: SUNY Press) [*3* 30; *4* 8; *5* 9, 20, 101; *6* 3, 74, 136, 140, 147, 167, 170, 223, 239; *8* 17, 31, 202, 205, 206, 211, 217, 219, 221, 228, 235, 236, 248, 258, 268, 273, 310, 333, 383, 388; *9* 15, 113, 214, 271, 277, 295, 304, 319, 334; *10* 69, 163, 168, 169, 179, 201, 202, 236, 245, 252]

Katz, J., & Hartnett, R. T. 1976. *Scholars in the Making: The Development of Graduate and Professional Students* (Cambridge MA: Ballinger) [*8* 53]

Kauffman, S. A. 1993. *The Origins of Order: Self-Organization and Selection in Evolution* (Oxford: Oxford University Press) [*9* 312; *10* 251]

Kauffman, S. A. 1995. *At Home in the Universe: The Search for Laws of Complexity* (London: Viking Press) [*9* 312; *10* 251]

Keat, R., & Urry, J. 1975. *Social Theory as Science* (London: Routledge & Kegan Paul) [*5* 171]

Keller, E. F. 1988. Feminist perspectives on science studies. *Science, Technology & Human Values* 13:235–61 [*3* 55; *5* 31; *6* 224; *8* 22; *9* 283; *10* 103]

Keller, E. F. 1991. Feminism and science. In *The Philosophy of Science*, edited by Boyd, R., Gasper, P., & Trout, J. D. (Cambridge MA: MIT Press) pp. 279–88 [*10* 103]

Kempson, R. M. 1988. The relation between language, mind and reality. In *Mental Representations: The Interface between Language and Reality*, edited by Kempson, R. M. (Cambridge: Cambridge University Press) pp. 3–26 [*10* 117]

Kendall, D. G. 1975. Recovery of structure from fragmentary information. *Philosophical Transactions of the Royal Society* 279:547–82 [*8* 313]

Kern, L. H., Mirels, H. L., & Hinshaw, V. G. 1983. Scientists' understanding of propositional logic: An experimental investigation. *Social Studies of Science* 13:131–46 [*6* 128; *8* 252, 298]

Keynes, J. M. 1936. *The General Theory of Employment, Interest and Money* (London: Macmillan) [*7* 83]

Kiang, N. 1995. How are scientific corrections made? *Science and Engineering Ethics* 1:347–56 [*9* 9]

Kitcher, P. 1991. Explanatory unification. In *The Philosophy of Science*, edited by Boyd, R., Gasper, P., & Trout, J. D. (Cambridge MA: MIT Press) pp. 329–45 [*10* 217]

Kitcher, P. 1993. *The Advancement of Science: Science without Legend, Objectivity without Illusions* (Oxford: Oxford University Press) [*1* 3; *5* 64, 141; *6* 127, 167, 174, 222; *8* 15, 93, 127, 268, 273, 328, 419; *9* 38, 40, 57, 118, 217, 240, 245, 333; *10* 70, 124, 179, 217]

Klausner, S. Z. 1983. Social knowledge for social policy. In *Realizing Social Science Knowledge*, edited by Holzner, B., Knorr, K. D., & Strasser, H. (Vienna: Physica-Verlag) pp. 93–121 [*7* 66, 73, 74, 81; *9* 324]

Klemke, E. D., Hollinger, R., & Kline, A. D., eds. 1980. *Introductory Readings in the Philosophy of Science* (Buffalo NY: Prometheus) [*5* 162; *8* 265, 336; *9* 242]

Knorr, K. D. 1975. The nature of scientific consensus and the case of the social sciences. In *Determinants and Controls of Scientific Development*, edited by Knorr, K. D., Strasser, H., & Zilian, H. G. (Dordrect: Reidel) pp. 227–56 [*9* 136]

Knorr, K. D. 1978. The nature of scientific consensus and the case of the social sciences. *International Journal of Sociology* 8:113–45 [*9* 131, 135, 136]

Knorr-Cetina, K. D. 1981. *The Manufacture of Knowledge: An Essay on the Constructivist and Contextual Nature of Science* (Oxford: Pergamon) [*3* 6, 44; *4* 63; *5* 92, 171, 196; *7* 21; *8* 31, 34, 236; *9* 83]

Knorr-Cetina, K. D. 1981a. Social and scientific method, or what do we make of the distinction between the natural and social sciences. *Philosophy of the Social Sciences* 11:335–60 [*5* 32, 92, 171; *6* 18; *8* 31, 237; *10* 21]

Knorr-Cetina, K. D. 1983. The Ethnographic Study of Scientific Work. In *Science Observed: Perspectives on the Social Study of Science*, edited by Knorr-Cetina, K. D., & Mulkay, M. (London: Sage) pp. 115–40 [*8* 181; *10* 77]

Knorr-Cetina, K. D. 1995. How superorganisms change: Consensus formation and the social ontology of high energy physics experiments. *Social Studies of Science* **25**:119–47 [*4* 17, 75; *5* 74; *7* 25; *8* 4, 160, 169, 177; *9* 21, 43, 125, 247]

Knorr-Cetina, K. D., & Amann, K. 1990. Image dissection in natural scientific inquiry. *Science Technology & Human Values* **15**:259–83 [*6* 30]

Knorr-Cetina, K. D., & Mulkay, M. 1983. Introduction. In *Science Observed: Perspectives on the Social Study of Science*, edited by Knorr-Cetina, K. D., & Mulkay, M. (London: Sage) pp. 1–18 [*3* 4; *9* 243]

Koestler, A. 1964. *The Act of Creation* (New York NY: Dell) [*9* 286]

Kolata, G. B. 1986. Anthropologists suggest cannibalism is a myth. *Science* **232**:1497–1500 [*7* 58; *9* 53]

Körner, S. 1966. *Experience and Theory* (London: Routledge & Kegan Paul) [*6* 102, 242; *10* 67]

Körner, S. 1970. *Categorial Frameworks* (Oxford: Blackwell) [*6* 111; *8* 96, 107; *10* 56, 153]

Körner, S. 1974. On the subject matter of philosophy. [*8* 96; *10* 99, 146]

Kosslyn, S. M. 1990. Les images mentales. *La Recherche* **11**:156–63 [*8* 399]

Kuhn, T. S. 1962. *The Structure of Scientific Revolutions* (Chicago IL: University of Chicago Press) [*1* 23; *6* 10, 66; *7* 33; *8* 18, 74, 84, 90, 104, 126, 128, 139, 195, 199, 359; *9* 255, 261, 266]

Kuhn, T. S. 1978. *Black-Body Theory and the Quantum Discontinuity, 1894–1912* (Oxford: Oxford University Press) [*9* 233]

Kuhn, T. S. 1984. *Scientific Development and Lexical Change, Thalheimer Lectures* (Baltimore MD: Johns Hopkins University Press) [*5* 134, 141; *6* 123; *9* 266, 326; *10* 173]

Lakatos, I. 1970. Falsification and the methodology of scientific research programmes. In *Criticism and the Growth of Knowledge*, edited by Lakatos, I., & Musgrave, A. (Cambridge: Cambridge University Press) pp. 91–195 [*8* 298; *9* 190]

Lakatos, I. 1978. *The Methodology of Scientific Research Programmes* (Cambridge: Cambridge University Press) [*8* 74, 120, 257, 287; *9* 109, 112, 190, 205; *10* 147]

Lakatos, I., & Musgrave, A., eds. 1970. *Criticism and the Growth and Knowledge* (Cambridge: Cambridge University Press) [*8* 86]

Lakoff, G. 1987. *Women, Fire and Dangerous Things: What Categories Reveal about the Mind* (Chicago IL: Chicago University Press) [*8* 413]

Langley, P., Simon, H. A., Bradshaw, G. L., & Zytkow, J. M. 1987. *Scientific Discovery: Computational Explorations of the Creative Processes* (Cambridge MA: MIT Press) [*8* 14, 208, 216, 271, 389, 425; *10* 9]

Langmuir, I. 1953. Pathological Science. Unpublished tape recording, transcribed and privately circulated by R. N. Hall.

Langton, C. G., ed. 1995. *Artificial Life: An Overview* (Cambridge MA: MIT Press) [*9* 299]

Latour, B. 1987. *Science in Action* (Milton Keynes: Open University Press) [*2* 19; *3* 7; *7* 24; *8* 26, 170, 339; *9* 27, 48, 88, 98, 153, 293; *10* 148]

Latour, B., & Woolgar, S. 1979. *Laboratory Life: The Social Construction of Scientific Facts* (London: Sage) [*3* 6, 44; *4* 63, 72; *5* 196; *7* 52; *8* 75, 135, 171, 249, 338, 353; *9* 50, 80, 164]

Laudun, L. 1977. *Progress and its Problems* (London: Routledge & Kegan Paul) [*2* 10; *7* 6; *8* 7, 73]

Laudun, L. 1990. *Science and Relativism: Some Key Controversies in the Philosophy of Science* (Chicago: University of Chicago Press) [*7* 19, 56; *8* 100, 105; *9* 85; *10* 93, 179]

Laudun, L. 1991. A confutation of convergent realism. In *The Philosophy of Science*, edited by Boyd, R., Gasper, P., & Trout, J.D. (Cambridge MA: MIT Press) pp. 223–46 [*10* 184]

Law, J., & Williams, R.J. 1982. Putting facts together: A study of scientific persuasion. *Social Studies of Science* 12:535–8 [*9* 81]

Lear, J. 1985. Shame: review of Charles Taylor. *London Review of Books* (19 Sept.):22 [*5* 154]

Lehman, M.M. 1977. Human thought as an ingredient of system behaviour. In *The Encyclopaedia of Ignorance: Life Sciences & Earth Sciences*, edited by Duncan, R., & Weston-Smith, M. (Oxford: Pergamon) pp. 347–54 [*6* 194]

Leinfellner, W. 1984. Evolutionary Causality, Theory of Games and Evolution of Intelligence. In *Concepts and Approaches in Evolutionary Epistemology*, edited by Wuketits, F. (Dordrecht: Reidel) pp. 233–77 [*9* 293]

Lemaine, G. 1980. Science normale et science hypernormale: les stratégies de différenciation et les stratégies conservatrices dans la science. *R. franç. sociol.* 21:499–527 [*8* 23, 35, 55, 62, 131, 132, 382; *9* 220, 273, 313]

Lemaine, G., Clémençon, M., Gomis, A., Pollin, B., & Salvo, B. 1977. *Stratégies et choix dans la recherche: À propos des travaux sur le sommeil* (Paris: Mouton) [*6* 229; *8* 32, 120, 196, 388; *9* 71, 274, 286]

Lemaine, G., MacLeod, R., Mulkay, M., & Weingart, P., eds. 1976. *Perspectives on the Emergence of Scientific Disciplines* (The Hague: Mouton) [*4* 46; *8* 190]

Levi, P. 1989. *Other People's Trades* (Harmondsworth: Penguin) [*5* 133]

Levy, S. 1992. *Artificial Life: The Quest for a New Creation* (London: Penguin) [*9* 299]

Leydesdorff, L. 1991. In search of epistemic networks. *Social Studies of Science* 21:75–110 [*6* 75]

Lindblom, C.E. 1990. *Inquiry and Change: The Troubled Attempt to Understand and Shape Society* (New Haven & London: Yale University Press) [*6* 90, 168; *7* 14, 107, 125; *8* 14, 152, 289, 323, 358, 374; *9* 45, 94, 259; *10* 2, 47, 124]

Lindblom, C.E., & Cohen, D.K. 1979. *Usable Knowledge: Social Science and Social Problem Solving* (New Haven CT: Yale University Press) [*7* 67]

Linstone, H.A., & Turoff, M., eds. 1975. *The Delphi Method: Techniques and Applications* (Reading MA: Addison-Wesley) [*8* 267]

Lloyd, G.E.R. 1979. *Magic, Reason and Experience* (Cambridge: Cambridge University Press) [*6* 123; *9* 72; *10* 167]

Loasby, B.J. 1989. *The Mind and Method of the Economist: A Critical Appraisal of Major Economists in the 20th Century* (Aldershot: Elgar) [*8* 103]

Lodge, D. 1975. *Changing Places* (London: Penguin) [*2* 48]

Longino, H. 1990. *Science as Social Knowledge: Values and Objectivity in Scientific Inquiry* (Princeton NJ: Princeton University Press) [*5* 141; *7* 55; *8* 19, 165; *10* 112, 173, 223, 248]

Lorenz, K. 1941. Kants Lehre vom Apriorischen im Lichte gegenwärtiger Biologie. *Blätter für Deutsche Philosophie* 15:94–125 [*5* 114]

Lorenz, K. 1973 (1977). *Die Ruckseite des Spiegels: Versuch einer Naturgeschichte menschlichen Erkennis*. English Translation edn (New York NY: Harcourt-Brace) [*9* 315]

Lucas, J.R. 1986. Philosophy and Philosophy of. Unpublished lecture to British Academy. 29 Oct. 1986.

Lucas, J.R. 1990. Reason and reality. In *Harré and his Critics: Essays in Honour of Rom Harré, with his Commentary on them*, edited by Bhaskar, R). pp. 41–47 [*10* 173]

Lucas, J. R. 1995. Prospects for realism in quantum mechanics. *International Studies in the Philosophy of Science* **9**:225–34 [*9* 228; *10* 163, 173, 210]

Luce, R. D., & Narens, L. 1987. Measurement scales on the continuum. *Science* **236**:1527–32 [*5* 38]

Luria, W. R. 1973. *The Working Brain* (London: Penguin) [*8* 416]

Lyons, J. 1970. *Chomsky* (London: Fontana/Collins) [*6* 130]

Lyons, J. 1981. *Language and Linguistics* (Cambridge: Cambridge University Press) [*6* 36; *8* 393; *10* 71]

MacIntyre, A. 1981. *After Virtue* (London: Duckworth) [*3* 32]

MacRoberts, M. H., & MacRoberts, B. R. 1984. The negational reference: or the art of dissembling. *Social Studies of Science* **14**:91–4 [*9* 111]

Maddox, J. 1995. Duesberg and the new view of HIV. *Nature* **373**:189 [*9* 56]

Magee, B. 1973. *Popper* (London: Collins) [*8* 287; *9* 82, 173, 174, 177]

Magyar, G. 1981. How trustworthy are experimental facts? *European Journal of Physics* **2**:244–9 [*5* 72]

Mahoney, M. J. 1979. Psychology of the scientist: an evaluative review. *Social Studies of Science* **9**:349–75 [*9* 11, 29, 218]

Mann. 1983. *The Student Encyclopaedia of Sociology* (London: Morgan) [*3* 11]

Mansfield, E. 1991. The social rate of return from academic research. *Research Policy* [*3* 128]

Margolis, J. 1991. *The Truth about Relativism* (Oxford: Blackwell) [*6* 7; *10* 146]

Markus, G. 1987. Why is there no hermeneutics of natural science? *Science in Context* **1**:5–54 [*3* 67; *5* 33, 94, 141, 171, 179; *6* 253; *7* 3, 123; *8* 63, 129, 147, 175, 380; *9* 22, 124, 130, 146, 148, 160, 325; *10* 222]

Marmo, G., & Vitale, B. 1980. Quality, form and globality: An assessment of catastrophe theory. *Fundamenta Scientiae* **1**:35–54 [*6* 212]

Marshall, E. 1990. Science beyond the Pale. *Science* **249**:14–16 [*8* 280]

Martin-Rovet, D., & Carlson, T. 1995. The international exchange of scholars. *Minerva* **33**:71–94, 171–91 [*2* 48]

Masterman, M. 1970. The nature of a paradigm. In *Criticism and the Growth of Knowledge*, edited by Lakatos, I., & Musgrave, A. (Cambridge: Cambridge University Press) pp. 59–90 [*8* 115]

Maynard Smith, J., & Szathmáry, E. 1995. *The Major Transitions in Evolution* (Oxford: Freeman) [*10* 256]

Mayr, E. 1982. *The Growth of Biological Thought* (Cambridge MA: Harvard University Press) [*6* 54, 158; *9* 127, 288]

Mayr, E. 1988. *Towards a New Philosophy of Biology: Observations of an Evolutionist* (Cambridge MA: Harvard University Press) [*5* 17, 87; *6* 50, 55, 62, 64; *8* 291; *9* 127; *10* 244, 248]

McCormmach, R. 1982. *Night Thoughts of a Classical Physicist* (Cambridge MA: Harvard University Press) [*4* 19]

McLuhan, M. 1962. *The Gutenberg Galaxy* [*6* 43]

Mead, G. H. 1938. *The Philosophy of the Act* (London: University of Chicago Press) [*5* 25, 64, 89, 120, 138, 181, 184; *6* 249; *8* 327; *10* 56, 57, 166, 186]

Mead, M. 1976. Towards a human science. *Science* **191**:903 [*5* 158]

Meadows, A. J. 1974. *Communication in Science* (London: Butterworth) [*3* 20]

Medawar, P. 1964. Is the scientific paper fraudulent? *The Saturday Review* (Aug. 1):42–3 [*5* 178]

Medawar, P. B. 1967. *The Art of the Soluble* (London: Methuen) [*5* 79; *6* 4; *8* 12, 205, 342; *9* 6]

Medawar, P. B. 1969. *Induction and Intuition in Scientific Thought* (Philadelphia PA: American Philosophical Society) [*10* 15]

Menard, H. W. 1971. *Science: Growth and Change* (Cambridge MA: Harvard University Press) [*9* 209]

Merleau-Ponty, M. 1962. *Phenomonology of Perception*, translated by Smith, C., English translation edn (London: Routledge & Kegan Paul) [*5* 127]

Merton, R. K. 1938 (1970). *Science Technology and Society in Seventeenth Century England* (New York NY: Harper) [*3* 21]

Merton, R. K. 1942 [1973]. The Normative Structure of Science. In *The Sociology of Science: Theoretical and Empirical Investigations*, edited by Storer, N. W. (Chicago IL: University of Chicago Press) pp. 267–78 [*3* 18; *5* 13; *9* 1]

Merton, R. K. 1973. *The Sociology of Science* (Chicago IL: University of Chicago Press) [*3* 29, 36, 53, 54, 58, 63, 66, 75, 78, 79, 87, 92, 93, 95, 97, 117; *4* 62; *5* 28, 200; *7* 1, 21, 35, 102; *8* 13, 223; *9* 1, 7, 197, 242]

Mervis, J. 1993. Expert testimony. *Science* **261**:22 [*10* 170]

Meyer-Krahmer. 1997. Science-based technologies and interdisciplinarity: Challenges for firms and policy. In *Systems of Innovation: Technologies, Institutions and Organizations*, edited by Edquist, C. (London: Pinter) pp. 298–317 [*4* 71; *8* 60]

Michalos, A. C. 1980. Philosophy of science: Historical, social and value aspects. In *A guide to the Culture of Science, Technology and Medicine*, edited by Durbin, P. T. (New York NY: Free Press) pp. 197–281 [*5* 1, 165]

Midgley, M. 1989. *Wisdom, Information and Wonder* (London: Routledge & Kegan Paul) [*4* 54; *7* 95; *8* 60; *9* 66, 91, 184, 267; *10* 44, 91, 137, 146]

Midgley, M. 1992. *Science as Salvation* (London: Routledge) [*2* 26; *8* 257, 319, 420, 428; *9* 184, 310, 331; *10* 41, 121, 123, 132, 147, 184, 202, 220, 245]

Midgley, M. 1992a. Can science save its soul? *New Scientist* (1 Aug.):24–7 [*5* 30; *7* 77; *9* 238; *10* 132, 248]

Miller, A. L. 1984. *Imagery in Scientific Thought: Creating 20th Century Physics* (Boston MA: Birkhauser) [*8* 404, 406]

Millikan, R. 1984. *Language, Thought and Other Biological Categories* (Cambridge MA: MIT Press) [*5* 16, 52, 146, 148; *6* 59, 66, 115, 244; *8* 430; *9* 215, 222; *10* 28, 187, 218]

Mitchell, E. W. J. 1986. Opportunities and needs for basic and strategic research in the next decade. (Royal Society) [*2* 25]

Mitroff, I. I. 1974. *The Subjective Side of Science* (Amsterdam: Elsevier) [*3* 61; *8* 384; *9* 27, 42]

Mokyr, J. 1990. *The Lever of Riches* (New York NY: Oxford University Press) [*7* 32; *8* 32, 290; *9* 216, 252]

Monod, J. 1970 (1972). *Chance and Necessity*. English translation edn (London: Collins) [*10* 62]

Morell, V. 1993. Anthropology; Nature-culture battleground. *Science* **261**:1798–1801 [*9* 54]

Morrell, J., & Thackray, A. 1981. *Gentlemen of Science: Early Years of the British Association for the Advancement of Science* (Oxford: Oxford University Press) [*3* 24, 130]

Mukerji, C. 1989. *A Fragile Power: Scientists and the State* (Princeton NJ: Princeton University Press) [*2* 40; *3* 8, 132; *4* 16, 43, 60, 61, 66; *5* 28, 49, 51, 59, 63, 89, 92, 97, 128; *6* 66, 78, 84, 199, 231, 234, 237, 251; *7* 16, 20, 44, 49, 104, 108, 115; *8* 31, 42, 47, 54, 136, 163, 184, 218, 346, 418; *9* 14, 16, 33, 64, 146, 151]

Mulkay, M. 1979. *Science and the Sociology of Knowledge* (London: George Allen & Unwin) [*3* 31, 37, 41; *5* 32, 86; *6* 226; *8* 109, 112, 214, 304; *9* 112, 114, 117, 162; *10* 56, 154]

Mulkay, M. 1985. *The Word and the World: Explorations in the Forms of Sociological Analysis* (London: George Allen & Unwin) [*5* 99; *9* 17, 24, 60, 81, 117]

Mulkay, M. 1991. *Sociology of Science: A Sociological Pilgrimage* (Milton Keynes: Open University Press) [*3* 64; *5* 96, 99; *9* 8, 20, 24, 114, 126, 331; *10* 87]

Munz, P. 1993. *Philosophical Darwinism: On the Origin of Knowledge by Means of Natural Selection* (London: Routledge) [*5* 115, 148; *6* 102, 125, 136; *8* 248, 415; *9* 205, 279, 291, 326; *10* 46, 61, 65, 68, 69, 80, 124, 125, 148, 173, 209, 214, 219, 248]

Myers, G. 1990. *Writing Biology: Texts in the Social Construction of Scientific Knowledge* (Madison WI: U. of Wisconsin Press) [*8* 341, 372, 380; *9* 156]

Myers, G. 1991. Politeness and certainty. *Social Studies of Science* 21:37–74 [*3* 46; *9* 162]

Naess, A. 1968. *Scepticism* (London: Routledge & Kegal Paul) [*8* 257; *9* 2; *10* 147]

Naess, A. 1972. *The Pluralist and Possiblist Aspect of the Scientific Enterprise* (London: Allen & Unwin) [*4* 1; *5* 10, 137; *6* 3, 202, 216; *8* 8, 113, 295, 304, 310; *9* 235, 271; *10* 143, 154, 175]

Nagashima, N. 1973. Japan. In *Modes of Thought*, edited by Horton, R., & Finnegan, R. (London: Faber) pp. 92–165 [*6* 175]

Nagel, E. 1961. *The Structure of Science: Problems with the Logic of Scientific Explanation* (London: Routledge & Kegan Paul) [*5* 154; *6* 195, 200; *8* 255; *9* 215; *10* 24, 238, 248]

Nagel, E. 1979. *Teleology Revisited and other Essays in the Philosophy and History of Science* (New York NY: Columbia University Press) [*5* 48; *8* 303, 329; *9* 204]

Nagel, T. 1986. *The View from Nowhere* (Oxford: Oxford University Press) [*6* 80]

NAS. 1972. *Physics in Perspective*. (Washington DC: US National Academy of Sciences) [*2* 21]

Nelson, R. R. & Winter, S. 1982. *An Evolutionary Theory of Economic Change* (Cambridge MA: Belknap) [*9* 321]

Neurath, O. 1932. *Protocol Statements* [*8* 376]

Nicod, J. 1923 (1969). *Geometry and Induction* (London: Routledge & Kegan Paul) [*5* 137; *10* 39]

Nicoll, A., & Brown, P. 1994. HIV beyond reasonable doubt. *New Scientist* (15 Jan. 1994):24–8 [*9* 55]

Nowotny, H. 1989 (1994). *Time: The Modern and Postmodern Experience*. English translation edn (Cambridge: Polity Press) [*10* 54]

Nowotny, H., & Felt, U. 1997. *After the Breakthrough: The Emergence of High-Temperature Superconductivity as a Research Field* (Cambridge: Cambridge University Press) [*3* 52; *4* 16, 24, 52, 70; *5* 22, 199; *7* 101; *8* 190]

Nowotny, H., & Rose, H., eds. 1979. *Counter-movements in the Sciences* (Dordrecht: Reidel) [*9* 238]

Oakeshott, M. 1933. *Experience and its Modes* (Cambridge: Cambridge University Press) [*1* 9; *5* 15, 25]

Odlyzko, A. 1995. *The Decline of Unfettered Research*. Unpublished essay [*2* 37; *8* 162, 210]

OECD. 1980. The Measurement of Scientific and Technical Activities. (Paris: OECD)

Oeser, E. 1984. The Evolution of Scientific Method. In *Concepts and Approaches in Evolutionary Epistemology*, edited by Wuketits, F. (Dordrecht: Reidel) pp. 149–84 [*8* 414; *9* 165, 180; *10* 60, 68, 153]

O'Hear, A. 1996. Self-conscious belief. In *The Certainty of Doubt: Tributes to Peter Munz*, edited by Fairburn, M., & Oliver, W.H. (Wellington, NZ: Victoria University Press) pp. 336–51 [*10* 41, 122, 126]

Oldfield, R.C., & Marshall, J.C., eds. 1968. *Language* (London: Penguin Books) [*6* 134]

Oppenheim, P., & Putnam, P. 1991. Unity of science as a binding hypothesis. In *The Philosophy of Science*, edited by Boyd, R., Gasper, P., & Trout, J. D. (Cambridge MA: MIT Press) pp. 405–27 [*10* 234]

Oreskes, K., Scrader-Frechette, K., & Belitz, K. 1994. Verification, validation and confirmation of numerical models in the earth sciences. *Sciences* 263:641–6 [*6* 209]

Orlans, H. 1976. The advocacy of social science in Europe and America. *Minerva* 14:6–32 [*7* 77]

OTA. 1986. *Research Funding as an Investment: Can We Measure the Returns?* (Washington DC: Office of Technology Assessment)

Parisi, G. 1993. *Physics World* (Sept.):42–7 [*10* 248]

Pasteur, L. 1854. Chance favours only the prepared mind.

Peters, R. S. 1970. In *Explanation in the Behavioural Sciences*, edited by Borger, R., & Cioffi, E. (Cambridge: Cambridge University Press) p. 33 [*10* 48]

Phillips, R. 1981. *Mushrooms and other Fungi of Great Britain and Europe* (London: Pan Books) [*6* 47]

Piaget, J. 1954. Language and thought from the genetic point of view. *Acta Psychologica* 10:88–98 [*6* 131; *8* 407; *10* 26]

Piaget, J., & Inhelder, B. 1963. Mental images: Intellectual operations and their development. In *Intelligence*, edited by Fraisse, P., & Piaget, J. (London: Routledge & Kegan Paul) [*8* 406; *10* 26]

Pickering, A. 1984. *Constructing Quarks: A Sociological History of Particle Physics* (Edinburgh: Edinburgh University Press) [*6* 201, 226; *8* 31, 106, 236, 335; *8* 350, 371; *9* 114, 251; *10* 164]

Pickering, A. 1995. *The Mangle of Practice: Time, Agency and Science* (Chicago IL: U. of Chicago Press) [*6* 11; *8* 32; *10* 141, 181, 221]

Pinch, T. 1986. Controversies in science. *Physics Bulletin* 37:417–20 [*9* 59, 65]

Pinch, T. J. 1979. Normal explanations of the paranormal: the demarcation problem and fraud in parapsychology. *Social Studies of Science* 9:329–48 [*9* 211]

Pinch, T. J. 1993. Generations of SSK. *Social Studies of Science* 23:363–73 [*9* 201, 243]

Pirsig, R. M. 1974. *Zen and the Art of Motorcycle Maintenance* (London: Bodley Head) [*8* 260; *10* 173]

Pivcevic, E. 1970. *Husserl and Phenomenology* (London: Hutchinson) [*5* 139, 145, 153; *6* 239; *8* 253, 348; *9* 183; *10* 11, 42, 58, 147]

Plotkin, H. C. 1994. *The Nature of Knowledge: Concerning Adaptations, Instinct and the Evolution of Intelligence* (Harmondsworth: Penguin) [*5* 114; *8* 415; *9* 279, 298; *10* 65]

Plott, C. R. 1986. Laboratory experiments in economics: The implications of posted-price institutions. *Science* 232:732–8 [*5* 61]

Polanyi. 1969. *Knowing and Being* (London: Routledge & Kegan Paul) [*3* 15; *5* 49; *6* 28; *8* 5, 9, 308]

Polanyi, M. 1958. *Personal Knowledge* (London: Routledge & Kegan Paul) [*Pref.* 1; *2* 36; *3* 1; *5* 11, 49, 94, 106, 124, 127, 128, 137, 147; *6* 34, 66; *8* 67, 308, 392, 405, 422, 431; *9* 248; *10* 22, 114, 123, 160, 207, 236, 248]

Polkinghorne, J. 1985. *Creation and the Structure of the Physical World* (London: King's College) [*8* 279; *10* 139, 165]

Popper, K. 1945. *The Open Society and its Enemies* (London: Routledge and Kegal Paul) [*5* 137; *7* 75]

Popper, K. R. 1935 (1959). *The Logic of Scientific Discovery*. English translation edn (London: Hutchinson) [*Pref.* 2; *1* 13; *6* 110; *8* 198, 256, 281, 286; *9* 92, 230, 280, 317; *10* 15, 157]

Popper, K. R. 1963 (1968). *Conjectures and Refutation: The Growth of Scientific Knowledge*. 3rd edn (New York NY: Harper Torchbooks) [*1* 22; *2* 10; *4* 1; *5* 86; *6* 110, 116; *8* 183, 256, 281, 284, 286, 288; *9* 13, 92, 185, 222, 246, 280, 317; *10* 6, 7, 65, 216]

Popper, K. R. 1972. *Objective Knowledge: An Evolutionary Approach* (Oxford: Oxford University Press) [*5* 18; *9* 170, 172, 174, 176, 185, 223, 230, 277; *10* 15, 81, 187, 217]

Porter, T. M. 1992. Quantification and the accounting ideal in science. *Social Studies of Science* **22**:633–51 [*5* 39]

Porter, T. M. 1995. *Trust in Numbers: The Pursuit of Objectivity in Science and Public Life* (Princeton NJ: Princeton University Press) [*5* 43, 72, 82, 88, 180; *6* 24, 114; *7* 40, 66, 115; *9* 224; *10* 182, 210]

Price, D. J. de S. 1961. *Science since Babylon* (New Haven CT: Yale University Press) [*4* 45, 47]

Price, D. J. de S. 1963 (1986). *Little Science, Big Science – and Beyond*. Enlarged republication edn (New York NY: Columbia University Press) [*3* 19; *4* 42; *8* 57, 381]

Price, D. K. 1979. Endless frontier or bureaucratic mess. In *Limits of Scientific Inquiry*, edited by Holton, G., & Morison, R. S. (New York NY: Norton) pp. 75–92 [*4* 76]

Putnam, H. 1975. *Mathematics, Matter and Method* (Cambridge: Cambridge University Press) [*6* 144]

Putnam, H. 1978. *Meaning and the Moral Sciences* (London: Routledge & Kegal Paul) [*5* 171; *6* 162]

Putnam, H. 1981. *Reason, Truth and History* (Cambridge: Cambridge University Press) [*9* 269]

Putnam, H. 1991. The 'corroboration' of theories. In *The Philosophy of Science*, edited by Boyd, R., Gasper, P., & Trout, J. D. (Cambridge MA: MIT Press) pp. 131–7 [*5* 11; *7* 10; *8* 310]

Quine, W. V. O. 1953. *From a Logical Point of View* (New York NY: Harper & Rowe) [*8* 295, 376]

Radnitzky, G. 1968. *Anglo-Saxon Schools of Metascience* (Goteborg: Akademiforlaget) [*8* 149]

Randall, J. L. 1975. *Parapsychology and the Nature of Life* (London: Souvenir Press) [*9* 240]

Ravetz, J. R. 1971. *Scientific Knowledge and its Social Problems* (Oxford: Clarendon Press) [*3* 39; *4* 67; *5* 63, 82, 93, 172; *7* 96; *8* 32, 90, 133, 201, 431; *9* 32, 105, 115, 155, 277; *10* 20, 112]

Regis, E. 1991. *Great Mambo Chicken and the Transhuman Condition* (London: Penguin) [*9* 238]

Reid, C. 1970. *Hilbert* (London: George Allen & Unwin) [*9* 3]

Remmling, G. W., ed. 1973. *Towards the Sociology of Knowledge* (London: Routledge & Kegan Paul) [*10* 33]

Rescher, N. 1978. *Scientific Progress* (Pittsburgh PA: University of Pittsburg Press) [*8* 20, 138, 158; *9* 335]

Rescher, N. 1980. The ethical dimension in scientific research. In *Introductory Readings in the Philosophy of Science*, edited by Klemke, E. D., Hollinger, R., & Kline, A. D. (Buffalo NY: Prometheus) pp. 238–53 [*7* 125]

Rescher, N. 1988. *Rationality* (Oxford: Oxford University Press) [*6* 127, 153, 252; *8* 174, 257, 261; *9* 137; *10* 12, 113, 116, 122, 140, 147]

Restivo, S. 1983. *The Social Relations of Physics, Mysticism and Mathematics* (Dordrecht: Reidel) [*6* 153]

Restivo, S. 1990. The social roots of pure mathematics. In *Theories of Science in Society*, edited by Cozzens, S. E., & Gieryn, T. F. (Bloomington IN: Indiana University Press) pp. 120–43 [*6* 140, 228]

Restivo, S. 1994. *Science, Society and Values: Towards a Sociology of Objectivity* (Bethlehem
PA: Lehigh University Press) [2 *17, 35;* 3 *10, 38, 136, 137;* 4 *4, 15, 25, 63;* 5 *3, 29, 141,
202;* 6 *255;* 7 *39, 51, 54, 125;* 8 *31, 39, 91, 150, 153, 154, 158, 162, 171, 311, 362;* 9 *28, 93,
137, 141, 198, 239, 251, 334;* 10 *77, 126, 148, 173, 186, 191*]

Riedl, R. 1984. *Biology of Knowledge: The Evolutionary Basis of Reason*, translated by
Foulkes, P., 3rd edn (Chichester: Wiley) [5 *114*]

Rip, A. 1994. The republic of science in the 1990s. *Higher Education* **28**:3–23 [3 *133;* 4 *59,
64, 81;* 7 *39;* 8 *171*]

Roberts, M. J. 1974. On the nature and condition of social science. *Daedalus* **130**:47–64
[6 *160*]

Roberts, R. R. 1989. *Serendipity: Accidental Discoveries in Science* (New York NY: Wiley) [8
227]

Root-Bernstein, R. S. 1989. *Discovering* (Cambridge MA: Harvard University Press) [5 *85;*
6 *196;* 8 *118, 185, 212, 222, 226, 239, 299, 315;* 9 *196, 277, 282*]

Rorty, R. 1986. The contingency of language. *London Review of Books* (17 April):3–6 [6
220, 352; 9 *274;* 10 *173, 185*]

Rorty, R. 1996. Something to steer by. *London Review of Books* (20 Jan.):7–8 [7 *59*]

Rose, H . 1997. Science wars: My enemy's enemy is – only perhaps – my friend. In
Science Today: Problem or Crisis, edited by Levinson, R., & Thomas, J. (London:
Routledge) pp. 51–66 [7 *93*]

Rose, J., & Ziman, J. 1964. *Camford Observed* (London: Gollancz) [2 *49*]

Rose, S. 1992. *The Making of Memory: From Molecules to Mind* (London: Bantam Press) [5 *63;*
6 *225;* 8 *9;* 9 *30, 165;* 10 *259*]

Rose, S. 1997. *Lifelines: Biology, Freedom, Determinism* (London: Penguin) [6 *92, 218;* 10 *248*]

Rosenberg, N., & Nelson, R. 1994. American universities and technical advance in
industry. *Research Policy* **23**:323–48 [2 *22, 29;* 4 *70;* *91*]

Rosenmayer, L. 1983. Sociology – deciphering 'Zeitgeist', creating consciousness or
cooperating to 'solve problems'. In *Realizing Social Science Knowledge*, edited by
Holzner, B., Knorr, K. D., & Strasser, H. (Vienna: Physica-Verlag) pp. 125–46 [6
16; 7 *77, 86;* 8 *141, 324*]

Roth, P., & Barrett, R. 1990. Deconstructing quarks. *Social Studies of Science* **20**:579–746
[8 *352*]

Rothschild, V. 1972. Forty-five varieties of research and development. *Nature*
239:373–8 [2 *23*]

Rouse, J. 1987. *Knowledge and Power: Towards a Political Philosophy of Science* (Ithaca NY:
Cornell University Press) [5 *37, 46, 59, 61, 95, 150, 171;* 6 *76;* 7 *8, 92, 100;* 8 *31, 236, 239,
289, 338, 353;* 9 *78, 114, 277;* 10 *50, 78, 109, 115, 175, 189, 198, 215, 221*]

Rudwick, M. J. S. 1976. The emergence of a visual language for geological science,
1760–1840. *History of Science* **14**:149–95 [6 *67*]

Rudwick, M. J. S. 1985. *The Great Devonian Controversy: The Shaping of Scientific Knowledge
among Gentlemanly Specialists* (Chicago IL: University of Chicago Press) [5 *69;* 6 *66;*
7 *52;* 8 *73;* 9 *16, 58, 84, 258*]

Ruse, M. 1996. *Monad to Man: The Concept of Progress in Evolutionary Biology* (Cambridge
MA: Harvard University Press) [9 *123, 327*]

Ryle, G. 1949. *The Concept of Mind* (London: Hutchinson) [5 *121;* 6 *129;* 9 *140*]

Salomon, J. J. 1970 (1973). *Science and Politics*. English translation of French original edn
(London: Macmillan) [2 *8; 38;* 4 *53, 58, 74;* 10 *260*]

Sapp, J. 1987. *Beyond the Gene* (New York NY: Oxford University Press) [9 *284*]

Saviotti, P. P. 1996. *Technological Evolution, Variety and the Economy* (Cheltenham: Edward Elgar) [*8* 103; *9* 321]

Schäfer, W., ed. 1983. *Finalization in Science: The Social Orientation of Scientific Progress* (Dordrecht: D. Reidel) [*7* 125; *8* 37, 123, 124, 143, 145, 146, 159, 172, 175, 177; *9* 93]

Schaffner, A. C. 1994. The future of science journals: Lessons from the past. *Information Technology & Libraries* 13 (Dec) [*5* 194; *9* 20, 138]

Schmaus, W. 1981. Fraud and sloppiness in science. *Perspectives on the Professions* 1 (3/4):1–4 [*9* 194]

Schmaus, W., Segerstrale, U., & Jessuph, D. 1992. A Manifesto, & Welcome to our new allies. *Social Epistemology* 6:243–65, 315–20 [*1* 28; *5* 14, 182; *7* 36; *8* 321; *9* 197, 202, 211; *10* 9, 84, 86, 110, 158, 247]

Schon, D. A. 1963. *Displacement of Concepts* (London: Tavistock) [*6* 220]

Schütz, A. 1945 (1967). *Collected Papers I. The Problem of Social Reality* (The Hague: Martinus Nijhoff) [*5* 140]

Schütz, A., & Luckmann, T. 1974. *The Structures of the Life-World*. English translation edn (London: Heinemann) [*5* 140; *6* 239; *10* 11]

Searle, J. R. 1995. *The Construction of Social Reality* (London: Penguin) [*8* 116; *9* 221; *10* 37, 45, 173, 188, 190, 193, 204]

Secord, P. F. 1990. 'Subjects' versus 'persons' in social psychological research. In *Harré and his Critics: Essays in Honour of Rom Harré, with his Commentary on them*, edited by Bhaskar, R. pp. 165–88 [*5* 18; *10* 241]

Segerstrale, U. 1987. Debates involving science. *Science* 238:1296–7 [*9* 99]

Segerstrale, U. 1990. The murky boundary between scientific intuition and fraud. *International Journal of Applied Psychology* 5:11–20 [*9* 188, 200]

Segerstrale, U. 1991. The social context of scientific epistemology. Paper read at Affiliated Conference on Social Epistemology and Social Theory of Science: 9th International Conference on Logic, Methodology and Philosophy of Science, Aug. 12–15, at Uppsala [*3* 26; *7* 5, 45, 121, 123; *9* 35, 62]

Segerstrale, U. 1992. Beleaguering the cancer establishment. *Science* 255;613–15 [*7* 57; *8* 13, 293, 325, 333; *9* 236; *10* 238, 247]

Shapin, S. 1994. *A Social History of Truth: Civility and Science in Seventeenth Century England* (Chicago IL: University of Chicago Press) [*3* 22, 68; *4* 3; *5* 11, 64, 69, 71, 74, 141; *8* 306; *9* 116, 137; *10* 18, 42]

Shapin, S., & Schaffer, S. 1985. *Leviathan and the Air Pump: Hobbes, Boyle and the Experimental Life* (Princeton NJ: Princeton University Press) [*2* 3; *3* 16, 22, 26; *4* 3; *5* 26, 46, 63, 68, 80, 81, 86, 177; *7* 18, 46; *8* 294, 349; *9* 26, 36, 41, 154]

Shepard, R. N. 1992. The perceptual organization of colours. In *The Adapted Mind: Evolutionary Psychology and the Generation of Culture*, edited by Barkow, J. H., Cosmides, L., & Tooby, J. (New York NY: Oxford University Press) pp. 495–532 [*6* 39; *10* 71]

Shils, E. 1982. Knowledge and the sociology of knowledge. *Knowledge: Creation, Diffusion, Utilization* 4:7–32 [*1* 11; *2* 32; *8* 107; *9* 244; *10* 102, 124, 142]

Shinn, T. 1987. Geometrie et langage: La production des modèles en sciences sociales et en sciences physiques. *Bulletin de Methodologie Sociologique* 16:5–38 [*6* 199, 230]

Shinn, T., & Whitley, R., eds. 1985. *Expository Science: Forms and Functions of Popularisation*. Vol. 9, *Sociology of the Sciences Yearbook* (Dordrecht: D. Reidel) [*9* 150]

Shotter, J. 1990. Rom Harré: Realism and the turn to social constructionism. In *Harré and his Critics: Essays in Honour of Rom Harré, with his Commentary on them*, edited by Bhaskar, R. pp. 206–23 [*10* 173]

Shrum, W. 1984. Scientific specialities and technical systems. *Social Studies of Science* **14**:63–90 [*8* 46, 63, 161, 167]

Shweder, R. 1992. 'Why do men barbecue?' and other postmodern ironies of growing up in the decade of ethnicity. *Daedelus* **121**:279–309 [*5* 160]

Simmel, G. 1895. On the relationship between the theory of selection and epistemology. *Archiv für systematische Philosophie* **1**:34–45 [*10* 66, 113]

Simon, H. 1956. *Psychological Review* **63**:129 [*9* 318]

Sinclair-de-Zwart, H. 1969. Developmental psycholinguistics. In *Studies in Cognitive Development*, edited by Elkind, & Flavell (Oxford: Oxford University Press) [*8* 409]

Sindermann, C.J. 1982. *Winning the Games Scientists Play* (New York NY: Plenum) [*9* 9, 12, 45, 47, 197]

Sinding, C. 1996. Literary genres and the construction of knowledge in biology: Semantic shifts and scientific change. *Social Studies of Science* **26**:43–70 [*3* 46; *9* 324]

Sismondo, S. 1993. Some social constructions. *Social Studies of Science* **23**:515–70 [*7* 37; *8* 308, 347, 361, 363, 370, 375, 378; *10* 173, 190, 193]

Skair, L. 1973. *Organised Knowledge: A Sociological View of Science and Technology* (London: Hart-Davis MacGibbon) [*3* 9; *8* 68]

Slaughter, M.M. 1982. *Universal Languages and Scientific Taxonomy in the Seventeenth Century* (Cambridge: Cambridge University Press) [*6* 141]

Slezak, P. 1989. Scientific discovery by computer as empirical refutation of the strong programme. *Social Studies of Science* **19**:563–600 [*1* 5; *5* 104; *6* 214; *7* 85; *8* 219, 246]

Smit, W.A. 1995. Science, technology and the military. In *Handbook of Science and Technology Studies*, edited by Jasanoff, S., Mrkle, G.E., Petersen, J.C., & Pinch, T. (Thousand Oaks CA: Sage) pp. 598–626 [*2* 15]

Smith, P.M., & Turrey, B.B. 1996. The future of the behavioural and social sciences. *Science* **271**:611–12 [*6* 51; *7* 111]

Smith, R.M. 1997. Still blowing in the wind: The American quest for a democratic, scientific political science. *Daedalus* **126** (1):253–287 [*7* 78]

Sokolowski, R. 1988. Natural and artificial intelligence. *Daedalus* **117** (1):45–64 [*9* 165]

Solomon, J. 1992. *Getting to Know about Energy in School and in Society* (London: Falmer Press) [*6* 235, 245; *8* 340]

Spiegel-Rösing, I., & de Solla Price, D., eds. 1977. *Science, Technology and Society* (London: Sage) [*2* 13]

Stamp, E. 1981. Why can accounting not become a science like physics? *ABACUS* **17**:13–27 [*5* 39]

Star, S.L. 1983. Simplification in scientific work: an example from neuroscience research. *Social Studies of Science* **13**:205–28 [*8* 24]

Star, S.L. 1985. Scientific work and uncertainty. *Social Studies of Science* **15**:391–428 [*5* 89; *9* 215]

Star, S.L., & Griesemer, J.R. 1989. Instiutional ecology, 'translations' and boundary objects. *Social Studies of Science* **19**:387–420 [*6* 61]

Steen, L.A. 1988. The science of patterns. *Science* **240**:611–16 [*6* 58, 149]

Steinberger, J. 1993. What do we learn from neutrinos? *Science* **259**:1872–6 [*8* 334]

Stevens, R.J. 1988. Human nature and the nature of science. *The American Biology Teacher* **50**:354–61 [*6* 194, 211; *10* 140, 214, 248]

Stoneham, A.M. 1995. National or Notional Laboratories. Zeneca Lecture of Royal Society.

Strawson, P.F. 1960. *The Bounds of Sense* (London: Methuen) [*10* 108, 173]

Studer, K.E., & Chubin, D.E. 1980. *The Cancer Mission* (Beverley Hills CA: Sage) [**3** *141;* **8** *159, 172*]

Suppé, F., ed. 1974. *The Structure of Scientific Theories* (Urbana IL: U. of Illinois Press) [**6** *109, 159;* **10** *215, 244*]

Suttie, I.D. 1935. *The Origins of Love and Hate* (London: Kegan Paul) [**8** *369*]

Sutton, J.R. 1984. Organizational autonomy and professional norms in science. *Social Studies of Science* **14**:197–224 [**8** *46, 161, 167*]

Swinnerton-Dyer, P. 1995. The importance of academic freedom. *Nature* **373**:186–8 [**7** *90;* **8** *33*]

Taubes, G. 1996. Electronic journals etc. *Science* **271**:764–8 [**5** *195*]

Taylor, C. 1985. *Human Agency and Language: Philosophical Papers I* (Cambridge: Cambridge University Press) [**3** *43;* **5** *141, 146, 154, 189;* **6** *125, 219;* **7** *70;* **10** *1, 28, 30, 38, 84, 85, 173*]

Thagard, P. 1993. *Computational Philosophy of Science* (Cambridge MA: MIT Press) [**1** *24;* **6** *52, 186, 226;* **8** *89, 254, 322, 386, 425;* **9** *339;* **10** *1*]

Thom, R. 1972 (1975). *Structural Stability and Morphogenesis*. English translation edn (Readin MA: Benjamin) [**6** *132, 228*]

Thouless, R.H. 1971. *Straight and Crooked Thinking* (London: Pan Books) [**9** *86*]

Tibbetts, P., & Johnson, P. 1985. The discourse and *praxis* models in recent reconstructions of scientific knowledge generation. *Social Studies of Science* **16**:739–50 [**8** *173*]

Tilley, N. 1980. *The Logic of Laboratory Life*. (Unpublished draft.) [**8** *110, 134;* **9** *5, 21, 44;* **10** *56*]

Tooby, J., & Cosmides, L. 1992. The psychological foundations of culture. In *The Adapted Mind: Evolutionary Psychology and the Generation of Culture*, edited by Barkow, J.H., Cosmides, L., & Tooby, J. (New York NY: Oxford University Press) pp. 19–36 [**10** *73*]

Topitsch, E. 1980. The end of finalisation: The retreat of a spurious ideal. *Minerva* **18**:179–95 [**8** *137, 176;* **9** *28*]

Toulmin, S. 1953. *The Philosophy of Science* (London: Hutchinson) [**6** *66, 73, 89, 101;* **8** *310, 431*]

Toulmin, S. 1972. *Human Understanding, Vol. 1* (Oxford: Oxford University Press) [**3** *16;* **5** *182, 189;* **6** *97, 118, 131, 145, 162;* **8** *32, 79, 91, 95, 380, 396, 423;* **9** *37, 117, 258, 270, 276, 287, 292, 295, 338, 340, 341;* **10** *10, 51, 74, 98, 109, 122, 221*]

Toulmin, S. 1980. The intellectual authority and the social context of the scientific enterprise: Holton, Rescher and Lakatos. *Minerva* **18**:652–7 [**6** *167, 242;* **8** *120;* **9** *237, 271;* **10** *221*]

Toulmin, S. 1990. *Cosmopolis: The Hidden Agenda of Modernity* (Chicago IL: University of Chicago Press) [**4** *85;* **6** *17, 95, 163, 167, 185;* **7** *65, 88;* **8** *191, 310;* **9** *212;* **10** *133, 135, 259*]

Traweek, S. 1988. *Beamtimes and Lifetimes: The World of High Energy Physicists* (Cambridge MA: Harvard University Press) [**2** *40;* **3** *38, 48, 70;* **5** *44, 60, 75, 92;* **8** *4, 28, 59, 220, 302, 360;* **9** *106;* **10** *122, 177, 203*]

Trenn, T.J. 1974. Review of Holton on Themata. *Philosophy of Science* **41**:415–18 [**6** *222*]

Tricker, R.A.R. 1965. *The Assessment of Scientific Speculation* (London: Mills & Boon) [**5** *127;* **8** *268, 270, 272*]

Trouvé, J.M. 1992. The evolution of science and the 'tree of knowledge'. In *New Horizons in the Philosophy of Science*, edited by Lamb, D. (Avebury: Aldershot) pp. 87–114 [**10** *217*]

Turkle, S. 1982. The subjective computer: A study in the psychology of personal computation. *Social Studies of Science* **12**:173–206 [*5* 159]

Turnbull, D. 1989. *Maps are Territories: Science is an Atlas* (Geelong Vic.: Deakin University Press) [*6* 66, 69, 81, 87, 91, 93]

Turnbull, D. 1994. Cartography and science: mapping the construction of knowledge spaces. Paper read at Making Space, March 1994, at Canterbury [*8* 174; *10* 218]

Turnbull, D. 1995. Rendering turbulence orderly. *Social Studies of Science* **25**:9–34 [*6* 61]

Turner, S. P. 1990. Forms of patronage. In *Theories of Science in Society*, edited by Cozzens, S. E., & Gieryn, T. F. (Bloomington IN: Indiana University Press) pp. 185–211 [*3* 12, 129, 134; *4* 80; *8* 51]

Valée, J. P. 1989. The search for knowledge: Science versus religion. *Journal of the Royal Astronomical Society of Canada* **83**:8–23 [*10* 136]

Varela, F. J., Thompson, E., & Rosch, E. 1991. *The Embodied Mind: Cognitive Science and Human Experience* (Cambridge MA: MIT Press) [*5* 120; *6* 35, 39; *8* 375]

Varma, R., & Worthington, R. 1995. Immiseration of industrial scientists in corporate laboratories in the United States. *Minerva* **33**:235–338 [*8* 155]

Vincenti, W. G. 1990. *What Engineers Know and How They Know It: Analytical Studies from Aeronautical History* (Baltimore MD: The Johns Hopkin University Press) [*2* 7; *7* 103; *10* 249]

Vollmer, G. 1984. Mesocosm and Objective Knowledge. In *Concepts and Approaches in Evolutionary Epistemology*, edited by Wuketits, F. (Dordrecht: Reidel) pp. 69–121 [*5* 117, 141, 243; *8* 326, 336, 415; *9* 300; *10* 66, 68, 69, 173, 177]

von Weizsäcker, C. F. 1952. *The World View of Physics* (London: Routledge & Kegan Paul) [*10* 236]

Vygotsky, L. S. 1962. *Thought and Language* (Cambridge MA: MIT Press) [*5* 187]

Waddington, C. H. 1941 (1948). *The Scientific Attitude*. 2nd edn (West Drayton: Penguin) [*3* 30; *7* 42]

Wagner-Döbler, R. 1997. Self-organization of scientific specialization and diversification: A quantitative case study. *Social Studies of Science* **27**:147–170 [*8* 58]

Waters, C. K. 1990. Confessions of a Creationist. In *Evolution, Cognition and Realism: Studies in Evolutionary Epistemology*, edited by Rescher, N. (Lanham MD: University Press of America) pp. 79–90 [*9* 295, 301, 339]

Watkins, J. 1970. Imperfect rationality. In *Explanation in the Behavioural Sciences*, edited by Borger, R., & Cioffi, E. (Cambridge: Cambridge University Press) pp. 147–237 [*6* 172]

Watson, H. 1990. Investigating the social foundations of mathematics: Natural number in culturally diverse forms of life. *Social Studies of Science* **20**:283–312 [*6* 152; *10* 107]

Watson, J. D. 1968. *The Double Helix* (London: Weidenfeld & Nicolson) [*4* 21; *10* 48]

Watson-Verran, H., & Turnbull, D. 1995. Science and other indigenous knowledge systems. In *Handbook of Science and Technology Studies*, edited by Jasanoff, S., Markle, G., Pinch, T., & Petersen, J. (Thousand Oaks CA: Sage Publications) pp. 115–39 [*5* 172]

Weber, M. 1918 (1948). Science as a vocation. In *From Max Weber*, edited by Gerth, H. H., & Mills, C. W. (London: Routledge & Kegan Paul) pp. 129–56 [*3* 135; *4* 83; *5* 109]

Weinberg, A. 1967. *Reflections on Big Science* (Oxford: Pergamon) [4 39; 8 41]

Weinberg, S. 1972. Science and trans-science. *Minerva* **10**:209–222 [7 119]

Weingart, P. 1989. The end of academia? The social reorganization of knowledge production. In *The University within the Research System – An International Comparison*, edited by Battaglini, A., & Monaco, F. (Baden-Baden: Nomos) [4 22, 34, 43, 71, 81; 7 89; 8 46]

Weingart, P. 1995. Interdisciplinarity: Institutional responses to change in the world of science. (Unpublished) [5 194; 8 179, 357; 10 241, 247]

Weinstein, D. 1981. *Scientific Fraud and Scientific Ethics* (Chicago IL: Centre for the Study of Ethics in the Professions, Illinois Institute of Technology) [9 191]

Weiss, C. H. 1983. Three terms in search of reconceptualization: knowledge, utilization and decision making. In *Realizing Social Science Knowledge*, edited by Holzner, B., Knorr, K. D., & Strasser, H. (Vienna: Physica-Verlag) pp. 201–18 [7 71, 84]

Weizenbaum, J. 1976. *Computer Power and Human Reason* (San Francisco CA: Freeman) [5 155; 6 206]

Whewell, W. 1840. *The Philosophy of the Inductive Sciences. II* (London) [8 274]

Whitbeck, C. 1995. Truth and trustworthiness in research. *Science and Engineering Ethics* **1**:403–16 [5 74]

Whitley, R. 1984. *The Intellectual and Social Organization of the Sciences* (Oxford: Clarendon Press) [3 28; 5 191; 8 77]

Whorf, B. 1941. In *Language, Culture and Personality*, edited by Spier, L. (University of Utah Press) [6 36; 10 106]

Wigner, E. P. 1969. In *The Spirit and Uses of the Mathematical Sciences*, edited by Saaty, T. L., & Weyl, F. J. (New York NY: McGraw Hill) pp. 123–31 [6 158]

Wilder, R. L. 1967. The role of intuition. *Science* **156**:605–10 [8 429]

Williams, B. 1985. *Ethics and the Limits of Philosophy* (London: Fontana) [5 148, 153]

Winch, P. 1958. *The Idea of a Social Science* (London: Routledge & Kegan Paul) [5 163, 189]

Wolpert, L. 1992. *The Unnatural Nature of Science* (London: Faber & Faber) [1 2; 2 18; 3 138; 5 58; 6 139, 240; 9 225]

Woolgar, S. 1983. Irony in the social study of science. In *Science Observed: Perspectives on the Social Study of Science*, edited by Knorr-Cetina, K. D., & Mulkay, M. (London: Sage) pp. 239–66 [10 186]

Wuketits, F. M. 1984. Evolutionary Epistemology: A Challenge to Science and Philosophy. In *Concepts and Approaches in Evolutionary Epistemology*, edited by Wuketits, F. (Dordrecht: Reidel) pp. 1–33 [8 415; 10 65]

Wynne, B. 1995. Public understanding of science. In *Handbook of Science and Technology Studies*, edited by Jasanoff, S., Markle, G. E., Petersen, J. C., & Pinch, T. (Thousand Oaks CA: Sage) pp. 361–88 [5 172]

Yearley, S. 1994. Understanding science from the perspective of the sociology of scientific knowledge: an overview. *Pubic Understanding of Science* **3**:245–58 [5 10; 8 310]

Yonezawa, F. 1993. Computer-simulation methods in the study of non-crystalline materials. *Science* **260**:635–40 [6 208]

Yoon, C. Y. 1993. Country creatures, great and small. *Science* **260**:620–2 [6 54]

Zahler, R. S., & Sussmann, H. J. 1977. Claims and accomplishments of applied catastrophe theory. *Nature* **269**:759–63 [6 212]

Zhao, K.-H. 1989. Physics nomenclature in China. *American Journal of Physics* **58**:449–51 [6 247]

Ziman, J.M. 1960. *Electrons and Phonons: The Theory of Transport Phenomena in Solids* (Oxford: Clarendon Press) [*9* 161]

Ziman, J.M. 1968. *Public Knowledge: The Social Dimension of Science* (Cambridge: Cambridge University Press) [**Pref.** 4; *3* 20, 42, 51, 64; *4* 11; *5* 79, 82, 105, 149, 178, 193; *7* 3; *8* 121, 262, 359; *9* 4, 10, 78, 102, 163; *10* 135]

Ziman, J.M. 1978. *Reliable Knowledge: An Exploration of the Grounds for Belief in Science* (Cambridge: Cambridge University Press) [**Pref.** 5; *3* 51; *4* 11; *5* 42, 82, 98, 136, 149, 164, 175; *6* 29, 45, 66, 73, 75, 79, 82, 87, 89, 178, 183, 211, 215, 227; *7* 15, 34, 105; *8* 110, 331, 366, 397, 402, 403, 413, 417, 429; *9* 10, 63, 78, 102, 128, 179, 181, 193; *10* 113, 173, 176]

Ziman, J.M. 1980. *Teaching and Learning about Science and Society* (Cambridge: Cambridge University Press) [*3* 142; *4* 56]

Ziman, J.M. 1981. *Puzzles, Problems and Enigmas: Occasional Pieces on the Human Aspects of Science* (Cambridge: Cambridge University Press) [**Pref.** 3; *2* 46; *3* 45, 69; *4* 12; *5* 4; *6* 199, 214; *8* 4, 16, 200, 359; *9* 110, 161, 197, 249; *10* 129]

Ziman, J.M. 1981a. What are the options? Social determinants of personal research plans. *Minerva* **19**:1–42 [*2* 9]

Ziman, J.M. 1983. The collectivisation of science. *Proceedings of the Royal Society B* **219**:1–19 [*4* 38]

Ziman, J.M. 1984. *An Introduction to Science Studies* (Cambridge: Cambridge University Press) [**Pref.** 6; *3* 146; *4* 38, 73; *5* 149; *8* 283]

Ziman, J.M. 1985. Pushing back frontiers – or redrawing maps. In *The Identification of Progress in Learning*, edited by Hägerstrand, T. (Cambridge: Cambridge University Press) pp. 1–12 [*4* 49; *8* 71, 183; *9* 250]

Ziman, J.M. 1987. *Knowing Everything about Nothing: Specialization and Change in Scientific Careers* (Cambridge: Cambridge University Press) [*2* 47; *3* 141; *4* 16, 68; *5* 109, 191; *8* 57, 76, 81, 120]

Ziman, J.M. 1991. Academic science as a system of markets. *Higher Education Quarterly* **45**:41–61 [*2* 14]

Ziman, J.M. 1994. *Prometheus Bound: Science in a Dynamic Steady State* (Cambridge: Cambridge University Press) [**Pref.** 7; *2* 6, 11, 24; *4* 13, 16, 23, 37, 72, 84; *8* 43, 151, 234; *9* 336]

Ziman, J.M. 1995. *Of One Mind: The Collectivization of Science* (Woodbury NY: AIP Press) [**Pref.** 8; *2* 9; *3* 143; *4* 26, 38, 49, 55, 65; *5* 44, 191; *7* 22, 60, 97, 112; *8* 29, 38, 43, 71, 76, 81, 263; *9* 147, 149, 294]

Ziman, J.M. 1996. Darwin and/or Lamarck. Selection and/or design: Technological innovation as an evolutionary process. *Times Higher Education Supplement* [*9* 287]

Ziman, J.M. 1996a. Science studies from Pole to Pole: A Hooker-Restivo connection? *Minerva* **34**:309–18 [*1* 8; *8* 157; *10* 262]

Ziman, J.M. 1996b. Is science losing its objectivity? *Nature* **382**:751–4 [*3* 10; *4* 2; *7* 99]

Ziman, J.M., ed. 2000. *Technological Innovation as an Evolutionary Process* (Cambridge: Cambridge University Press) [*9* 281, 287, 296, 306, 309; *10* 192]

Ziman, J.M., Sieghart, P., & Humphrey, J. 1986. *The World of Science and the Rule of Law* (Oxford: Oxford University Press) [*3* 139]

Zuckerman, H. 1977. *Scientific Elite: Nobel Laureates in the United States* (New York NY: The Free Press) [*7* 48; *8* 58]

Zuckerman, H. 1979. Theory choice and problem choice in science. *Sociological Inquiry* **48**:65–95 [*8* 2, 21, 66, 213, 231; *9* 7, 220].

Index

385

Printed in the United States
By Bookmasters